The Ascendancy of the Scienti

The Ascendancy of the Scientific Dictatorship

An Examination of Epistemic Autocracy, From the 19th to the 21st Century

Paul & Phillip Collins

ISBN : 1-4196-3932-3

To order additional copies, please contact us.
BookSurge, LLC
www.booksurge.com
1-866-308-6235
orders@booksurge.com

The Ascendancy of the Scientific Dictatorship

Special Thanks to:

Michael Corbin
Terry Melanson
Derek & Sharon Gilbert
Messian Dread
Professor Mary Rucker
Judge Joseph Palmer
Joan D'Arc
Al Hidell
Andrei Mignea
Linda Collins
Tracy Collins
Shane Smith
Brother John Matthews
Fran Tarr
Vera Strode
&
Our Precious Lord and Savior,
Jesus Christ

Recommended Web Resources:

Conspiracy Archive
<www.conspiracyarchive.com>

A Closer Look
<www.4acloserlook.com>

Peer Into Darkness
<www.peeringintodarkness.com>

The Architecture of Modern Political Power
<www.mega.nu>

Michael Heiser, Ph.D.
<www.thedivinecouncil.com>

Why an Updated, Expanded, Revised Edition?

When we published the first edition of *The Ascendancy of the Scientific Dictatorship*, we automatically knew that more could be said on the topic. New information was being made available even as the book was being printed. Excellent investigative researchers like Michael Corbin and Terry Melanson were already pointing us in new directions. So much of the information synchronized with our own research that we knew a new, vastly expanded examination of the scientific dictatorship was warranted.

Moreover, previously unavailable pieces of information prompted us to rethink some of our initial contentions. In particular, new evidence for the abiotic oil production led us to refine our arguments concerning Peak Oil Theory. While the manipulated supply of oil may, in fact, be declining, it appears as though the world supply has yet to be fully tapped. In addition, later interviews on *A Closer Look*, which is one of the most important radio talk shows today, elucidated connections between the National Science Foundation and the politics of scientism. Although we had touched on convergent technology programs in the first edition, we soon found our examination of the topic was terribly incomplete.

However, while many things changed, other things remained the same. In particular, our examination of the scientific dictatorship continued from a Christian perspective. Of course, this was one aspect of the first edition that earned us scathing criticism. In most instances, we found that such criticisms were premised upon anti-Christian biases and scientistic presuppositions. We make no apologies for our faith and it remains a permanent fixture of this new edition.

At any rate, it is our sincere hopes that this new edition of *The Ascendancy of the Scientific Dictatorship* will thoroughly elucidate the ideational continuums underpinning the ongoing conspiracy to establish world oligarchy.

Yet, in holding scientific research and discovery in respect, as we should, we must also be alert to the equal and opposite danger that public policy could itself become the captive of a scientific-technological elite."

Eisenhower's Farewell Address to the Nation January 17, 1961

Dedication

Chapter One:

The Birth of Modern Scientific Dictatorships

Science

Knowledge

S ince the dawn of ancient civilization, the ruling class manipulated humanity largely through religious institutions and mysticism. However, as the mists of antiquity receded and contemporary history advanced, the elite's theocratic power structure underwent an epistemic transformation. The result of this transformation was the emergence of a "scientific dictatorship." The history and background of this "scientific dictatorship" is a conspiracy, created and micro-managed through the historical tide of Darwinism. In turn, Darwinism finds its origins with Freemasonry, a philosophical scion of the Ancient Mystery cults that began in Mesopotamia roughly 6000 years ago.

The Question of Conspiracy

The conspiracy to establish a scientific dictatorship is predominantly a conspiracy of ideas. It is a continuity of thought, consciously promulgated within conspiratorial circles and disseminated on the popular level under numerous appellations. At the level of public consumption, the conspiracy tends to become more of an unconscious ideational contagion. In short, the conspiracy "started the ball to rolling" and humanity has been doing the rest since. However, conscious agents of the conspiratorial tradition remain to facilitate the "scientific dictatorship's" gradual migration from abstraction toward tangible enactment. To suggest otherwise is to invoke the hopelessly flawed accidentalist view of history.

The accidentalist perspective of history strains the law of averages considerably. In *None Dare Call it Conspiracy*, Gary Allen observes the occurrence of "32,496 consecutive coincidences" in American history between 1930 and 1970 (9). These included depressions, wars, and racial tensions (9). Accidentalist historians argue that, given all of the variables and complexities of the human condition, such events are to be expected. In less academic language, "Hey, these things just happen." Fair enough. However, Allen correctly argues that such a succession of coincidences "stretches the law of averages a bit" (9).

Given the law of averages, one could expect at least half of these events to produce a favorable outcome for the nation (Allen 8). Surely, if all of America's problems resulted from the simple ineptitude of poor leadership, then their errors would intermittently benefit the citizenry

(Allen 8). Clearly, they have not. This suggests some conscious force at work in political and social dynamics.

Understandably, conspiracy research has not always commanded the serious attention of scholars and orthodox historians. It has been largely relegated to the realm of fantasy, the providence of sensationalistic journalists and socially maladjusted individuals that are prone to paranoid delusions. Certainly, many of those who promulgate so-called "conspiracy theories" fall into one or both of these two categories. Serious researchers labor in the shadows of such people and, as a result, are marginalized.

However, conspiracy is not a fantasy or a joke. It is a natural cognitive function. The mistake made by many researchers is not their recognition of some conspiratorial design intrinsic to the unfolding of history, but their attempts to explain the complexities of that design with simplistic unifying theories. Typically, such unifying theories presuppose the existence of a "master plan." Although there are common conspiratorial objectives and discernible continuities of thought, such a "master plan" does not exist. It is a fiction, justifiably satirized by characters like Imperious Leader of *Rocky and Bulwinkle* fame.

In *The Architecture of Modern Political Power*, Daniel Pouzzner effectively dispels of the myth of a "master plan" and offers a reasonable portrait of the conspiratorial body known as the Establishment:

> The establishment program is not quite a traditional conspiracy. As in many, its members do not all know each other, have sometimes conflicting conceptions of what is to be done, and have sometimes conflicting agendas. From here, the distinctions mount. It is a largely "open" conspiracy, in that much of its membership, structure, methods, and operations, are matters of public record, however scattered and obscure. Its manner of coordination is atypical. (36)

Various researchers have enumerated the machinations comprising the Establishment: the Council on Foreign Relations, the Trilateral Commission, the Bilderberg Group, etc. Yet, many have distorted or embellished upon the nature of these organizations. Ignoring the mystique that many sensationalists bestow upon these organizations, Pouzzner accurately characterizes them as "a network of affiliations and alliances, some strong and some weak, some advertised and some secret" (78). While there is no "master plan," there is a master objective. All of these groups share "a common goal of world rule by oligarchy"

(78). However, their projects in establishing an oligarchical world government exhibit "varying degrees of coordination, coherency, and internal contention" (78).

Pouzzner makes it abundantly clear that this somewhat diffuse network constitutes a conspiracy, "but a largely open one, and one of humans, hence neither monolithic nor unerring" (78). As a "conspiracy of humans," the conspiracy is replete with all of the inherent weaknesses of humanity. Pouzzner explains: "Moreover, the core of the establishment has nothing approaching absolute authority" (78). Pouzzner identifies the most powerful conspiratorial entities as "the first-tier international bankers and the intelligence apparatus they largely control" (78). Yet, even they must occasionally select their orders cautiously and state them in more euphemistic language (78).

Although diffusely connected by the "common goal of world rule by oligarchy," the Establishment has never exhibited a stable unity. Pouzzner elaborates:

> The myriad interlocking subconspiracies one encounters while exploring this compilation are arranged in interlocking hierarchies. There is no clean command hierarchy in general; in fact there is a degree of incoherency and fluctuation in the command topology. Subconspiracies are linked by conspirators who are members of multiple subconspiracies, and these crucial links between pairs of subconspiracies have explicit knowledge of the existence and role of each of those subconspiracies. (78)

Cardinal Stefan Wyszynski divided the various elite factions into "three Internationales" (Martin 21). Given the perpetually shifting command topology of the conspiracy, these categories are not absolute. However, they do establish some loose demarcations, allowing one to develop some general idea of the separate and occasionally conflicting interests comprising the conspiracy. Vatican insider Malachi Martin enumerates these "Internationales":

> There exist on this earth, Wyszynski used to say, only three Internationales. The "Golden Internationale" was his shorthand term for the financial powers of the world—the Transnationalist and Internationalist globalist leaders of the West.

The "Red Internationale" was, of course, the Leninist-Marxist
Party-State of the Soviet Union. . .
The third geopolitical contender—the Roman Catholic
Church; the "Black Internationale". . . (21)

According to Wyszynski, the "Golden Internationale" was further
divided into tow factions: the Transnationalists and the Internationalists
(18). The Transnationalists are, according to Martin, "[e]ntrepeneurial in
their occupations" and endeavor to "exercise their entrepreneurship on
a worldwide basis" (18) Meanwhile, the Internationalists are those who
are proficient in international politics (18). Their overall objective is to
draw the national governments of the world into an interdependent
union (18). Synopsizing the globalist theme underpinning both of these
factions, Martin states:

> In the current competition to establish and head a one-world
> government, Transnationalists and Internationalists can be
> said for all practical purposes to act as one main contender.
> The Genuine Globalists of the West. Both groups are products
> par excellence of the system of democratic capitalism. Both
> are closely intertwined in their membership that individuals
> move easily and with great effect from an Internationalist to a
> Transnationalist role and back again. (18)

Likewise, the enormous conspiratorial body constituting the
crusade to establish a global government could also be considered a
single conspiracy. Pouzzner recapitulates:

> Though it is obviously not monolithic, omnipotent,
> omniscient, or unerring, it is nonetheless obligatory to
> consider the collection of subconspiracies as one single, huge
> conspiracy, protected from itself by compartmentation. It
> is a huge, global network of secretive manipulation, and it
> lurks behind most decisions of political, social, or economic
> consequence. (78)

Likewise, this examination of the scientific dictatorship shall
approach the concept as a single conspiracy. Thus, from now on,
factionalism should be taken as a given. None of the references
to a conspiracy contained herein should be construed to mean a
"monolithic, omnipotent, omniscient, or unerring" plot. Instead, they

should be understood as references to a somewhat diffuse constellation of organizations and groups acting as conduits for elitist interests. The only invariant among these factions is some philosophy of collectivism, which exhibits certain degrees of variance determined by the ideological propensities of the organization's members.

The legitimizing "science" of this emergent dictatorship is couched in the occult doctrines of Freemasonry. However, it is not the contention of this researcher that Masonry is some extremely elaborate conspiratorial network or that all of its members are actively involved in a conspiracy. In *None Dare Call It Conspiracy*, Gary Allen correctly warns against such claims, which are premised upon "racial or religious bigotry" (10-11). In fact, the vast majority of Masons inhabiting the lower echelons of the Lodge are consciously deceived by those occupying the higher layers of organizational strata. Thirty-third Degree Freemason Albert Pike confessed as much in *Morals and Dogma*, the Masonic "bible." First published in 1871, *Morals and Dogma* presents the mandate for secrecy even among fellow Masons:

> The Blue Degrees are but the outer court or portico of the Temple. Part of the symbols are displayed there to the Initiate, but he is intentionally misled by false interpretations. It is not intended that he shall understand them; but it is intended that he shall imagine he understands them. Their true explication is reserved for the Adepts, the Princes of Masonry. The whole body of the Royal and Sacerdotal Art was hidden so carefully, centuries since, in the High Degrees, as that it is even yet impossible to solve many of the enigmas which they contain. (819)

Pike expresses no regrets about this deception. In fact, he candidly voices his approval:

> It is well enough for the mass of those called Masons, to imagine that all is contained in the Blue Degrees; and whoso attempts to undeceive them will labor in vain, and without any true reward violate his obligations as an Adept. Masonry is the veritable Sphinx, buried to the head in the sands heaped round it by the ages. (819)

This revelation should thoroughly demolish the notion that every Mason is a conspiratorial agent. Not every member is privy to the esoteric doctrines or occult secrets of the higher degrees.

In addition, it must be understood that Freemasonry is not the locus of conspiratorial power. However, secret societies such as Masonry typically act as organizational conduits for the implementation of the Establishment's clandestine agendas. Pouzzner elaborates:

> Unknown to the vast majority of Bilderberg, TLC, CFR, COA, and RIIA affiliates of record, at least in its specifics, is a parallel apparatus of covert action, closely intertwined with the intelligence community (Central Intelligence Agency, National Security Agency, etc.), law enforcement community (Department of Justice, Secret Service, American Society for Industrial Security, etc.), certain select secret societies (Skull & Bones, etc.), criminal and labor (Scientology, Unification Church, Nightstalkers and Delta Force, SEALs, etc.), and the middle and lower ranks of various industries (including the top leadership of some front and niche firms, and organizations such as International Executive Services Corporation). This apparatus protects and facilitates the activities of the affiliates, with the subtly menacing promise of accountability. (37)

Freemasonry qualifies as part of this "parallel apparatus of covert action." It and other secret societies function as implementation instruments for the various elite factions of the Establishment. Yet, in regards to the involvement of secret societies in this "parallel apparatus," Pouzzner makes a crucial point:

> Those who are, by design, essentially hidden from public view--including most of the intelligence and law enforcement communities, and obviously, secret societies and criminal syndicates--are directly exposed to the coercive and conspiratorial character of the apparatus, but have no political capital with which to threaten it. (38)

Given its distinct lack of political capital, Freemasonry's influence within the conspiracy is contingent upon the strategic placement of its members and the pervasive nature of the ideas it promulgates. The latter shall be the focus of this discussion.

The Epistemological Cartel

In *The Architecture of Modern Political Power*, Daniel Pouzzner outlines the tactics employed by the elite to maintain their dominance. Among them is "[o]stensible control over the knowable, by marketing institutionally accredited science as the only path to true understanding" (75). Thus, the ruling class endeavors to discourage independent reason while feigning illusory power over human knowledge. This tactic of control through knowledge suppression and selective dissemination is reiterated in the anonymously authored document "Silent Weapons for Quiet Wars":

> Energy is recognized as the key to all activity on earth. Natural science is the study of the sources and control of natural energy, and social science, theoretically expressed as economics, is the study of the sources and control of social energy. Both are bookkeeping systems. Mathematics is the primary energy science. And the bookkeeper can be king if the public can be kept ignorant of the methodology of the bookkeeping. All science is merely a means to an end. The means is knowledge. The end is control. (203)

The word "science" is derived from the Latin word *scientia*, which means "knowing." An elite monopoly of the knowable, which is enforced through institutional science, could be characterized as an "epistemological cartel." The ruling class has bribed the "bookkeepers" (i.e., natural and social scientists). Meanwhile, the masses practically deify the "bookkeepers" of the elite, and remain "ignorant of the methodology of the bookkeeping." The unknown author of "Silent Weapons for Quiet Wars" provides an eloquently simple summation: "The means is knowledge. The end is control. Beyond this remains only one issue: Who will be the beneficiary?" (203).

In *Brave New World Revisited*, Aldous Huxley more succinctly defined this epistemological cartel:

> The older dictators fell because they could never supply their subjects with enough bread, enough circuses, enough miracles, and mysteries.
> Under a scientific dictatorship, education will really work with the result that most men and women will grow up to love their

servitude and will never dream of revolution. There seems to
be no good reason why a thoroughly scientific dictatorship
should ever be overthrown. (116)

This is the ultimate objective of the elite: an oligarchy legitimized
by arbitrarily anointed expositors of "knowledge" or, in Huxley's own
words, a "scientific dictatorship."

The New Theocracy
How did the "scientific dictatorship" of the twentieth century
begin? In earlier centuries, the ruling class controlled the masses
through more mystical belief systems, particularly Sun worship. Yet,
this would all change. In *Saucers of the Illuminati*, Jim Keith documents
the shift from a theocracy of the Sun to a theocracy of "science":

> Since the Sun God (and his various relations, including sons
> and wives) were, after several thousands years of worship,
> beginning to fray around the edges in terms of believability,
> and a lot commoners were beginning to grumble that this
> stuff was all made up, the Illuminati came up with a new and
> improved version of their mind control software that didn't
> depend upon the Sun God or Moon Goddess for ultimate
> authority. (78)

Priests and rituals were soon supplanted by a new breed of
"bookkeepers" and a new "methodology of bookkeeping." Keith
elaborates:

> As the Sun/Moon cult lost some of its popularity, "Scientists"
> were quick to take up some of the slack. According to their
> propaganda, the physical laws of the universe were the ultimate
> causative factors, and naturally, those physical laws were only
> fathomable by the scientific (i.e. Illuminati) elite. (78-79)

Catholic scholar Rama Coomaraswamy identifies nominalism as
one of the chief catalysts for this shift. Coomaraswamy explains the
concept of nominalism:

> While nothing under the sun is new, we shall initiate our sad
> tale with William of Ockham.
> Born in 1290, Ockham is one of the earliest of those who

misunderstood the nature of the soul. He not only denied free will, he also denied that the Intellect was capable of forming universal concepts. He and his followers--usually labeled "nominalists"--claimed that all ideas were really images, that is, impressions on the imagination originating in sensual perception. The error--it is one shared by virtually all modern "philosophers" and psychologists--is that nominalists confound the individualized image of the imagination with the concept or idea which resides in the Intellect. According to St. Thomas, the difference between images and ideas consists in the fact that images are representations of things in their singularity, particularity and concreteness, whereas ideas are representations of things in their universality. Despite his denial of "universals," Ockham continued to believe in God. But he held such belief to have no objective character and the nature of his faith was "blind". I would ask you to remember Faith requires our assent to what the intellect tells us is Truth and it is the nature of this faculty to "see." The acceptance of nominalism precludes such "vision" and inevitably results in a bifurcation between what can be observed and measured, and what is believed. (No pagination)

According to Coomaraswamy, nominalism's epistemological rigidity heralded an overwhelming preoccupation with sense certainty and the ontological plane of the physical universe. Thus, scientific inquiry would eventually be limited to quantifiable entities. Coomaraswamy elaborates:

It [Nominalism] is but a short step to envisioning the measurable as the totality of reality, and the relegating of concepts such as the "good" and the "beautiful"--to say nothing of Revelation--concepts beyond measurement and hence seen as having no objective measurable reality--to the realm of private and subjective convictions where they become whatever we feel or want them to be. (No pagination)

Coomaraswamy contends that nominalist epistemology resulted in the obliteration of metaphysics and the emergence of a scientific deism of sorts:

Man, by his very nature seeks to know the truth, the nature

and purpose of his existence. Nominalism, precludes this possibility. Denying the intellect, it denies that man can abstract from the things of this world and penetrate their underlying reality; being dependent upon phenomena, its only certainties are statistical approximations. Obviously experiential knowledge has its place and function, but once it is declared to be the only legitimate source of knowledge, man is deprived of the absolute and has no access to the nature of his being. Metaphysics is destroyed, sacred knowledge is nullified, and man is forced to turn from Revelation and Intellection to individualism and rationalism. Cut off from what is "above", He must turn to what is "below". It was Descartes who epitomized this deviation in his *Cogito ergo sum*. The individual consciousness of the thinking subject (or more precisely, his ephemeral ego) was proclaimed to be the source of all reality and truth; the knowing subject—-man-- was henceforth bound to the realm of reason as applied to phenomena and separated from both Intellection and Revelation. It is a short step from this to the radical doubt of Hume and the agnosticism of Kant. (No pagination)

Thus, human reason mind was apotheosized and God became a nebulous irrelevancy. This paradigm shift facilitated the rise of the elite's new theocracy. The official state-sanctioned religion of this theocracy was 'scientism': the belief that the investigational methods of the natural sciences should be ecumenically imposed upon all fields of inquiry. In his article "The Shamans of Scientism," Michael Shermer describes scientism as:

a scientific worldview that encompasses natural explanations for all phenomena, eschews supernatural and paranormal speculations, and embraces empiricism and reason as the twin pillars of a philosophy of life appropriate for an Age of Science. (No pagination)

This form of epistemological imperialism is not to be confused with legitimate science. Researcher Michael Hoffman makes this distinction in his book *Secret Societies and Psychological Warfare*:

Science, when practiced as the application of man's God-given talents for the production of appropriate technology on a human scale, relief of misery and the reverential exploration and appreciation of the glory of Divine Providence as revealed in nature, is a useful tool for mankind. Scientism is science gone mad, which is what we have today. (Hoffman 49)

Concerning this important distinction, Rama Coomaraswamy states:

Traditional man, placing science in a hierarchal relationship to the totality of truth, sees no conflict between what is demonstrable by measurement and what he knows from Revelation. His attitude towards the "modern scientistic outlook" with its claim to the totality of truth and its refusal to recognize any moral master is, however, quite another matter. In no way can he give his assent to irrational postulates such as progress, evolution, and the perfectability of man qua man— ideas which have their origin in man's collective subconscious rather than in God. If any conflict exists, it is not between science and faith properly understood, but between modern and traditional attitudes. (No pagination)

Convinced that their outlook encompasses the "totality of truth," the shamans of scientism are overtly hostile towards supernatural explanations. According to their criteria, all inquiry must be restricted to this ontological plane of existence. Shermer succinctly voices this so-called "modern attitude":

. . .cosmology and evolutionary theory ask the ultimate origin questions that have traditionally been the province of religion and theology. Scientism is courageously proffering naturalistic answers that supplant supernaturalistic ones and in the process is providing spiritual sustenance for those whose needs are not being met by these ancient cultural traditions. (No pagination)

Scientism's bestowal of metaphysical primacy upon the ontological plane of the physical universe precludes the knowledge humanity attains through abstraction. Simultaneously, it exalts concepts like "progress, evolution and the perfectability of man" with a religious reverence. While scientific materialists and their fellow travelers (e.g., behaviorists, physicalists, functionalists, secular humanists, Marxists, etc.) relegate texts such as the Biblical Eden account to mere myth, an Edenic motif remains firmly embedded within their own Weltanschauung.

In the beginning of this secular mythology, Eden was a singularity, which was eventually divided into countless pluralities by the Big Bang. According to the myth, the reconstitution of Eden is achieved through evolution, which invariably requires the assistance of Man (spelled with a capital M to signify humanity's potential to achieve apotheosis through the evolutionary process). Man unites evolution with the science of "progress," which is bodied forth through biological methodologies(e. g., eugenics, population control, etc.) and social methodologies (e.g., communism, fascism, and other forms of sociopolitical Utopianism). As evolution is guided down the desired course, Man returns to the singularity (i.e., a world government and a unified consciousness). Thus, Eden is reborn. However, Eden is confined to this ontological plane and immortality is attainable only through the continuity of the species.

If elements of this mythology sound familiar, it is because it is certainly nothing new. It is derivative of ancient occult cosmologies, particularly Gnosticism. The only difference is that the scientistic version stipulates that its Eschaton resides entirely within this physical universe. However, the scientistic myth resembles a religion in every way. This is a reality the shaman of scientism cannot deny, even though their scientistic hubris prevents them from acknowledging it. Shermer candidly delineates the scientist's new role as a mythmaker:

> . . .because of language we are also storytelling, mythmaking primates, with scientism as the foundational stratum of our story and scientists as the premier mythmakers of our time. (No pagination)

Just as the ancient theocratic orders required their myths of the Sun God and Moon Goddess, the contemporary theocratic orders require new myths. However, the new gods and goddesses are designed according to scientistic parameters. Now, they are impersonal, naturalistic entities that pervade the fabric of the physical universe. They work in accordance with human reason and can even be harnessed by man, thus ensuring his primacy over this ontological plane. Of course, a being that can command the natural forces would be tantamount to a god. Like the religious myths of antiquity, the scientistic myths promise such an apotheosis. Rene Guenon comments on the scientistic myth:

Thus it comes about that there has grown up in the "scientistic" mentality. . .a real "mythology": most certainly not in the original and transcendant meaning applicable to the traditional "myths," but merely in the "pejorative" meaning which the word has acquired in recent speech. (151)

Indeed, many scientific concepts are based upon assumptions that are mythic in character. Many contemporary scientists would scorn the commoner for invoking unseen forces to explain visible phenomena. However, scientists are guilty of the same sin. Guenon cites one example:

Endless examples could be cited: one of the most striking and "immediate," so to speak, is the "imagery" of atoms and the multiple elements of various kinds into which they have lately become dissociated in the most recent physical theories(the effect of this is of course that they are no longer in any sense atoms, which literally means "indivisibles," though they go on being called by that name in the face of logic). "Imagery" is the right word, because it is no more than imagery in the minds of the physicists; but the "public at large" believes firmly that real "entities" are in question, such as could be seen and touched by anyone whose senses were sufficiently developed or who had at his disposal sufficiently powerful instruments of observation; is not that "mythology" of a most ingenuous kind? (151-52)

How does this differ from the religious "imagery" invoked by priests and magicians to mesmerize audiences in the past? There is no difference. Yet, in this chronocentric period of human history, there is no scarcity of contempt for the thinkers of antiquity. The dominant scientistic culture still pours "scorn on the conceptions of the ancients at every opportunity" (152). Hoffman further elaborates on the folly of scientism:

The reason that science is a bad master and dangerous servant and ought not to be worshipped is that science is not objective. Science is fundamentally about the uses of measurement. What does not fit the yardstick of the scientist is discarded. Scientific determinism has repeatedly excluded some data from its measurement and fudged other data, such as Piltdown Man, in order to support the self-fulfilling nature of its own agenda, be it Darwinism or "cut, burn and poison" methods of cancer "treatment." (Hoffman 49)

Because of its preoccupation with quantification, science tends to overlook the qualitative aspect of things. In relation to the human condition, the qualitative encompasses abstract concepts such as liberty and dignity. Because such concepts are immune to quantification, they are automatically precluded by science. Thus, an exclusively scientific approach reduces the complex human creature into a grossly oversimplified biological automaton. Concerning this preoccupation with quantification, Rama Coomaraswamy states:

Science deals with measurable phenomena. Its laws resume past experience and its closest approximation to truth is by means of statistical averages. Such a methodology can never establish absolute or objective certainties but only predict that what has happened in the past will probably occur in the future. When the scientist departs from the measurable, when he reasons or speculates about the facts he has gathered, he defines the results as a "working hypothesis" or a "theory." As more facts become known, theories are modified and even radically changed. The conclusions of science are never stable, but rather can be described as a constantly changing "consensus." They are "objective" only in so far as they can be quantitatively demonstrated, but they are never "universal" in the sense that they are absolute or applicable throughout time and space. Those who doubt this have but to look at the innumerable and rapidly changing cosmological theories proposed for our consideration over the last 50 years. Needles to say, those who adhere to the traditional viewpoint can have no argument with measurable fact. (No pagination)

This is the folly of scientism. In ecumenically imposing science upon all fields of inquiry, scientism ignores *qualia*. In such a climate of epistemological rigidity, human liberty and dignity are either limited or absolutely obliterated. Scientism comprises the ideational nucleus of all modern scientific dictatorships.

It must be understood that this new institution of knowing is a form of mysticism like its religious precursors. A society governed exclusively by science amounts to little more than a secular theocracy. Coomaraswamy explains:

> Unfortunately modern man sees science, not as a specialized kind of knowledge about the material world we live in, but as an almost "mystical" concept encompassing his most cherished convictions; his belief in evolution, progress, and that all reality is subjective, measurable, and centered on man qua man. For him, what science cannot measure and explain with its limited methodology simply doesn't exist--all that is knowable is encompassed within its aegis. For modern man the scientist has replaced the priest, and when he speaks--even if it be outside the realm of his competence--his words are imbued with quasi-divine authority. Everything modern man believes in--be it hygiene, socialism or modern psychology--is described as "scientific," an adjective which seemingly endows its subject with the quality of truth and objectivity. It is to Science that we are directed in seeking a solution to our every problem. Modern man often proclaims his belief in science, and well he should, for science, or rather scientism, has become his religion. He accepts its fuzzy "dogmas," not because they are rational or intellectually compelling, but because he feels they are true. Such a faith is "visceral" and "blind." It defies definition and can be described as an "immenantist awareness finding its source in the subconscious". (No pagination)

Indeed, Science (spelled with a capital S to signify its virtual apotheosis) is the new religion and scientists have become its priesthood. Shermer candidly recapitulates:

> . . .we are, at base, a socially hierarchical primate species. We show deference to our leaders, pay respect to our elders and follow the dictates of our shamans; this being the Age of Science, it is scientism's shamans who command our veneration. (No pagination)

Contemporary science is predicated upon empiricism, the epistemological stance that all knowledge is derived exclusively through the senses. Yet, an exclusively empirical approach relegates cause to the realm of metaphysical fantasy. This holds enormous ramifications for science. What is perceived as *A* causing *B* could be merely a consequence of circumstantial juxtaposition. Although temporal succession and spatial proximity are axiomatic, causal connection is not. Affirmation of causal relationships is impossible. Given the absence of causality, all of a scientist's findings must be taken upon faith. Ironically, science relies on the affirmation of such cause and effect relationships.

Certainly, one could argue that empiricism is probabilistic in nature and, therefore, acknowledges probable causal connections. Yet, this researcher would contend that the probabilistic nature of empiricism is still mystical in character because the scientist must invest some measure of faith in the probability of certain outcomes.

For instance, consider the law of gravity. In the absence of causal connection, the law of gravity would no longer qualify as a law in any way. It would be premised upon faith in the probability of an object dropping to earth, an outcome that is still uncertain in a world without causality. In addition, scientists with more dogmatic propensities insist so strongly on the alleged reality of certain theories that one would question whether probability weighed upon their empirical observations at all.

For example, consider the insistence of some neo-Darwinians that a particular anatomical structure is definitely a vestigial organ and, thus, constitutes proof of evolution. This is a statement of certainty, which is impossible to make in a universe ruled by probability. Empiricism, probabilistic though it may be, requires some degree of faith and qualifies as a form of mysticism.

Returning to Pouzzner's previous statement, "ostensible control over the knowable" is achieved through the promulgation of "institutionally accredited science" (Pouzzner 75). Now, the elite had to meet two requirements to insure their epistemological dominance: a science specifically designed for their needs and an institution to accredit and disseminate it.

Epistemic Autocracy Defined

Epistemic autocracy is, essentially, the primacy of one epistemology that is politically and socially expedient to a ruling elite. Throughout the 20th century, the dominant epistemology has been empiricism. Empiricism works in tandem with the nominalist contention that "all ideas were really images, that is, impressions on the imagination originating in sensual perception" (Coomaraswamy, no pagination). Nominalism rejects man's ability to "abstract from the things of this world and penetrate their underlying reality," thus destroying metaphysics (no pagination). With metaphysics destroyed, anti-metaphysical views prevail. According to Rene Guenon, such anti-metaphysical views are:

> . . .known more especially in their philosophical aspect by such names as "pantheism," "immanentism" and "naturalism," all of which are closely interrelated, and many people would doubtless recoil before such a consequence if they could know what it is they are really talking about. (288)

Guenon argues that this climate of anti-metaphysical thought furthers the objectives of a movement called the "counter-initiation," which:

> works with a view to introducing its agents into "pseudo-initiatic" organizations, using the agents to "inspire" the organizations, unperceived by the ordinary members and usually also by the ostensible heads, who are no more aware than the rank-and-file of the purpose they are really serving; but it is well to add that such agents are in fact introduced in a similar way and wherever possible into all the more exterior "movements" of the contemporary world, political or otherwise, and even as was mentioned earlier, into authentically initiatic or religious organizations, but only when their traditional spirit is so weakened that they can no longer resist so insidious a penetration. (293-94)

Freemasonry, which Pike candidly characterizes as "the veritable Sphinx, buried to the head in the sands heaped round it by the ages," qualifies as one of these "'pseudo-initiatic' organizations." Masonry's core doctrines and principles are "unperceived by the ordinary members." The organization promulgates an internal culture of obscurantism, which keeps the "rank-and-file" unaware of the "purpose they are really serving." The deception is compounded by the fact that higher initiates are encouraged to conduct semiotic warfare on lower initiates. Lower initiates must go through a labyrinthine series of rituals, the meanings of which are veiled by esoteric symbols. In so doing, the Lodge unconsciously indoctrinates its members into "pantheism," "immanentism," and "naturalism." According to Guenon, these concepts:

. . .quite literally amount to an "inversion" of spirituality, to a substitution for it of what is truly its opposite, since they inevitably lead to its final loss, and this constitutes "satanism" properly so-called. Whether it be conscious or unconscious in any particular case makes little difference to the result; it must not be forgotten that the "unconscious satanism" of some people, who are more numerous than ever in this period in which disorder has spread into every domain, is really in the end no more than an instrument in the service of the "conscious satanism" of those who represent the "counter-initiation." (288-89)

With the "counter-initiation" flourishing in this age of anti-metaphysical thought, society is witnessing the rise of what could be dubbed the "Satanic State." Empiricism is part of the anti-metaphysical philosophical climate that is facilitating this rise. Empiricism's rejection of causality allows the power elite to rationalize the philosophically impossible position of atheism. Rejection of causality invariably leads to the rejection of a First Cause, which is God. The popularization of atheism has been accompanied by secularization. However, as sociologist William Sims Bainbridge makes clear, secularization does not represent the complete obliteration of religion. Instead, it represents the opening stage of an occult counterculture movement:

Secularization does not mean a decline in the need for religion, but only a loss of power by traditional denominations. Studies of the geography of religion show that where the churches become weak, cults and occultism explode to fill the spiritual vacuum. ("Religions for a Galactic Civilization," no pagination)

With the Prime Mover philosophically dethroned, a new moral authority could be erected. Of course, this moral authority would be designed according to the precepts of the ruling class doctrine. Without exception, the newly designed deity promoted by the elite has always been the omnipotent State. Astride the state is the apotheosized "Reason" of man. This is secular humanism, which is really Luciferianism disseminated on the popular level (see Chapter Three: Luciferianism: The Religion of Apotheosis for further explication).

It comes as little surprise that the scientific dictatorship, which initially supplanted the dominant ecclesiastical authorities of history, will be invariably transmogrified back into a theocratic power structure. Empiricism, which is one of its core doctrines, amounts to little more than academically dignified occultism. Although ostensibly dissimilar with the more mystical epistemologies of the past, the doctrine of sense certainty was no less mystical in character. This becomes evident in empiricism's ultimate rejection of causality. In the absence of causation, all scientific findings must be taken upon faith. In effect, the scientist becomes the veritable priest, bestowing validity upon whatever conclusions he or she deems fit. Those conclusions that contradict the epistemic orthodoxy are deemed heretical and summarily disregarded.

The elite's epistemic autocracy was firmly established through the British Royal Society, a Masonic institution that would exercise considerable power over scientific knowledge for many years to come.

The British Royal Society

The new secular church and clergy of the elite originated within the walls of the British Royal Society. Established in 1660 under the complete name of the Royal Society of London for Improving Natural Knowledge, it remains the world's oldest scientific institute ("Royal Society"). The creators of the Royal Society were also members of the Masonic Lodge. According to Baigent, Leigh, and Lincoln in *Holy Blood, Holy Grail*:

> Virtually all the Royal Society's founding members were Freemasons. One could reasonably argue that the Royal Society itself, at least in its inception, was a Masonic institution—derived, through Andrea's Christian Unions, from the "invisible Rosicrucian brotherhood." (144)

Jim Keith makes it clear that the Masonic Lodge "has been alleged to be a conduit for the intentions of a number of elitist interests" (*Casebook on Alternative Three*, 20). In service to the elite, the Royal Society Freemasons would re-sculpt epistemological notions and disseminate propaganda. Jim Keith provides a brief summation of the Royal Society's role in years to come: "The British Royal Society of the late seventeenth century was the forerunner of much of the media manipulation that was to follow" (*Saucers of the Illuminati*, 79).

Certainly, one could argue that the Royal Society itself was not entirely conspiratorial in character. One might suggest that the early Royal Society members were mainly naive Baconians who believed in an oversimplified epistemology of empirical science devoid of intention, devoid of hypotheses (Newton's *hypothesis non fingo*). No doubt, many of the later Royal Society members fell into this category.

However, the Royal Society possessed an inner circle that was slightly less altruistic. In 1864, Freemason and Royal Society member T.H. Huxley established the X Club, a group of nine men wielding "personal influence on almost every famous scientist in the world, as well as on many distinguished radicals" (Taylor 189). Interestingly enough, Adrian Desmond and James Moore characterize the X Club as "a sort of masonic Darwinian lodge, invisible to outsiders" (526). The X Club's invisibility to outsiders suggests an organizational culture of obscurantism. It is possible that, as an appendage of the Masonic Royal Society, the X Club inherited the Craft's tradition of secrecy. The Masonic pedigree of the Club's founder, T.H. Huxley, reinforces this contention.

With the exception of Herbert Spencer, its members "were all presidents and secretaries of learned societies" (Taylor 189). Although George Lyell and Charles Darwin were not members, their ideas enjoyed the X Club's greatest respect (Taylor 189). Dining together before each Royal Society meeting, the X Club formulated tactics for controlling the scientific press (Taylor 189). From 1864 to 1884, Huxley and his associates "literally 'governed'" British science in this fashion(Taylor 189). Herein was the "media manipulation" mentioned earlier by Jim Keith.

The case of *Bathybius haeckelii* most effectively illustrates the conspiratorial character of the X Club. In 1868, T.H. Huxley claimed to have discovered infinitesimal creatures in samples of mud removed from the North Atlantic (Taylor 187). Having been stored in alcohol, the "primitive organisms" in Huxley's sample were obviously dead (Taylor 187). Nevertheless, he believed that the protoplasmic matter was the missing *Monera* in Ernst Haeckel's phylogenetic tree (Taylor 187). Dubbing the microscopic species *Bathybius haeckelii*, Huxley claimed that they were the missing transitional forms between inorganic matter and organic life(Taylor 187-88). This constituted a victory for abiogenesis, which was probably a concept derived from the occult doctrines of Masonry's Kabalistic heritage (to be discussed shortly).

In 1875, Huxley's claim was thoroughly refuted by a chemist aboard the HMS Challenger (Taylor 188). Observing that the addition of seawater to alcohol created an "amorphous precipitate of sulphate lime," the chemist discovered that *Bathybius* was really gypsum (Taylor 188). It was merely a lifeless rock, not primordial life. However, Huxley and his Masonic colleagues were unwilling to relinquish their epistemic primacy. The X Club exerted its considerable influence over the scientific media and almost completely suppressed the discovery (Taylor 189). An obscure report did appear in the *Quarterly Journal of the Microscopical Science* (Taylor 189). Yet, no public statement concerning the fiasco was made (Taylor 189).

Worse still, Haeckel's unrevised and unabridged *History of Creation* remained in circulation for fifty years (Taylor 189-90). Until 1923, this book featured pictures of *Bathybius* and presented the lifeless rock as the evolutionary segue for primordial life (Taylor 452). There is a name for the perpetuation of a lie. It is called "propaganda." Hyperbolic though it may sound, Huxley and his nine other colleagues represented a cabal of propagandists. Simply stated, the X Club was a media cartel. It also controlled the Royal Society (Taylor 189). This organizational model--a small, elite group presiding over a larger body of oblivious adherents--resembled the Masonic directorial framework.

In fact, the Royal Society may have served as a recruiting ground for Masonic groups that were currently involved in subversive activity throughout the European continent. Among one of these groups was the infamous Illuminati (which shall be examined in greater detail later). The first real whistleblower to give outsiders a glimpse at the inner workings of the Illuminati was Professor John Robison (Griffin 44). Robison was the Secretary General of the Royal Society at Edinburgh (Griffin 44). Believing that Robison would make an excellent addition to the secret society, an illuminist invited Robison to join the Illuminati (Griffin 44).

Robison pretended to be receptive, while secretly believing that the organization had nefarious intentions (Griffin 44-46). Thus, he was able to win the group's trust and was entrusted with several documents that were never intended for public consumption (Griffin 46). This lead to Robison's own 1797 book, *Proofs of a Conspiracy*, which revealed many of the Illuminati's secrets (Griffin 46). While condemning the Illuminists, Robison never suspected the involvement of either Masonry or the Royal Society in a subversive activities. He was only partially correct. Not every Royal Society mason may have been a conspirator. However, not every Royal Society mason was a naive Baconian either.

Moreover, the Baconian tradition itself was not one of mere scientific inquiry. It was replete with the secrecy and elitism inherent to conspiracies. Sir Francis Bacon was a member of a secret society called the Order of the Helmet (Howard 74). The organization's name was derived from Pallas Athene, the Greek goddess of wisdom who

was portrayed wearing a helmet (Howard 74). Although regarded as
an innovator of science by orthodox academia, Bacon's studies mostly
embraced occultism. In his youth, Bacon was "a student of Hermetic,
Gnostic, and neo-Platonist philosophy and had studied the Cabbala"
(Howard 74).

Allegedly, Bacon was also a Grand Master of the secret Rosicrucian
Order (Howard 74). The Rosicrucians were closely associated with
Freemasonry (Howard 50). In fact, a Rosicrucian poem written in
1638 voices the organization's close ties with the Lodge (Howard 50).
It reads, "For what we pessage is not in grosse, for we brethen of the
Rosie Crosse, we have the Mason's Word and second sight, things to
come we can foretell aright. . ." (qutd. in Howard 50). In other words,
Rosicrucians knew the "inner secrets of Freemasonry and possessed the
psychic power to predict the future" (Howard 50).

In 1627, Bacon published a novel entitled *The New Atlantis* (Howard
74). The pages of Bacon's book were adorned with Freemasonic symbols,
such as "the compass and square, the two pillars of Solomon's temple and
the blazing triangle, and the eye of God, indicating his association with
the secret societies who supported his Utopian concepts" (Howard 75).
The novel "describes the creation of the Invisible College advocated in
Rosicrucian writings" (Howard 74). This Rosicrucian mandate for an
"Invisible College" was realized with the formation of the Royal Society
in 1660 (Howard 57).

Author Frank Fischer provides a most elucidating description of
Bacon's "Utopian concepts":

> For Bacon, the defining feature of history was rapidly becoming
> the rise and growth of science and technology. Where Plato
> had envisioned a society governed by "philosopher kings,"
> men who could perceive the "forms" of social justice, Bacon
> sought a technical elite who would rule in the name of
> efficiency and technical order. Indeed, Bacon's purpose in *The
> New Atlantis* was to replace the philosopher with the research
> scientist as the ruler of the utopian future, New Atlantis was a
> pure technocratic society. (66-67)

A technocratic society, or Technocracy, can be defined as follows:

> Technocracy, in classical political terms, refers to a system of
> governance in which technically trained experts rule by virtue

of their specialized knowledge and position in dominant political and economic institutions. (Fischer 17)

Oxford Professor Carroll Quigley also wrote about a dictatorship of "experts," suggesting that a cognitive elite "will replace the democratic voter in control of the political system" (Quigley 866). Of just such a democracy of "experts," Freemason H.G. Wells stated:

> The world's political organization will be democratic, that is to say, the government and direction of affairs will be in immediate touch with and responsive to the general thought of the educated whole population. (26)

Literary critic and author W. Warren Wagar comments on this statement:

> Read carefully. He did not say the world government would be elected by the people, or that it would even be responsive to the people--just to those who were "educated." (Wells, *The Open Conspiracy: H.G. Wells on World Revolution*, 26)

Bacon's Utopian vision--a technocratic world government ruled by "experts," particularly scientists--was a "scientific dictatorship." There is evidence to suggest that this elitist vision continued within the Royal Society. Darwin's maternal grandfather, Josiah Wedgwood, was "one of the technocrats inside Birmingham's elite industrial circle, the 'Lunar Society'" (Desmond and Moore 7). This elite circle of technocrats was active from about 1764 to 1800 and its prominent influence "continued long afterwards under the banner of The Royal Society" (Taylor 55). The Royal Society would retain the elitist character of its progenitor. Membership in the organization was considered "largely a privilege of the wealthy, well-connected, scientific elite" (Desmond and Moore 279).

The technocratic character of the Royal Society is most effectively illustrated by the "science" it vigorously promoted: Darwinism. A central feature of Darwin's evolutionary theory is natural selection. Ian Taylor observes that "the political doctrine implied by natural selection is elitist, and the principle derived according to Haeckel is 'aristocratic in the strictest sense of the word'" (411).

Saint-Simon: Father of Technocracy

Bacon's most immediate successor in the development of technocratic theory was Henri de Saint-Simon. Fischer states: " . . .Saint-Simon's work can be interpreted as a prescription for Bacon's prophecy" (69). E.H. Carr characterizes Saint-Simon as "the precursor of socialism, the precursor of the technocrats, and the precursor of totalitarianism" (2). Saint-Simon's philosophy was pure scientism and his vision for a Utopian society was premised entirely upon scientistic precepts. Fischer explains:

> In his [Saint-Simon's] view, a new unity based upon an all-encompassing ideology had to be forged. Only a belief in science and technology could replace the divisive ideologies prevalent at the time, particularly those of the church. In short, priests and politicians—-the older rulers of Europe—- had to be supplanted by scientists and technicians. (69)

To achieve such an end, Saint-Simon proposed a scientific dictatorship called the "Administrative State" (69). This new form of governance would eradicate "competing political interests" and supplant them with a system of "expert management" (69). "Scientists and technicians" would constitute this apolitical system of bureaucracy (69). Of course, Saint-Simon's Utopian vision was inherently anti-democratic. An apolitical system precludes democratic functions such as voting and representative governance. Because science is predominantly a system of quantification, a society governed under its principles would have to conform to the rigid parameters of reductionist epistemology. Of course, humanity's irreducible complexity does not readily lend itself to a reductionistic criterion of governance. Nevertheless, Saint-Simon clung to his religious conviction that humanity needed to "abandon mass democracy and, in turn, politics" (Segal 63).

Saint-Simon's technocratic philosophy eventually became a revolutionary ideology devoted to the transformation of society into a scientific dictatorship. According to James H. Billington, the Saint-Simonian movement underwent two developmental phases, the first of which being the "scientistic period" (211). Billington states:

> The scientistic phase of Saint-Simonian thinking grew directly out of the activities of the first people to call themselves "ideologists." Destutt de Tracy, who first popularized the term "ideology" in 1796-67, suggested in the first part of his

Elements of Ideology in 1801 that traditional metaphysics must be superseded by "ideology," a new method of observing facts, inferring consequences, and accepting nothing not suggested directly by sensation. Building on the tradition of Locke, and of Condillac's *Treatise on Sensation* of 1775, de Tracy maintained that all thinking and feeling were physical sensations in the strictest sense of the word. (211-12)

Automatically, astute readers will identify the theme of sense certainty, which is a hallmark of epistemic autocracy. Out of this rigid epistemology would arise the radical concept of a "science of man," which seems to have found expression in the modern social sciences. Billington explains:

Henri de Saint-Simon extended this radical empiricism into the altogether new field of social relations. Having spent eleven months in prison during the Reign of Terror, expecting death at every moment, Saint-Simon had a deep fear of revolution. He dreamed of founding a new science of man as a means both to overcome disorder and to remove the overgrowth of false political rhetoric, which concealed the real, material questions of society. Thus, ironically, this aristocrat of the ancien regime seeking to provide (in the words of one of his titles) *the means for bringing an end to the revolution*, ended up popularizing the most revolutionary of all modern ideas: there cane be a science of human relations. (212)

It certainly is ironic. Saint-Simon's theoretical "*means for bringing an end to the revolution*" would contribute to the philosophical foundations of the various revolutionary movements promoting a scientific dictatorship. These movements would include every radical socialist group of modernity. Commenting on Saint-Simon's revolutionary resume, Billington characterizes Saint-Simon as:

a father of socialism as well as sociology, and a John the Baptist of revolutionary ideology, crying out in the Wilderness of the Napoleonic restoration eras for a new historicism and moral relativism. (214)

Yet, Saint-Simon still hoped to see revolution eventually stifled. The means to such an end, in his mind, was the scientific regulation of human behavior. In other words, a scientific dictatorship. This was also the sentiment of Aldous Huxley. Although they are separated by vast gulfs of times, Saint-Simon and Huxley both share one core contention: Under the efficient management of a scientific dictatorship, people would "grow up to love their servitude and will never dream of revolution."

Saint-Simon's technocratic philosophy was accepted by the adherents of Destutt de Tracy, which were known as "ideologists" (Billington 212). It is with the ideologists that the "scientistic period" of the Saint-Simonian movement began (211). The group responsible for the management and control of the ideologists was known as *ideologiste* (212). The members comprising this group contended that "the key to diagnosing and curing the ills of humanity lay in an objective understanding of the physiological realities that lay behind all thinking and feeling." (212)

This interpretation of society would be integral to the Saint-Simonian vision of a scientific dictatorship. It also presaged the primacy of physiology in the modern social sciences (e.g., materialist, physicalist, and functionalist philosophies of mind). Once enshrined in the halls of psychiatry and psychology, this physiological tradition would birth such atrocities as lobotomies, pharmacological manipulation, and electroshock "treatment." Convinced that all disorders are symptomatic of physiological abnormalities in the brain, many contemporary doctors continue to employ crude physical methods to "cure" psycho-spiritual conditions. This is the technocratic heritage that the social sciences owe to Saint-Simon. (See Chapter Three: <u>Maschinenmenschen: From Autonomous to Automaton</u> for further explication)

Saint-Simon presented the following strategy for the effective transformation of society into a fully functional Technocracy:

> Scientific indoctrination and intellectual unity were to be provided in a new "positive encyclopedia" on which he [Saint-Simon] worked from 1809 to 1813. His *Essay on the Science of Man* in 1813 suggested that every field of knowledge moved successively from a conjectural to a "positive" stage, and that the sciences reached this stage in a definite order. Physiology had now moved into a positive stage, just as astrology and alchemy had previously given way to astronomy and chemistry. Now the science of man must move towards the positive stage and completely reorganize all human institutions. (213)

This physiological approach to governance is a theme echoed by various socialist totalitarian regimes. It provided the theoretical groundwork for Marxism. Billington explains:

> Believing that the scientific method should be applied to the body of society as well as to the individual body, Saint-Simon proceeded to analyze society in terms of its physiological components: classes. He never conceived of economic classes in the Marxian sense, but his functional class analysis prepared the way for Marx. (213)

Saint-Simon's physiological analysis of society also inspired the scientific dictatorship of Nazi Germany. Ernst Haeckel, the famous evolutionist responsible for Hitler's introduction to social Darwinism, openly espoused this physiological view. He contended that each cell of an organism, "though autonomous, is subordinated to the body as a whole; in the same way in the societies of bees, ants, and termites, in the vertebrate herds, and in the human state, each individual is subordinate to the social body of which he is a member" (qutd. in Keith, *Casebook on Alternative Three* 157). Herein is the central theme of all socialist totalitarian regimes: the subordination of the individual to the collective. Yet, there is always an "elite" that occupies the developmental capstone of the physiological state. For Haeckel, it was the mythical Aryan that exhibited "symmetry of all parts, and that equal development, which we call the type of perfect human beauty" (qutd. in Keith, *Casebook on Alternative Three* 85).

The Saint-Simonian vision for a scientific dictatorship was accompanied by a new scientistic religion. Dubbed the "New Christianity," this scientistic faith offered "morality without metaphysics" and "technology without theology" (214). This faith "represented the culmination of the *ideologiste* attempt to supplant all religion by absorbing it into a progressive scheme of secular evolution" (215). The mission to replace religion with "secular evolution" has been the mission of every scientific dictatorship of the 20th century. Again, both communist Russia and Nazi Germany stand as prime examples. (See Chapter One: The Rise of Modern Scientific Dictatorships for further explication)

Saint-Simon hoped that his "New Christianity" would divorce governance from politics, resulting in an apolitical system of "expertise." Billington writes:

Political authority was to be replaced by social authority in his [Saint-Simon's] technocratic utopia. It was to be administered by three chambers: Inventions run by engineers, Review run by scientists, and Execution run by industrialists. A Supreme college was to draw up physical and moral laws, and two even higher academies, Reasoning and Sentiment, were to be filled by a new breed of propagandistic writer and artist. (215)

The ideologiste's plan for the installation of the "New Christianity" was something of a precursor to the scheme of socialist theoretician Antonio Gramsci. Gramsci's program involved a subtle form of semiotic deception. Traditional religious institutions were to be gradually eviscerated through socialist propaganda and inculcation. However, the standard religious iconography was to be left in place. As God slowly vanished, the omnipotent state was apotheosized. Likewise, the Saint-Simonian program entailed a Fabian strategy for the ritualistic enthronement of Technocracy. The "New Christianity" was meant to be an outgrowth of religion's eventual subsumption under science. Billington elaborates:

> In his commentary of 1802 Francois Dupis's *The Origins of All the Cults of Universal Religion* de Tracy suggested that past religions were not simply senseless superstition, but rather a kind of scientific baby talk: the generalized expression in imprecise language of the scientific thought of the age. Religious ritual was, moreover, socially necessary to dramatize scientific principles for still-ignorant people. Saint-Simon viewed his *New Christianity* as just such a necessity for the masses. His death left it unclear whether this faith was designed to provide the moral basis for the new social order or merely an interim faith until the masses were educated to accept a totally scientistic system. (215)

As is evidenced by later plans such as Gramsci's, the Saint-Simonian vision of a "totally scientistic society" accompanied by a purely scientistic religion survived. A new theocratic order had been born and the infallible scientist constituted its priesthood. This religion is still alive, as is evidenced by Shermer's contention that "it is scientism's shamans who command our veneration."

As a tangible enactment of Bacon's "Invisible College," the British Royal Society qualifies as a part of this scientistic tradition. As Fischer previously stated, Bacon's *New Atlantis* found its prescription in the work of Saint-Simon. With the creation of the Royal Society, the New Atlantis drew closer to incarnation. The vision always remained the same: a totally scientistic society.

Sociology: The Science of Control

Auguste Comte was the "principal disciple" of Saint-Simon (Fischer 70). Building upon the scientistic concepts of his mentor, Comte developed ideas that:

> proved to be very influential in the rise of modern sociology. Many call him the father of the discipline, a fact that also underscores the technocratic origins of modern social science itself. (71)

The technocratic character of sociology is illustrated by the field's inherent scientism. In sociology, scientism assumes the alias of positivism, Comte's philosophical conviction that:

> . . .all sciences (including the social sciences) should be based on rigorous observation and the scientific calculation of the mathematical laws that governed the world. (Dowbiggin 11)

In positivism, one immediately discerns the sort of radical empiricism that Saint-Simon extended to the emergent field of social relations. One can also identify scientism's characteristic preoccupation with quantifiable entities. This sort of thinking would underpin the belief systems of every contemporary socialist totalitarian:

> Twentieth-century liberals' statist and corporatist bent, as well as their confidence in reform, government interventionism, and technocratic elites, can be traced back to the Comtean tradition of the previous century. (11)

Comte's positivism contended that societies experienced three developmental stages: religious, metaphysical, and scientific (Thio 9). Comte asserted that the religious and metaphysical stages were marked by a "reliance on the superstition and speculation" (9). Of course, such thinking is consistent with materialism, which precludes the existence of supra-sensible entities. It also synchronizes with nominalism, which promulgates anti-metaphysics by rejecting man's inherent ability of abstraction. All that remains is the ontological plane of the physical universe. The stage is set for the metaphysical claim of "self-creation" and its corresponding Gnostic claim of "self-salvation." This theme underpins the philosophy of almost every contemporary sociopolitical utopian.

Remaining true to Saint-Simon's scientistic heritage, Comte bestowed epistemological primacy upon the scientific stage. This last developmental stage would witness the rise of a technocracy where "sociologists would develop a scientific knowledge of society and would guide society in a peaceful, orderly evolution" (9). Comte called this new social order a "sociocracy" and promoted it as "religion of humanity" (Fischer 71). Astute readers will automatically discern echoes of Saint-Simon's "New Christianity." That a new theocratic order would be preceded by a secular one is hardly a consequence. Again, secularization is merely a segue for the installation of a new religion and a new religious authority. According to Comte's utopian vision, the social scientist constituted the inner priesthood of the new religious authority. Fischer explains:

> Sociologists were to identify the principles of this new faith and to implement them through a "sociolatry." The sociolatry was to entail a system of festivals, devotional practices, and rites designed to fix the new social ethics in the minds of the people. In the process, men and women would devote themselves not to God (deemed an outmoded concept) but to "Humanity" as symbolized in the "Grand Being" and rendered incarnate in the great men of history. (71)

Thus, guided by the ecclesiastical authority of sociology, humanity would continue its evolution onward and upwards. In the "Grand Being," one may discern echoes of the Masonic concept of the Great Architect. It also reiterates the monistic ideas of Jung, Hegel, Wells, and others. All of these represent variants of evolutionary pantheism. Like classic pantheism, evolutionary pantheism depicts God as an immanent force pervading the fabric of the physical universe. However, evolutionary pantheism marries immanentism with Darwinian transformism. The resulting religion is a scientistic faith in progress. God becomes the emergent deity of man, who gradually migrates towards apotheosis through the process of evolution. Not surprisingly, sociologists like Emile Durkheim and Herbert Spencer advanced similar monistic theories.

Sociology's inherent scientism reached fanatical levels during the 1920s. This period witnessed the rise of "scientific" sociology, which was also called "objectivism" (Bannister 3-4). Predictably, objectivism was preoccupied with quantifiable entities and empirical observations. "Objective" sociology restricted its inquiries entirely to "the observable externals of human behavior, thus carrying to a logical conclusion the strict inductionism of sensationalist psychology" (3). Following Destutt de Tracy's example, sociologists were increasingly supplanting traditional metaphysics with an "ideology" premised on sense certainty. Robert Bannister expands on this anti-metaphysical approach:

> Epistemologically, it [objectivism] rested on the conviction that experience is the sole source of knowledge; ontologically, on a distinction between objects accessible to observation (about which knowledge is possible) and those not accessible (about which there can be no knowledge). (3)

Naturally, the reduction of man's every thought and emotion to physical sensations is accompanied by the physiological interpretation of society. After all, if the purely sensate man is but a microcosm of the social whole, then it is reasonable to assume that the macrocosm is governed by the same physiological principles. Thus, human civilization became the "social organism." The individual became little more than a cell, biologically subordinated to the physiological whole. Obviously, this collectivistic depiction of society underpins the philosophy of every contemporary scientific dictatorship.

Of course, the animal of society requires a zookeeper. This was precisely the role that the sociologist was designed to fill. Sociology is the art of social engineering. For the social scientist, man is a reactive animal to be conditioned. Deceased philosopher Michel Foucault feared just such behavioral tyranny. Foucault correctly observed the tendency of the social sciences to arbitrarily divide knowledge into rigid categories that distort reality. In so doing, the social scientist inoculates society with the very maladies that he or she claims to treat.

According to Foucault, the social sciences were instrumental in the exercise of *bio-power*. Bio-power is a form of social control through the imposition of technocratic organizational frameworks upon society. The technocratic character of these frameworks becomes evident in their ostensible emphasis upon efficiency and rationality. While institutions that exercise bio-power peddle rhetoric about productivity and effective resource allocation, the underlying motive for their operational protocol is the control of the individual. Sociology, in effect, attempts to scientifically legitimize this system of control. Hubert Dreyfus and Paul Rabinow describe this unholy alliance:

> The advance of bio-power is contemporary with the appearance and proliferation of the very categories of anomalies—-the delinquent, the pervert, and so on—-that technologies of power and knowledge were supposedly designed to eliminate. The spread of normalization operates through the creation of abnormalities which it then must treat and reform. By identifying anomalies scientifically, the technologies of bio-power are in a perfect position to supervise and administer them.
> This effectively transforms into a technical problem—-and thence into a field for expanding power—-what might otherwise be construed as a failure of the whole system of operation. Political technologies advance by taking what is essentially a political problem, removing it from the realm of political discourse, and recasting it in the neutral language of science. Once this is accomplished the problems have become technical ones for specialists to debate. . .When there was resistance, or failure to achieve its stated aims, this was construed as further proof of the need to reinforce and extend the power of the experts. A technical matrix was established. By definition, there ought to be a way of solving any technical problem. Once this matrix was established, the spread of bio-power was assured, for there was nothing else to appeal to; any other standard could be shown to be abnormal or to present merely technical problems. (195-96)

In effect, the social sciences created a self-perpetuating system of technocratic control. The disease and the purported "cure" were created in the same laboratory. The social engineers cultivated criminality and then presented themselves as the only solution. Once sociology was dignified as a legitimate blueprint for the management of the state, the dialectic of criminal deviance followed by police state policies was empowered. Oligarchs have a vested interest in maintaining this dialectic. Thus, the ruling class of today has re-engineered America's judicial system so that it actually cultivates criminality (see Chapter Three: Cultivating Criminality for further explication).

At first, bio-power was exercised in the military and prisons. However, it eventually metastasized, becoming formally institutionalized in schools, factories, hospitals, and other social organizations. Accompanying the extension of bio-power was the diffusion of disciplinary technologies. The most prevalent of these was Jeremy Bentham's Panopticon, which Foucault characterized as "the diagram of a mechanism of power reduced to its ideal form. . .a figure of political technology that may and must be detached from any specific use" (205). The power elite's appropriation of Bentham's panoptic schema as a defining feature of their societal construct illustrates the continuity of the Enlightenment's scientistic tradition throughout the ruling class conspiracy (see Chapter Two: The Report From Iron Mountain for further explication).

While social engineering is hardly an American invention, the concept gained substantial political and social capital in the United States. Lester Ward, who is considered the founder of American sociology, believed that sociology was far more than a "fact-gathering" enterprise (Bannister 13). He contended that "its goal is a radical 'sociocracy,' not the palliatives that pass for social reform" (13). Ward's "radical 'sociocracy'" began to take shape shortly after World War II. Many American social scientists were also alumni of the Office of Strategic Services, which plagiarized and refined Nazi psychological warfare techniques. With their Nazi counterparts vanquished, OSS social scientists diffused themselves throughout civilian institutions and commandeering several strategically sensitive positions. These included positions in the mass media and tax-exempt foundations (see Chapter Three: The Social Scientific Dictatorship for further explication). With social engineers firmly entrenched within America's informational infrastructure, a "radical 'sociocracy'" began to ascend in the West.

The Gnostic Division Between Science and Theology
Before the advent of the British Royal Society, science (i.e., the study of natural phenomena) and theology (i.e., the study of God) were inseparable. The two were not separate repositories of knowledge, but natural correlatives. In *Confession of Nature*, Gottfried Wilhelm Leibniz established the centrality of God to science. According to Leibniz, the proximate origins of "magnitude, figure, and motion," which constitute the "primary qualities" of corporeal bodies, "cannot be found in the essence of the body" (de Hoyos).

Linda de Hoyos reveals the point at which science finds a dilemma:

> The problem arises when the scientist asks why the body fills this space and not another; for example, why it should be three feet long rather than two, or square rather than round. This cannot be explained by the nature of the bodies themselves, since the matter is indeterminate as to any definite figure, whether square or round. For the scientist who refuses to resort to an incorporeal cause, there can be only two answers. Either the body has been this way since eternity, or it has been made square by the impact of another body. "Eternity" is no answer, since the body could have been round for eternity also. If the answer is "the impact of another body," there remains the question of why it should have had any determinate figure before such motion acted upon it. This question can then be asked again and again, backwards to infinity. Therefore, it appears that the reason for a certain figure and magnitude in bodies can never be found in the nature of these bodies themselves. (No pagination)

The same can be established for the body's cohesion and firmness, which left Leibniz with the following conclusion:

> Since we have demonstrated that bodies cannot have a determinate figure, quantity, or motion, without an incorporeal being, it readily becomes apparent that this incorporeal being is one for all, because of the harmony of things among themselves, especially since bodies are moved not individually by this incorporeal being but by each other. But no reason can be given why this incorporeal being chooses one magnitude, figure, and motion rather than another, unless he is intelligent and wise with regard to the beauty of things and powerful with regard to their obedience to their command. Therefore such an incorporeal being be a mind ruling the whole world, that is, God. (de Hoyos, no pagination)

This argument refutes the scientistic claim that the physical universe constitutes the "totality of reality." Guenon recapitulates:

The truth is that the corporeal world cannot be regarded as being a whole sufficient to itself, nor as being isolated from the totality of universal manifestation: on the contrary, whatever the present state of things may look like as a result of "solidification," the corporeal world proceeds entirely from the subtle order, in which it can be said to have its immediate principle, and through that order as intermediary it is attached successively to formless manifestation and finally to the non-manifested. If it were not so, its existence could be nothing but a pure illusion, a sort of fantasmagoria behind which there would be nothing at all, which amounts to saying that it would not really exist in any way. That being the case, there cannot be anything in the corporeal world such that its existence does not depend directly on elements belonging to the subtle order, and beyond them, on some principle that can be called "spiritual," for without the latter no manifestation of any kind is possible, on any level whatsoever. (213-14)

Of course, this conclusion was antithetical to the doctrine of the scientific dictatorship, which contended that "the physical laws of the universe were the ultimate causative factors" (Keith, *Saucers of the Illuminati*, 78-79). Metaphysical naturalism (i.e., nature is God) had to be enthroned. Meanwhile, God's presence in the corridors of science had to be expunged. To achieve this, the Royal Society created a Gnostic division between science and theology, thus insuring the primacy of matter in the halls of scientific inquiry (Tarpley "How the Venetian System Was Transplanted into England"). The epistemological primacy of nominalism played no small role in establishing this Gnostic division. Coomaraswamy explains:

Having accepted the nominalist position, scientists soon began to consider the physical universe--the measurable world--as the totality of reality. All else was relegated to an ontological limbo. This, as Dr. Wolfgang Smith has pointed out, is not a scientific discovery, but a metaphysical assumption. Having taken this step man increasingly saw the phenomenological world, not as a reflection of God's beauty and goodness, but as a mechanical clock. (No pagination)

This mechanistic cosmology was exalted during the "age of Enlightenment" (Coomaraswamy, no pagination). Simultaneously,

Revelation and Intellection were summarily rejected (no pagination). Within this cosmological framework, man became an "autonomous entity" and God was relegated to irrelevant ambiguity. Such a deistic Weltanschauung was very amicable to Freemasonic philosophy (no pagination). Coomaraswamy comments: "And so we have the Masonic-Rousseauan concept of man whose 'dignity' lies in his 'independence'— he is his own authority and he creates his own culture" (no pagination).

As science became more preoccupied with observable phenomenon and the physical universe, it became less hospitable to metaphysical positions that reside outside the realm of quantification. Indeed, biases and presuppositions pervade the very fabric of the elite's epistemic monopoly. Academia itself has become the official church for this cult of epistemological selectivity. Christian philosopher Ravi Zacharias personally encountered the enormous prejudicial hurdles of scientism during a casual conversation with a few scholars, wherein one scientist makes a shocking confession:

> I asked them a couple of questions. "If the Big Bang were indeed where it all began, may I ask what preceded the Big Bang?" Their answer, which I had anticipated, was that the universe was shrunk down to a singularity.
> I pursued, "But isn't it correct that a singularity as defined by science is a point at which all the laws of physics break down?"
> "That is correct," was the answer.
> "Then, technically, your starting point is not scientific either."
> There was silence, and their expressions betrayed the scurrying mental searches for an escape hatch. But I had yet another question.
> I asked if they agreed that when a mechanistic view of the universe had held sway, thinkers like Hume had chided philosophers for taking the principle of causality and applying it to a philosophical argument for the existence of God. Causality, he warned, could not be extrapolated from science to philosophy.
> "Now," I added, "when quantum theory holds sway, randomness in the subatomic world is made a basis for randomness in life. Are you not making the very same extrapolation that you warned us against?"
> Again there was silence and then one man said with a self-deprecating smile, "We scientists do seem to retain selective

sovereignty over what we allow to be transferred to philosophy and what we don't." (64)

This "selective sovereignty," vigorously enforced by the epistemic dictatorship of the elite, effectively marginalized dissenters and consummated the apotheosis of the "bookkeepers." Hoffman explains:

> The cryptocracy has successfully harnessed to its own ends the huge potential for promoting secret political-occult agendas to the public, by presenting them as unassailable "objective scientific truth." Since the bogey of "science" instills in secularists a sort of blind reverence, opponents of political and occult agendas promoted through the propaganda of scientism are quickly stigmatized as "Neanderthal," especially with regard to their opposition to Darwinism, a dogma proved false by Norman Macbeth in his magisterial *Darwin Retried* and exposed as a cult by Gertrude Himmelfarb in *Darwin*. (Hoffman 49)

Suddenly, "ostensible control over the knowable" became the Divine Providence of god-like "bookkeepers." Meanwhile, their opponents became heretics and were "burned at the stake" (i.e., marginalized by academia and other secular institutions). Hoffman states:

> The doctrine of man playing god reaches its nadir in the philosophy of scientism which makes possible the complete mental, spiritual and physical enslavement of mankind through technologies such as satellite and computer surveillance; a state of affairs symbolized by the "All Seeing Eye" above the unfinished pyramid on the U.S. one dollar bill. (Hoffman 50)

Fettered by the epistemological totalitarianism of scientism, human thought must now operate within narrow parameters. Worse still, the enthronement of scientism might be facilitating the technological empowerment of an authoritarian elite. In short, the theocracy of science is Huxley's "scientific dictatorship."

Evolution and the Occult Doctrine of "Becoming"
With the British Royal Society acting as their headquarters of propaganda, the elite had created an institution to provide credibility

for their specially designed "science." Now, they needed to introduce
the "science." Recall that the founding members of the Royal Society
were all Freemasons. Thus, whatever "science" these men would design
would be derivative of Masonic doctrine. In *The Meaning of Masonry*,
W.L. Wilmshurst reveals the worldview underpinning the new Masonic
"science":

> This--the *evolution* of man into superman--was always the
> purpose of the ancient Mysteries, and the real purpose of
> modern Masonry is not the social and charitable purposes
> to which so much attention is paid, but the expediting of the
> spiritual evolution of those who aspire to perfect their own
> nature and transform it into a more god-like quality. And
> this is a definite science, a royal art, which it is possible for
> each of us to put into practice; whilst to join the Craft for
> any other purpose than to study and pursue this science is
> to misunderstand its meaning. (Wilmshurst, 47; emphasis
> added)

Later in the book, Wilmshurst reiterates this theme:

> Man who has sprung from earth and developed through the
> lower kingdoms of nature to his present rational state, has yet
> to complete his *evolution* by becoming a god-like being and
> unifying his consciousness with the Omniscient-to promote
> which is and always has been the sole aim and purpose of all
> Initiation. (Wilmshurst, 94; emphasis added)

Yet, Wilmshurst is not the only Masonic scholar to voice the
evolutionist sentiments of the Lodge's doctrine. In *Evrim Yolu*
(translated *The Way of Evolution*), Master Mason Selami Isindag makes
the following assertion:

> The most important characteristic of our school of morality
> is that we do not depart from the principles of logic and we
> do not enter the unknowns of theism, secret meanings or
> dogmas. On this basis we assert that the first appearance of
> life began in crystals under conditions that we cannot know
> or discover today. *Living things were born according to the law of
> evolution and slowly spread over the earth. As a result of evolution,*

today's human beings came to be and advanced beyond other animals both in consciousness and intelligence. (*Evrim Yolu,* 141; emphasis added)

In *Masonluktan Esinlenmeler* (translated *Inspirations from Freemasonry*), Isindag recapitulates this evolutionist theme:

> Masonry is not godless. But the concept of God they have adopted is different from that of religion. The god of Masonry is an exalted principle. *It is at the apex of the evolution.* By criticizing our inner being, knowing ourselves and deliberately walking in the path of science, intelligence and virtue, we can lessen the angle between him and us. Then, this god does not possess the good and bad characteristics of human beings. It is not personified. It is not thought of as the guide of nature or humanity. It is the architect of the great working of the universe, of its unity and harmony. It is the totality of all the creatures in the universe, a total power encompassing everything, an energy. Despite all this, it cannot be accepted that it is a beginning this is a great mystery. (*Masonluktan Esinlenmeler,* 73; emphasis added)

Isindag claims, "life began from one cell and reached its present stage as a result of various changes and evolutions" (*Masonluktan Esinlenmeler,* 78). Completing this evolutionary portrait of man, Isindag concludes:

> From the point of view of evolution, human beings are no different from animals. For the formation of man and his evolution there are no special forces other than those to which animals are subjected. (*Masonluktan Esinlenmeler,* 137)

Later in *Masonluktan Esinlenmeler,* Isindag reveals that an Eastern Masonic sect promulgated evolutionary thought:

> In the Islamic world there was a counterpart of Masonry called the *Ikhwan as-Safa'* [The Brethren of Purity]. This society was founded in Basra in the time of the Abbasids and published an encyclopedia composed of 54 large volumes. *Seventeen of these dealt with natural science and it contained scientific explanations that*

closely resembled those of Darwin. These found their way even to Spain and had an influence on Western thought. (*Masonluktan Esinlenmeler*, 274-75; emphasis added)

With God's effective exile from science, man's position as *imago viva Dei* (created in the image of the Creator) was summarily rejected. Now, Freemasonry could introduce its occult doctrine of "becoming," the belief in man's gradual evolution towards apotheosis. Masonic scholar Manly P. Hall synopsizes this doctrine:

Man is a god in the making, and as in the mystic myths of Egypt, on the potter's wheel, he is being molded. When his light shines out to lift and preserve all things, he receives the triple crown of godhood, and joins that throng of Master Masons, who in their robe of Blue and Gold, are seeking to dispel the darkness of night with the triple light of the Masonic Lodge. (54-55)

Thirty-third degree Mason J.D. Buck condenses this contention into one simple statement: "The only personal God Freemasonry accepts is humanity in toto . . . Humanity therefore is the only personal god that there is" (216). In short, the "apex of evolution" is man apotheosized. This was one of Illuminati founder Adam Weishaupt's "inner Areopagites: man made perfect as a god-without-God" (Billington 97). This religion is nothing new. Throughout the years, it has reappeared under numerous appellations. W. Warren Wagar enumerates this religion's numerous manifestations:

Nineteenth-and early twentieth-century thought teems with time-bound emergent deities. Scores of thinkers preached some sort of faith in what is potential in time, in place of the traditional Christian and mystical faith in a power outside of time. Hegel's *Weltgeist*, Comte's *Humanite*, Spencer's organismic humanity inevitably improving itself by the laws of evolution, Nietzsche's doctrine of superhumanity, the conception of a finite God given currency by J.S. Mill, Hastings Rashdall, and William James, the vitalism of Bergson and Shaw, the emergent evolutionism of Samuel Alexander and Lloyd Morgan, the theories of divine immanence in the liberal movement in Protestant theology, and du Nouy's telefinalism--all are exhibits in evidence of the influence chiefly of evolutionary thinking, both before and after Darwin, in Western intellectual history. The faith of progress itself--especially the idea of progress as built into the evolutionary scheme of things-is in every way the psychological equivalent of religion. (106-07)

Expanding on the religion of progress and its numerous permutations, Rama Coomaraswamy makes the following observation:

In point of fact, the idea of "progress," used in this sense, pre-dated Darwin by decades if not by centuries. One finds it used during the English Reformation where the "Recussants"--those who refused to abandon the Catholic faith--were described as "backward," while those who accepted the "established" state-enforced religion--were "progressive." The concept was further developed during the so-called "age of enlightenment" when people like Rousseau, Voltaire and Diderot dreamed of creating a perfect society without God. Kant embraced it in his "Idea of a Universal History on a Cosmopolitical Plan," a text in which he taught that history followed predetermined laws and revealed what be called "a regular stream or tendency" which demonstrated a "natural purpose" which would end in a "Universal civil society." Spencer spoke of the "law of progress" and defined evolution as "a change from an indefinite incoherent homogeneity to a definite coherent heterogeneity through continuous differentiations and integrations." He went on to teach that "the operation of evolution is absolutely universal. . .Whether it be in the development of the earth, in the development of life upon its surface, in the development of society, of government, of manufactures, of commerce, of language, of literature, science, art, this same advance from the simple to the complex, through successive differentiations, holds uniformly. . ." Hegel taught that humanity was driven ceaselessly upwards by an all-powerful, all-rational "It", and that the path of the ascent was an eternal, immutable, predestined, zigzag--his thesis and antithesis--always resulting in a higher synthesis. Evolutionary theory developed as a result of applying these ideas to biology. It provided a "scientific" basis for man's belief in progress and found ready acceptance in a world that sought to free itself from all divine sanction. From the time of Darwin, progress and evolution have become almost interchangeable terms that are mutually supportive and pervasive influences in our lives. (No pagination)

Masonic doctrine, particularly the occult process of "becoming," is derivative of the Ancient Mystery religion. This religion originated in Babylon and Egypt roughly 6000 years. It espoused a faith in the emergent deity of humanity, the rise of Man (spelled with a capital M to denote his purported divinity). Darwinism, which was actively promoted by the Freemasonic Royal Society, merely provided a conduit for this religion's transmission into the modern world.

According to *Mackey's Encyclopedia of Freemasonry*, Erasmus Darwin, grandfather of Charles Darwin, was the first to promulgate the concept of evolution:

> Dr. Erasmus Darwin (1731-1802) was the first man in England to suggest those ideas which later were to be embodied in the Darwinian theory by his grandson, Charles Darwin (1809-1882), who wrote in 1859 *Origin of Species*. (Qutd. in Daniel 34)

Freemason Erasmus Darwin, Charles' grandfather, "originated almost every important idea that has since appeared in evolutionary theory" (Darlington, "The Origin of Darwinism," 62). Erasmus was also the founder of the Lunar Society. According to author Ian Taylor, the Lunar Society was active from about 1764 to 1800 and its prominent influence "continued long afterwards under the banner of The Royal Society" (55). The group's name owed itself to the fact that members met monthly at the time of the full moon (55). The membership of this group boasted such luminaries as John Wilkinson (who made cannons), James Watt (who owed his notoriety to the steam engine), Matthew Boulton (a manufacturer), Joseph Priestly (a chemist), Josiah Wedgewood (who founded the famous pottery business), and Benjamin Franklin (55).

The Lunar Society was intimately tied to Freemasonry. Interestingly enough, in an article by Lord Richie-Calder, Lunar Society members were assigned the very esoteric appellation of "merchants of light." This was precisely the same description used for the hypothetical society presented in Sir Francis Bacon's *New Atlantis* (Taylor 55). In her examination of J.G. Findel's *History of Freemasonry*, Nesta Webster made the following observation: "Findel frankly admits that the New Atlantis contained unmistakable allusions to Freemasonry and that Bacon contributed to its final transformation" (Webster 120).

Researcher Ian Taylor adds:

> Webster pointed out that one of the earliest and most eminent precursors of Freemasonry is said to have been Francis Bacon, who is also recognized to have been a Rosicrucian; the Rosicrucian and Freemason orders were closely allied and may have had a common source. (Taylor 445)

Still, these are tenuous ties at best. Are there any sources that firmly establish a Darwinian/Freemasonic connection? *Mackey's Encyclopedia of Freemasonry* conclusively confirms a link:

> Before coming to Derby in 1788, Dr. [Erasmus] Darwin had been made a Mason in the famous Time Immemorial Lodge of Cannongate Kilwinning, No. 2, of Scotland. Sir Francis Darwin, one of the Doctor's sons, was made a Mason in Tyrian Lodge, No. 253, at Derby, in 1807 or 1808. His son Reginald was made a Mason in Tyrian Lodge in 1804. The name of Charles Darwin does not appear on the rolls of the Lodge but it is very possible that he, like Francis, was a Mason. (Qutd. in Daniel 34)

In 1794, Erasmus wrote a book entitled *Zoonomia*, which delineated his theory of evolution (Taylor 58). Being a Freemason, there is a distinct possibility that Erasmus cribbed liberally from the Lodge's occult doctrine of "becoming." Before Erasmus had penned his precursory notions of progressive biological development, Freemason John Locke (1632-1704) extrapolated the Hindu doctrine of reincarnation into the context of metaphysical naturalism and formulated a theory of evolution (Daniel 33-34). Researchers Paul deParrie and Mary Pride explains the evolutionary aspects of reincarnation:

> Ancient Babylonian and Hindu beliefs included the doctrine of evolution. The goddess Kali was designated, among other things, the goddess of "becoming" or evolution. Reincarnation, the spiritual form of evolution, was part of both of these religions. (deParrie and Pride, p. 27, 1988)

The British East India Company had imported the Hindu belief in reincarnation to England where it would be adopted by the British

Royal Society. A prominent member of the Royal Society, John Locke studied reincarnation extensively and, working with the occult doctrine as an extrapolative inspiration, developed his own evolutionary ideas. In fact, Locke's theory of evolution received the support of the male members of Darwin's family (Daniel 33-34). Two centuries later, this occult concept of "becoming" would be transmitted to Charles Darwin and *On the Origin of Species* would be published.

Darwinism has been one of the guiding inspirations underpinning the crusade to establish a socialist totalitarian world government. Whether it is of the Transnationalist or Internationalist pedigree, globalism exhibits the ideational thread of Darwinism. In *The Keys of This Blood*, Vatican Malachi Martin provides Pope John Paul's analysis of this Darwinian thread:

> From Pope John Paul's vantage point, the thing that seems to bind these two groups most closely in practical terms is that at heart, and philosophically speaking, both are sociopolitical Darwinists. Of course, the Pope doesn't for a moment imagine that such activists as these are likely to take time out from their total immersion in world affairs to formulate their basic group philosophy in the same way that the Humanists have. There is no Internationalist or Transnationalist equivalent of Professor Paul Kurtz's Humanist Manifesto II.
>
> Still, in John Paul's assessment, both of these globalist groups operate on the same fundamental assumptions about the meaning of human society today. Both agree on the face of it that the most important single trait that pervades the life of all nations is interdependence. And both agree that interdependence is a progressive function of evolutionary progress. Evolutionary, as in Darwin.
>
> In practical terms, both of these groups operate on the same working assumption Charles Darwin arbitrarily adopted to rationalize his feelings about mankind's physical origins and history. If it worked so well for Darwin, they almost seem to say, why not expand the idea of orderly progress through natural evolution to include such sociopolitical arrangements as corporations and nations? In this view, the most useful of Darwin's concepts is that of human existence as essentially a struggle in which the weakest perish, the fittest survive and the strongest flourish.
>
> When applied to sociopolitical arrangements, this Darwinist

process seems almost to dictate the Internationalist and Transnationalist one-world view of things. The continuing clash and contention in the world as it has been until now has resulted in a slow evolution of those who have survived from one stage of interdependent order to another. From time to time, natural "catastrophes" have intervened, forcing "nature" to take another path. But at each new stage, interdependence has become more important and more complex.

The greater the interdependence between groups, the higher the evolutionary stage, the more the balance achieved between interdependent groups results in the common good.

The view of the Internationalists and Transnationalists is that they are the ones who are equipped to bring mankind to the highest level of the sociopolitical evolution. Their effort is to bring together into one harmonious whole all those separate parts of our world that have not yet "evolved" into a natural cohesion for the common good. (Martin 314-15)

The oligarchical Dulles family is an example that supports Martin's above contention. Not only was this family made up of hardcore elitists, but they were also stalwart sociopolitical Darwinians. Researcher Anton Chaitkin elaborates:

In 1922 the liberal New York City preacher Harry Emerson Fosdick was attacked by fundamentalists for advocating Darwinian "survival of the fittest" evolutionary concepts from his Presbyterian pulpit; Fosdick was ensured by the Presbyterian General Assembly. At that time the father of Allen and John Foster Dulles, the Rev. Allen Macy Dulles, was Director of Apologetics at the Auburn Theological Seminary, a Presbyterian institution in upstate New York.

Reverend Dulles swung into action and organized a defense of Fosdick, circulating the "Auburn Affirmation" advocating "liberty of conscience" within the church. Reverend Dulles's Auburn Affirmation was signed by 1,293 Presbyterian ministers; the Reverend and his son, attorney John Foster Dulles, worked for a national liberal counterattack, and their faction took over the national leadership of the Presbyterian church on this issue. (560-61)

Of course, the vision of a socialist totalitarian world government synchronizes with the Masonic vision of a Baconian New Atlantis. In short, the final objective is a global scientific dictatorship and its legitimizing science is Darwinism. Indeed, Darwinism is both politically and socially expedient for the cause of globalism. Given the Masonic involvement in the inception and promulgation of Darwinism, its sociopolitical polyvalence may not have been an accident.

Metaphysical Naturalism and the Golem

Underpinning the concept of metaphysical naturalism is the notion that life originated with lifeless matter. This notion, dubbed "spontaneous generation," excludes the involvement of a supernatural Creator. Thus, nature became a god creating itself. Louis Pasteur, whose work established the Law of Biogenesis, provided the most succinct summation of this anthropomorphic mysticism:

> To bring about spontaneous generation would be to create a germ. It would be creating life; it would be to solve the problem of its origin. It would mean to go from matter to life through conditions of environment and of matter [lifeless material]. God as author of life would then no longer be needed. Matter would replace Him. God would need to be invoked only as author of the motions of the universe. (Dubos 395)

Like all of the false gods of antiquity, the voracity of this new deity was soon demolished. "Spontaneous generation" was proven impossible by the Law of Biogenesis. However, this fact did not stop certain "men of science" from chronically deifying nature. For instance, Charles Darwin unconsciously revealed his idolatrous impulses through statements like: "natural selection picks out with unerring skill the best varieties" (Hooykaas 18).

Evident in such statements is the idea that nature is sentient. After all, only a sentient being holds discriminative tastes and, therefore, "picks out" the recipients of its favor. Moreover, such statements reveal that "nature" itself is a sovereign deity acting as the ultimate arbiter of life and death. This meme has metastasized, presenting itself today as the Gaia Hypothesis. This hypothesis holds that the biosphere is a self-creating, self-sustaining, and self-regenerating entity (Lovelock 31-33). Central to this thesis is the contention that both the living and non-living are inseparable (Lovelock 31-33).

Although the concept of "spontaneous generation" was proven scientifically bankrupt years ago, many continue to resuscitate its corpse. Why does this theme of lifeless matter spontaneously generating life continue to emerge? The answer is because it has been with man for a very long time. It is derivative of the golem, an occult concept presented in the Hebraic Kabbalah. Thirty-third Degree Freemason Albert Pike revealed that "all the Masonic associations owe to it [the Kabbalah] their Secrets and their Symbols" (Pike 744). According to this occult text, the golem was an artificially created man generated from lifeless matter. The late Isaac Bashevis Singer, who studied the Kabbalah extensively, explained:

> "the golem is based on faith *that dead matter is not really dead, but can be brought to life.* What are the computers and robots of our time if not golems? The Talmud tells us of an interpreter by the name of Rava who formed a man by this mysterious power. We are living in an epoch of golem-making right now. The gap between science and magic is becoming narrower." (Qutd. in Hoffman 115; emphasis added)

Drawing upon the esoteric doctrines of their occult heritage, the Freemasonic members of the British Royal Society re-introduced the golem to the public mind under the guise of "metaphysical naturalism." Gradually, the corporeal machinations of nature supplanted the miraculous Creator. Master Mason Selami Isindag recapitulates this contention: "apart from nature there is no force that guides us, and is responsible for our thoughts and actions" (*Masonluktan Esinlenmeler*, 78).

Of course, the machinations of nature were only intelligible to anointed scientists of the epistemic dictatorship. Thus, the "bookkeepers" of the elite became the new expositors of "miracles." This virtual deification of the "bookkeepers" is evident in Singer's later statements regarding the golem:

> "I was interested in the golem from my early childhood. I was brought up in the home of a rabbi, and his sermons often spoke of miracles, by the Baal Shem Tov and other wonder rabbis. I realized early in my life that science and technology had actually created a civilization of miracles. Science is one long chain of miracles." (Qutd. in Hoffman 116)

Recall the words of Aldous Huxley in *Brave New World Revisited*: "The older dictators fell because they could never supply their subjects with enough bread, enough circuses, enough *miracles*, and mysteries" (Huxley 116; emphasis added). The new dictators do not intend to make the same mistake. With the effective enshrinement of metaphysical naturalism, the British Royal Society prepared to unleash their next golem. However, this golem would be an artificially created ape-man presented to the public imagination under the appellation of Darwinism.

The Veil of Materialism

While this new theocracy is veiled in secularism, it must be understood that the new state-sanctioned epistemology is a form of mysticism akin to its religious progenitor. This truth is illustrated by radical empiricism's rejection of causality, which stipulates the investment of faith in the purported results of scientific research. Likewise, the new state-sanctioned metaphysics is equally mystical in character. Accompanying radical empiricism is materialism, the metaphysical contention that matter holds primacy.

Naturalism works in tandem with materialism because it attempts to sustain the primacy of matter with the metaphysical claim of "self-creation" (i.e., abiogenesis). Of course, this claim suggests that living and dead matter are inseparable. Thus, living things are literally artificial entities that create themselves, an occult theme communicated through the Kabalistic myth of the golem. In a universe where materialistic metaphysics hold sway, the biosphere and the life it supports amount to one enormous golem. Accompanying this contention is the Gnostic doctrine of "self-salvation." If humanity is a god that created itself, then it is also responsible for its own salvation. Given these strange confluences of occult thought, materialism qualifies as little more than a new secular mysticism.

Not surprisingly, materialistic metaphysics pervade the fabric of many occult institutions. Even the acknowledgement of supra-sensible and incorporeal entities cannot hide the occultist's materialistic propensities. In fact, such propensities may have given rise to the occultist's mystical beliefs in the first place. Guenon explains:

> Without seeking for the moment to determine more precisely the nature and quality of the supra-sensible, in so far as it is actually involved in this matter, it will be useful to observe how far the very people who still admit it and think that they are aware of its action are in reality penetrated by materialistic

influence: for even if they do not deny all extra-corporeal reality, like the majority of their contemporaries, it is only because they have formed for themselves an idea of it which enables them in some way to assimilate it to the likeness of sensible things, and to do that is certainly scarcely better than to deny it. There is no reason to be surprised at this, considering the extent to which all the occultist, theosophist, and other schools of that sort are fond of searching assiduously for points of approach to modern scientific theories, from which they draw their inspiration more directly than they are prepared to admit; the result is what might logically be expected under such conditions. (153-54)

It comes as little surprise that occult Freemasonry and secular Marxism both share materialistic propensities. The late Malachi Martin examines the metaphysical commonalities between the two:

For both Marxists and Masons, however, different and opposed they may be politically, are at one in locating all of man's hopes and happiness in this worldly setting, without any intervention of a divine action from outside this cosmos and without appointing an otherworldly life as the goal of all human life and endeavor. Marxism and Masonry transcend, both of them, individuals and nations and human years and centuries. But it is rather an all-inclusive embrace, holding all close to the stuff and matter of the cosmos, not in any way lifting the heart and soul to a transhuman love and beauty beyond the furthest limit of dumb and dead matter. (534-35)

In a sense, materialism acts as a veil. The fact is that, although the occult theocracy of antiquity declined in power, it is still very much alive. It perpetuates itself through secularism. As sociologist William Sims Bainbridge makes clear, secularization actually represents the opening stage of an occult counterculture movement:

Secularization does not mean a decline in the need for religion, but only a loss of power by traditional denominations. Studies of the geography of religion show that where the churches become weak, cults and occultism explode to fill the spiritual vacuum. ("Religions for a Galactic Civilization," no pagination)

Thus, the thoroughly secularized society merely presages the emergence of a new theocratic order. The new ecclesiastical authority shall be occult in character, embracing what Guenon calls "neo-spiritualism" (155). The galvanizing mythology of this new theocratic order will most likely reflect the paradigmatic character of the Gnostic cosmology, depicting humanity as a collection of pluralities awaiting unification into a singularity through the sorcery of "science." As for the dominant religion, it will be Luciferianism, which was initially disseminated on the popular level as secular humanism. This is the anatomy of the emergent "Satanic state."

In addition to facilitating the rise of a new occult theocracy, materialism has also contributed to the enormous volumes of bloodshed witnessed by the 20th century. Arguably, contemporary regimes premised upon dialectical materialism have murdered far more people than any traditional theocracy premised upon a theistic faith. This is directly attributable to materialism's emphasis upon the primacy of matter. Materialistic metaphysics preclude the spirit, confining moral questions to the ontological plane of the physical universe. Severed from their ontological source, moral principles become tantamount to material phenomena. Thus, in a universe where materialism holds sway, it is reasonable to assume that evil is a purely corporeal entity that can be physically expunged. The ramifications of such an outlook are disturbing. In the article "What Evil Is and Why It Matters," Christian philosopher John Paul Jones reveals the consequences of this Weltanschauung:

> According to this [materialist] methodology, all we need do is find the material cause of evil and destroy it. After, all, since materialists assume all causes are material, they are logically obliged and conceptually predisposed to assume that evil is itself caused by material, physically destructible things or causes. (64)

The outgrowth of this paradigm is what Jones calls the "search and destroy" approach to dealing with evil (64). Jones expands on this approach:

> Consequently, those of a materialist mindset, whether

Christian or otherwise, are constantly engaged in campaigns to destroy the evil things or people they think are at the root of the problem. So we have, for example, the "war on drugs," the "war on guns," the "class war" and various genocides--all of which are known to cause more evil than they allegedly uproot, and today, as we witness the spread of eco-fascism in Europe that holds that we can solve the reputed environmental crisis by simply exterminating many millions of people, we also witness the approval of Chinese population control techniques, such as state-sanctioned abortion, infanticide, and forced sterilization. Strange fruits and bad apples, all. (64)

After years of war and waste, the materialist state is still incapable of expunging evil. This failure is directly attributable to materialism's misappropriation of matter as the totality of reality. In light of this metaphysical error, one is still left to ponder the source of evil. Yet, Biblical wisdom, which the materialist thoroughly rejects, may have already answered the question of evil. James 4:1-10 states:

From whence come wars and fighting among you? Come they not hence, even of your lusts that war in your members? Ye lust and have not; ye kill, and desire to have, and cannot obtain; ye fight and war, yet have not, because ye ask not. ye ask, and receive not, because ye ask amiss, that ye may consume it upon your lusts. Ye adulterers and adulteresses, know ye not that the friendship of the world is enmity with God? Whosoever therefore will be a friend of the world is the enemy of God? Do ye think that the scripture saith in vain, the Spirit that dwelleth in us lusteth to envy? But he giveth more grace. Wherefore he saith, God resisteth the proud, but giveth grace unto the humble. Submit yourselves therefore to God. Resist the devil, and he will flee from you. Draw nigh to God, and he will draw nigh to you. Cleanse your hands, ye sinners; and purify your hearts, ye doubleminded. Be afflicted, and mourn, and weep; let your laughter be turned to mourning, and your joy to heaviness. Humble yourselves in the sight of the Lord, and he shall lift you up.

Of course, such a conclusion is unthinkable to the materialist. It is interesting that Charles Fort believed:

> that man deliberately invented the dogma of materialism in order to shield himself from the evidence of what was being done to him by means of psycho-spiritual warfare methods hyped by "coincidence," symbolism and ritual. (Hoffman 68)

A metaphysical smoke screen currently obstructs humanity's view of the spiritual principles upon which so many of the world's dilemmas rest. It is the veil of materialism.

The Darwin Project

In the article "Toward a New Science of Life," *Executive Intelligence Review* journalist Jonathan Tennenbaum makes the following the statement concerning Darwinism:

> Now, it is easy to show that Darwinism, one of the pillars of modern biology, is nothing but a kind of cult, a cult religion. I am not exaggerating. It has no scientific validity whatsoever. Darwin's so-called theory of evolution is based on absurdly irrational propositions, which did not come from scientific observations, but were artificially introduced from the outside, for political-ideological reasons. (Tennenbaum, no pagination)

Given Darwinism's roots in occult Freemasonry and its expedient promotion of an emergent species of supermen (i.e., the elite), this is a fairly accurate assessment. Charles Darwin acted as the elite's apostle, preaching the new secular gospel of evolution. Darwinism could be considered a Freemasonic project, the culmination of a publicity campaign conducted by the Lodge. Evidence for this contention can be found in controversial *Protocols of the Wise Men of Sion.*

Although an examination of the *Protocols* and a critique of their authenticity are not the purposes of this essay, it is important to address the questions surrounding their origins. After all, the *Protocols* have been employed throughout history in numerous genocidal campaigns against the Jews. However, the authors of *Holy Blood, Holy Grail* provide evidence that the document may be Masonic in origin:

It can thus be proved conclusively that the *Protocols* did not issue from the Judaic congress at Basle in 1897. That being so, the obvious questions is whence they did issue. Modern scholars have dismissed them as a total forgery, a wholly spurious document concocted by anti-Semitic interests intent on discrediting Judaism. And yet the *Protocols* themselves argue strongly against such a conclusion. They contain, for example, a number of enigmatic references - references that are clearly not Judaic. But these references are so clearly not Judaic that they cannot plausibly have been fabricated by a forger, either. No anti-Semitic forger with even a modicum of intelligence would possibly have concocted such references in order to discredit Judaism. For no one would have believed these references to be of Judaic origin.

Thus, for instance, the text of the *Protocols* ends with a single statement. "Signed by the representatives of Sion of the 33rd Degree." Why would an anti-Semitic forger have made up such a statement? Why would he not have attempted to incriminate all Jews, rather than just a few - the few who constitute "the representatives of Sion of the 33rd Degree?" Why would he not declare that the document was signed by, say, the representatives of the international Judaic congress? In fact, the "representatives of Sion of the 33rd Degree" would hardly seem to refer to Judaism at all, or to any "international Jewish conspiracy." If anything, it would seem to refer to something specifically Masonic. And the thirty-third degree in Freemasonry is that of the so-called Strict Observance - the system of Freemasonry introduced by Hund at the behest of his "unknown superiors," one of whom appears to have been Charles Radclyffe. (Baigent, et al, 192-3)

Baigent, Leigh, and Lincoln conclude:

There was an original text on which the published version of the Protocols was based. This original text was not a forgery. On the contrary, it was authentic. But it had nothing whatever to do with Judaism or an "international Jewish conspiracy." It issued, rather, from some Masonic organization or Masonically oriented secret society that incorporated the word "Sion." (Baigent, et al, 194)

Given the Masonic language, one can completely discard the racist contention that the *Protocols* constitute evidence of an "international Jewish conspiracy." Nevertheless, the document holds some authenticity:

> The published version of the *Protocols* is not, therefore, a totally fabricated text. It is, rather, a radically altered text. But despite the alterations certain vestiges of the original version can be discerned. (Baigent, et al, 195)

The remnant of the original text strongly suggests Masonic origins. Having established the Masonic authorship of the *Protocols*, one may return to the issue at hand: Freemasonic involvement in the promotion of Darwinism. Consider the following excerpt from the *Protocols*, which reads distinctly like a mission statement:

> For them [the masses or cattle] let that play the principal part which we have persuaded them to accept as the dictates of science (theory). It is with this object in view that we are constantly, by means of our press, arousing a blind confidence in these theories. The intellectuals of the goyim [the masses or cattle] will puff themselves up with their knowledge and without any logical verification of it will put into effect all the information available from science, which our agentur specialists have cunningly pieced together for the purpose of educating their minds in the direction we want.
> Do not suppose for a moment that these statements are empty words: think carefully of the successes we arranged for *Darwinism*, Marxism, and Nietzsche-ism. (reprint in Cooper, 274-5; emphasis added)

In addition to establishing the Lodge's official sanction of Darwinism, this excerpt also reveals a direct relationship between Marxism, Nietzsche-ism, and evolutionary theory. This relationship shall be examined later.

It was the grandfather of Aldous Huxley, T.H. Huxley, who would act as the "official spokesman for the recluse Darwin" (White 268). Many years later, Aldous would propose a "scientific dictatorship" in *Brave New World Revisited*. Whether Aldous made this proposition on a whim or was penning a concept that had circulated within the Huxley family for years cannot be determined. Given the family's oligarchical tradition, the latter assertion remains a definite possibility. Yet, there may be a deeper Freemasonic connection, suggesting that the concept of a "scientific dictatorship" may have originated within the Lodge.

T.H. Huxley was a Freemason and, with no apparent achievements to claim as his own, was made a Fellow of the Royal Society at the age of 26 (White 267). T.H. Huxley tutored Freemason H.G. Wells, who would later teach Huxley's two grandsons, Julian and Aldous. Both Julian and Aldous were Freemasons (Daniel 147). Given this continuity of Freemasonic tutelage within the Huxley family, it is a definite possibility that the Huxlian concept of a "scientific dictatorship" is really Masonic. Considering Freemason H.G. Wells' endorsement of a "scientific dictatorship," which he considered a Technocracy, this is highly likely.

The rest is history. With the publicity campaigns of the Royal Society and the avid defense of evolution apologist T.H. Huxley, Darwin's theory would be disseminated and popularized. The seed had taken root and, in the years to come, numerous permutations of the elite's scientific dictatorship would emerge.

Darwinism Dismantled

Providing a complete and comprehensive delineation of the various concepts constituting Darwinism is a daunting task. The theory itself is a dense amalgam of "isms," thinly veiled occult concepts, philosophical doctrines, and ideologies. Again, Tennenbaum's statement that Darwinism "is based on absurdly irrational propositions, which did not come from scientific observations, but were artificially introduced from the outside, for political-ideological reasons" seems succinct and accurate. Yet, with what outside sources do these "absurdly irrational propositions" find their proximate origins?

One of the major influences on Darwin was Thomas Malthus, an Anglican clergyman who had received the blessings of French deist Jean-Jacques Rousseau and radical empiricist David Hume (Keynes 99). Malthus authored *Essay on the Principle of Population*, a treatise premised upon the thesis: "Population, when unchecked, increases in a geometrical ratio. Subsistence increases only in an arithmetic ratio"(qutd. in Taylor 61). Although Malthus articulated his observations in succinct mathematical equations, the labyrinthine and complex machinations comprising the natural order typically defy such overly simplistic reductionism. Nonetheless, Malthus concluded that society should adopt certain social policies to prevent the human population from growing disproportionately larger than the food supply.

Malthus' genocidal policies specifically targeted the poor. For instance, one of his proposals suggested the implementation of the following measures:

> Instead of recommending cleanliness to the poor, we should encourage contrary habits. In our towns we should make the streets narrower, crowd more people into the houses, and court the return of the plague. In the country, we should build our villages near stagnant pools, and particularly encourage settlement in all marshy and unwholesome situations. But above all, we should reprobate specific remedies for ravaging diseases; and those benevolent, but much mistaken men, who have thought they were doing a service to mankind by projecting schemes for the total extirpation of particular disorders. (Qutd. in Taylor 62-63)

Through the promotion of hygienically unsound practices amongst impoverished populations, Malthus believed that the "undesirable elements" of the human herd could be naturally culled by various maladies. The spread of disease could be further assisted through discriminative vaccination and zoning programs. Yet, amongst one of Malthus' most shocking proposals was his suggestion concerning children:

> We are bound in justice and honour formally to disclaim the right of the poor to support. To this end, I should propose a regulation be made declaring that no child born should ever be entitled to parish assistance. The [illegitimate] infant is comparatively speaking, of little value to society, as others will immediately supply its place' All children beyond what would be required to keep up the population to this [desired] level, must necessarily perish, unless room be made for them by the deaths of grown persons. (Qutd. in Taylor 63)

The dictum underpinning Malthus' logic would later be reiterated as "survival of the fittest." According to researcher Ian Taylor, the metastasis of this dictum "can be traced from Condorcet to Malthus, to Spencer, to Wallace, and to Darwin"(Taylor 65). Malthus' theory of carrying capacity was central to the Darwinian concept of natural selection:

Darwin gained a unique understanding from Malthus. Others such as the botanist Augustin de Candolle wrote of plants being "at war one with another." But nobody, Darwin said, conveys "the warring of the species" so strongly as Malthus. (Desmond and Moore 265)

The Malthusian Weltanschauung meshed comfortably with Darwin's Hobbesian contention that nature and society were a "war of all against all." This contention was edified by Malthus' statistical oversimplifications:

But it was Malthus's statistics that struck Darwin with a vengeance in his primed state. Malthus calculated that, with the brakes off, humanity could double in a mere twenty-five years. But it did not double; if it did the planet would be overrun. A struggle for resources slowed growth, and a horrifying catalogue of death, disease, wars, and famine checked the population. Darwin saw that an identical struggle took place throughout nature, and he realized that it could be turned into a truly creative force. (264-65)

It would be easy to characterize Malthus' impact on Darwin as an instance of bad science begetting more bad science. However, Darwin's contention that the Malthusian struggle between populations could be a "truly creative force" betrays his sociopolitical agenda. It is no mystery that Darwin was a proponent of free trade and unfettered competition (xxi). In all likelihood, Darwin probably believed that re-sculpting society along the contours of monopolistic capitalism would allow the ruling class to guide the "truly creative force" of natural selection.

Likewise, Malthus was an economist for the monopolistic East India Company (153). In service to this corporate entity, Malthus would argue that "[p]ublic charities—-the old poor relief—-only aggravated" overpopulation (153). It was the work of Darwin and Malthus that scientifically legitimized the genocidal "Poor Law Amendment Bill." This piece of legislation:

ended relief for all but the most destitute, those so ill or old that they would enter the abominable workhouses to receive food or money. Since their prison regimes were designed to discourage entry-—wives were torn from husbands to prevent breeding—-few were expected to apply for relief and a vast saving was guaranteed. (154)

Evident in this sordid portrait is the state's role as a weapon of biological warfare. Thanks to the enshrinement of socialist policies, Britain's population became biologically dependent upon the government. Meanwhile, the very same government designed public policies to cater to corporate interests (an unholy alliance that presaged Mussolini's corporatism). Once the population was made reliant upon socialist machinations for its subsistence, the state could selectively terminate welfare programs and "cull" specific segments of the "herd."

A similar operational model has been adopted by the sociopolitical Darwinians of today. This is painfully illustrated by the genocidal policies of the IMF and World Bank, two of the premiere Transnationalist machinations of modernity (see <u>Chapter One: The IMF and World Bank</u> for further explication). Modern globalists are merely following the lead of the British East India Company, which qualified as precursory Transnational corporation of sorts. Darwin and Malthus were the theoretical harbingers of this system.

In addition to Malthusianism, Darwinism's ingredients also include the occult religion of Gnosticism. Dr. Wolfgang Smith elaborates on the Gnostic features of Darwinism:

As a scientific theory, Darwinism would have been jettisoned long ago. The point, however, is that the doctrine of evolution has swept the world, not on the strength of its scientific merits, but precisely in its capacity as a Gnostic myth. It affirms, in effect, that living beings created themselves, which is in essence a metaphysical claim. . . Thus, in the final analysis, evolutionism is in truth a metaphysical doctrine decked out in scientific garb. In other words, it is a scientistic myth. And the myth is Gnostic, because it implicitly denies the transcendent origin of being; for indeed, only after the living creature has been speculatively reduced to an aggregate of particles does Darwinist transformism become conceivable. Darwinism, therefore, continues the ancient Gnostic practice of depreciating "God, the Father Almighty, Creator of Heaven and earth." It perpetuates, if you will, the venerable Gnostic tradition of "Jehovah bashing." And while this in itself may gladden Gnostic hearts, one should not fail to observe that the doctrine plays a vital role in the economy of Neo-Gnostic thought, for only under the auspices of Darwinist "self-creation" does the Good News of "self-salvation" acquire a semblance of sense. (Smith 242-43)

It is interesting to note that the British Royal Society, the Masonic institution responsible for the promulgation of Darwinism, rigorously imposed a division between science and theology upon the halls of scientific inquiry. Webster Tarpley characterizes this division as "literally Gnostic" (Tarpley "How the Venetian System Was Transplanted into England"). Indeed, the restriction of scientific research to the corporeal machinations of nature is redolent of Gnostic thinking. It is a distortion of Platonic metaphysics, the conceptual framework of which emphasizes a separation of the corporeal (the Becoming) and the incorporeal (the Being). This framework bears close resemblance to the traditional Christian Weltanschauung, which divides existence into the spiritual and the physical. However, Gnosticism rejected the Christian eschaton of heaven and hell. This is where the distortion begins.

According to Gnosticism, the physical universe is hell. Corporeal existence is a prison that fetters man through the demonic agents of space and time. However, through revelatory experience (*gnosis*), the sensate being of man could be transformed and this hell could become heaven. Possibly guided by this Gnostic philosophy, the Freemasonic Royal Society redirected scientific attention exclusively towards the material world. By focusing scientific efforts exclusively upon the temporal spatial realm, the members of the Royal Society probably hoped to see the eventual transformation of the irredeemable physical realm into the "immanentized eschaton" of an earthly heaven.

This was also the ultimate objective of Marxism, which was disseminated on the popular level as both fascism and communism. It is no coincidence that, historically, both the Nazis and the communists exhibited a religious adherence to the Gnostic myth of Darwinism. Smith writes:

In place of an Eschaton which ontologically transcends the confines of this world, the modern Gnostic envisions an End *within history*, an Eschaton, therefore, which is to be realized *within the ontological plane of this visible universe*." (238; emphasis added)

According to Vatican insider Malachi Martin, the Italian humanists who eventually created speculative Masonry "reconstructed the concept of *gnosis*, and transferred it to a thoroughly this-worldly plane" (519). Both Nazism and communism were birthed by organizational derivations of Masonry.

Given Gnosticism's derision for all things corporeal, it is extremely paradoxical that its adherents exhibit such a preoccupation with this material plane. Nonetheless, the Eschaton must manifest itself within the temporal spatial realm. Gnostic psychologist Carl Jung reiterates:

> According to [the alchemist] Basilius Valentinus, the earth (as *prima materia*) is not a dead body, but is inhabited by a spirit that is its life and soul. All created things, minerals included, draw their strength from this earth-spirit. This spirit is life. . .and it gives nourishment to all the living things it shelters in its womb. (329)

In light of Dr. Wolfgang Smith's intriguing observation, one could view the current rise of Gnosticism in the West as the natural corollary of Darwinism's unquestionable epistemological primacy in the West. The publication of *The DaVinci Code* and the release of the *Matrix* films are the most evident manifestations of this revival. The current resurgence of Gnostic thinking could represent the next stage of Darwinism's metastasis.

It comes as little surprise that Darwinism, which is premised upon metaphysical naturalism and materialism, is so compatible with Gnosticism. Both emphasize the primacy of this material plane. Had such a metaphysical doctrine remained confined to the realm of academic polemics, it may have been harmless enough. However, this was not to be the case. The Gnostic myth of Darwinism eventually migrated from the abstraction of speculative philosophy to other areas of study. With this migration, Darwinism enjoyed epistemological primacy. Julian Huxley elaborates:

> The concept of evolution was soon extended into other than biological fields. Inorganic subjects such as the life-history of stars and the formation of the chemical elements on the one hand, and on the other hand subjects like linguistics, social anthropology, and comparative law and religion, began to be studied from an evolutionary angle, until today we are enabled to see evolution as a universal and all-pervading process. (Qutd. in Newman 272)

Inevitably, the Gnostic myth of Darwinism subsumed social and political theory. The result was the sociopolitical Utopianism that underpinned all of the 20th century scientific dictatorships. Nazi Germany stands as a prime example of a Gnostic scientific dictatorship edified by Darwinism. In fact, Darwinian Sir Arthur Keith candidly admitted: "The German Fuhrer as I have consistently maintained, is an evolutionist; he has consciously sought to make the practice of Germany conform to the theory of evolution" (230). Darwinism's natural correlative, Gnosticism, was also present. Nazism was premised upon the occult doctrines of a Gnostic cult called Ariosophy, which promoted:

> [T]he rule of gnostic elites and orders, the stratification of society according to racial purity and occult initiation, the ruthless subjugation and ultimate destruction of non-German inferiors, and the foundation of a pan-German world-empire. Such fantasies were actualized with terrifying consequences in the Third Reich: Auschwitz, Sobibor and Treblinka are the hellish museums of Nazi apocalyptic, the roots of which lay in the millennial visions of Ariosophy. (Goodrick-Clarke, no pagination)

The Holocaust, which was an orgy of violence and death, represented Nazi Germany's efforts to "immanentize the Eschaton." In essence, Germany qualified as a Gnostic scientific dictatorship, edified by the "science" of Darwinism.

Communist Russia also exhibited all of the characteristics consistent with this profile. The *Encyclopedia of Religion* explains: "both Hegel and his materialist disciple Marx might be considered direct descendants of gnosticism" (576). In fact, Hegel is the ideological nexus where the Gnostic scientific dictatorships of Nazism and communism intersect. In *The Secret Cult of the Order*, Antony Sutton states: "Both Marx and Hitler have their philosophical roots in Hegel" (118). According to the *Encyclopedia of Religion*, the Gnostic Kabbalist named Christoph Oetinger significantly influenced Hegel's early work (576). From Hegel would spring two of the worst scientific dictatorships in history. Both of them were Gnostic at their core:

> In this century, with the presentation of traditional religious positions in secular form, there has emerged a secular Gnosticism beside the other great secular religions--the

mystical union of Fascism, the apocalypse of Marxist dialectic, the Earthly City of social democracy. The secular Gnosticism is almost never recognized for what it is, and it can exist alongside other convictions almost unperceived. (Webb 418)

As history has graphically demonstrated, the various religious crusades to "immanentize the Eschaton" are deadly serious. This truth is tangibly evidenced by the atrocities committed by the sociopolitical Utopians of secular Gnosticism. Both Auschwitz and the Soviet gulag are products of the same jihad. The secular theocracies that have waged this jihad have consistently been scientific dictatorships edified by Darwinism.

Martin explains that the humanist precursors to speculative Masonry desired "a special *gnosis*" (520). They believed that this "special gnosis" was a "secret knowledge of how to master the blind forces of nature for a sociopolitical purpose" (520). The subjugation and manipulation of nature is a theme consistently recapitulated by sociopolitical Utopians. For the sociopolitical Utopian, science represents a "special gnosis" designed to manipulate matter and reconfigure reality itself. It is an instrument for the re-sculpting of *prima materia* and "immanentizing the Eschaton." Raschke explains:

> The well-known maxim of Bacon, *nam et ipsa scientia potestas est* ("Knowledge itself is power"), is often commemorated as the credo of the new science, but it also suits quite precisely the magico-religious mentality of Gnosticism. (*The Interruption of Eternity* 49)

Technology has become the chief means of achieving Gnosticism's alchemical transformation of reality. Technology's potential for such an application is evident in the etymological origins of the appellation itself. It is derived from the Greek word *techne*, which means "craft." Simply defined, "crafting" is the skillful creation of something. Hence, expressions such as "outstanding craftsmanship" or a "master of the craft."

In the context of sociopolitical Utopianism, "crafting" is the skillful creation (or, more succinctly, re-sculpting) of reality itself. The "special *gnosis*" of science has provided the means through *techne*. Mark Pesce, co-inventor of Virtual Reality Modeling Language, elaborates upon *techne's* role in manipulating matter: "Each endpoint of techne has an expression in the modern world as a myth of fundamental direction-

—the mastery of matter. . ." (no pagination; emphasis added). This is the central precept of sociopolitical Utopianism: mastering reality itself.

Another "ism" constituting Darwinian evolution is scientific racism. This is one component of Darwin's theory that evolutionists have endeavored to either obfuscate or simply ignore. However, Darwin's emphasis upon race as an integral factor in humanity's evolution is virtually undeniable. The full title of Darwin's seminal tract on evolution was *On the Origin of the Species by Means of Natural Selection or the Preservation of Favoured Races in the Struggle for Life.*

This title prompts two disturbing questions. Exactly which race enjoys the "favour" of natural selection? More importantly, which race does not? In *The Descent of Man*, Darwin revealed natural selection's "favoured race" and the unfortunate people destined for extinction:

> At some future period, not very distant as measured by centuries, the civilised races of man will almost certainly exterminate, and replace the savage races throughout the world. At the same time the anthropomorphic apes... will no doubt be exterminated. The break between man and his nearest allies will then be wider, for it will intervene between man in a more civilised state, as we may hope, even than the Caucasian, and some ape as low as a baboon, instead of as now between the negro or Australian and the gorilla. (No pagination)

Given the possible Masonic origins of evolutionary thought, Darwin's racist Weltanschauung may have been inherited from the Craft. Masonry has had a long racialist tradition. According to its occult doctrine, which is racialist in character, it is the Anglo-Saxon who is divinely appointed to establish a global Masonic kingdom. This belief was thoroughly delineated in the September 1950 issue of *New Age Magazine*, the official journal of the Supreme Council, 33rd Degree Scottish Rite of Freemasonry. The author of the article, C. William Smith, opines:

> Looking back into history, we can easily see that the Guiding Hand of Providence has chosen the Nordic people to bring in and unfold the new order of the world. Records clearly show that 95 percent of the colonials were Nordics... Anglo-Saxons. Providence has chosen the Nordic race to unfold the "New Age" of the world. . .a "Novus Ordo Seclorum." (551)

For several years, the Lodge has practiced a policy of Masonic apartheid. The Prince Hall Grand Lodge, which permits black membership, is not considered a part of mainstream Freemasonry. The Scottish Rite and York networks are separate organizational entities from Prince Hall Masonry. In fact, mainstream Freemasonry considers the practices of black Freemasons as "spurious, illegitimate imitation" (Shaw and McKenney 29). Albert Pike, the Supreme Commander of the Southern Jurisdiction of Freemasonry, candidly decried the membership of African Americans in Masonry:

> I took my obligations from white men, not from negroes.
> When I have to accept negroes as brothers or leave Masonry,
> I shall leave it. (Qutd. in Darrah 319)

According to John J. Robinson, the black membership in the Lodge is "just a fraction of a fraction of one percent of total membership" (329). Even with the victories of the civil rights movement and the decline of racism, little has changed within the Lodge. William J. Whalen observes:

> By 1987, decades after most American institutions had accepted racial integration, only four of the forty-nine Grand Lodges could count even one black member in their jurisdictions. (23-25)

Of course, with the rise of racial integration, Freemasonry has been forced to recruit some black members. Lodges in New Jersey, Massachusetts, and New York have initiated a "handful of blacks" (23-25). However, Whalen makes it clear that "regular Freemasonry remains ninety-nine and forty-four hundredths percent white" (23-25). Conveniently, the "science" of Darwinism seems to legitimize such racist beliefs. Freemason T.H. Huxley, who was Darwin's most avid apologist, wrote:

> It may be quite true that some negroes are better than some white men; but no rational man, cognisant of the facts, believes that the average negro is the equal, still less the superior, of the average white man. And, if this be true, it is simply incredible that, when all his disabilities are removed, and our prognathous relative has a fair field and no favour, as well as

no oppressor, he will be able to compete successfully with his bigger-brained and smallerjawed rival, in a contest which is to be carried on by thoughts and not by bites. The highest places in the hierarchy of civilisation will assuredly not be within the reach of our dusky cousins, though it is by no means necessary that they should be restricted to the lowest. ("Emancipation-Black and White," 115)

The Freemasons in Darwin's own family may have passed this racist heritage onto him. While it is true that Darwin opposed slavery, he believed that blacks constituted an inferior race (Desmond and Moore xxi). According to Darwin, there were substantial dissimilarities intrinsic to the intellectual development of "men of distinct races" (*The Descent of Man* no pagination). Darwin asserted that the evolutionary chasm between apes and humans resided "between the negro or Australian and the gorilla," suggesting that blacks were closer to apes. The racism of such an assertion is axiomatic.

Perhaps one of the most significant constituent Weltanschauungs comprising Darwinism is Hegelianism. According to philosopher Georg Hegel, a pantheistic world spirit was directing "an ongoing developmental (evolutionary) process in nature, including humanity," which bodied itself forth as a "dialectical struggle between positive and negative entities" (381-82). This conflict always resulted in a "harmonious synthesis" (Taylor 381-82). The same dialectical framework is present in Darwinism.

In *Circle of Intrigue*, occult researcher Texe Marrs reveals the Hegelian structure intrinsic to Darwinian evolution. The organism (thesis) comes into conflict with nature (antithesis) resulting in a newly enhanced species (synthesis), the culmination of the evolutionary process (Marrs, *Circle of Intrigue,* 127). Of course, in such a world of ongoing conflict, violence and bloodshed are central to progress. Thus, Darwin's theory "gave credence to the Hegelian notion that human culture had ascended from brutal beginnings" (Taylor 386).

Yet, Darwinism's roots may go deeper than Hegelianism, returning to an esoteric source that has been there since the beginning. Hegel's ideas did not originate with himself, but Fichte (Sutton, *America's Secret Establishment*, 34). Who was Fichte? Antony Sutton reveals that he was a "Freemason, almost certainly Illuminati, and certainly promoted by the Illuminati" (Sutton, *America's Secret Establishment*, 34). In fact, Hegel's dialectical logic reiterates the Masonic dictum: *Ordo Ab Chao* (Order out of chaos). Again, it seems that the bedrock upon which Darwinism rests is Freemasonry, a channel for elitist interests.

The French Revolution: An Abortive Scientific Dictatorship

According to academia's officially sanctioned historians, the French Revolution was little more than a rebellion of the commoner against a corrupt aristocracy and religious institution. However, in *Essays on the French Revolution*, Lord Acton made an interesting observation:

> "The appalling thing in the French Revolution is not the tumult but the design. Through all the fire and smoke we perceive the evidence of calculating organization. The managers remain studiously concealed and masked; but there is no doubt about their presence from the first." (Qutd. in Reed 136)

Who were the "studiously concealed and masked managers" that orchestrated the French Revolution? In *Morals and Dogma*, Albert Pike revealed that it was Freemasonry that "aided in bringing about the French Revolution" (Pike 24). Indeed, the French Revolution represented the first full-scale attempt to tangibly enact the Masonic vision of a "scientific dictatorship."

The Lunar Society, which was the precursor to the Freemasonic Royal Society, was intimately connected to the revolutionary movement in France. Freemason Benjamin Franklin acted as the "shuttle diplomat between the French and English Utopian idealists" (Taylor 56). The son of James Watt was accused of being a French agent by Edmund Burke in the British House of Commons. Joseph Priestley had pledged his wholehearted support to the revolutionary French National Assembly (Taylor 56). Fellow Lunar Society member James Keir hosted a dinner to commemorate the fall of the Bastille. Most notably, Freemason and Lunar Society founder Erasmus Darwin actively supported the Jacobins (Taylor 56). Who were the Jacobins? William Hoar reveals that they were "agents of the Bavarian-bred Illuminati who operated out of the Club Breton" (*Architects of Conspiracy*, 2).

The French Revolution exhibited all of the hallmarks of a scientific dictatorship:

(1) A humanistic philosophy emphasizing man's evolutionary ascent towards apotheosis: After the Legislative Assembly rejected God as the object of man's worship and praise, the National Convention paraded a woman representing Athena from the convention hall to the chapel of Notre Dame. There, the Goddess of Reason took her place

on the high altar (Scott 306). In a Masonic context, this ritualistic enthronement of human reason could have represented the unification of man's consciousness with the Omniscient, which is the ultimate end of evolution (Wilmshurst 94). In other words, human reason became the ultimate source of moral precepts and man became God. Such ideas, which were the defining concepts of the Enlightenment, were adopted and perpetuated by Saint-Simon:

> In a sense, Saint-Simon was only reviving the Enlightenment vision of humanity advancing through successive stages to a scientific ordering of life (in Turgot's *Discourse on Universal History* of 1760); and of universal progress towards rational order (in Condorcet's Sketch for a *Historical Picture of the Progress of the Human Mind*, written in hiding shortly before his death in 1794). (Billington 213)

(2) A Malthusian depopulation campaign: Under the direction of Illuminist Robespierre, the new revolutionary government began carrying out a massive depopulation campaign that became known as the Terror. While Robespierre's goal of eliminating 15 million "useless eaters" was never realized, the Terror was successful in claiming the lives of some 300,000 Frenchmen, 297,000 of which were members of the lower and middle working classes. It should come as little surprise that Thomas Malthus was educated under the combined tutelage of two supporters of the French Revolution: Gilbert Wakefield and Lunar Society member Joseph Priestley (Taylor 59).

(3) A Hegelian framework: Recall the Hegelian structure intrinsic to evolution (Marrs 127). In hopes of accelerating France's evolution towards a "scientific dictatorship," the architects of the revolution promulgated a classic Hegelian dialectic: the bourgeoisie against the proletariat. The synthesis of these two polar extremes resulted in the subversion of individualism and the maintenance of class stratification.

Of course, the rest is history. The revolution swiftly degenerated into a bloodbath and many of the conspirators were slaughtered by the very mobs they had created. Yet, the esoteric symbol of this abortive scientific dictatorship remains. Long after she was enthroned in the cathedral of Notre Dame, Athena was transplanted upon new shores. Occult researcher Texe Marrs explains in *Dark Majesty*:

Today, statues of this Illuminist Goddess of Reason are found throughout the U.S.A.; one stands astride the U.S. Capitol building in Washington, D.C. Another is atop the dome of the Capitol building in Austin, Texas. Her statue has been erected in town squares and city parks. But the most fantastic idol of the Goddess of Reason, the most majestic statue of the pagan lady who bears the torch of light, who illuminates, uplifts, and frees mankind, is found in New York's harbor.

Towering above the shimmering but polluted waters, she holds in her outreached arm and hand a torch of fire and light. A gift of the Masonic Order, the modern inheritors of the Illuminati heritage, the Statue of Liberty was sculptured by Frederic Bartholdi, a member of the Masonic Lodge of Alsace-Lorraine in Paris, France. The statue is an esoteric idol of great significance to the secret societies plotting the New World Order. (*Dark Majesty*, 212)

Did the French Revolution truly end or did it simply change venues? The current cultural paradigm that pervades the corridors of power in America seems to affirm the latter. This is most effectively illustrated by the resurgence of the ideological force that inspired the Great Terror in revolutionary France: Jacobinism. In *America the Virtuous*, Claes G. Ryn observes the rise of the new Jacobinism throughout Western civilization:

Within today's Western democracies a new Jacobinism is exercising growing influence, especially in the United States. It is working to sever the remaining connections between the popular government and the traditional Western view of man and society. It employs an idiom somewhat different from that of the earlier Jacobinism, and it incorporates various new ideological and other ingredients, but it is essentially continuous with the old urge to replace historically evolved societies with an order framed according to abstract, allegedly universal principles, notably that of equality. Like the old Jacobinism, it does not oppose economic inequalities, but it scorns traditional religious, moral, and cultural preconceptions and social patterns that restrict or channel social and political advancement and economic activity . . . The new Jacobins are more accepting of existing society than were the old Jacobins, for they regard today's Western democracy as the result of great moral, social and political progress since the eighteenth century. They see it as an approximation of what universal principles require. (21)

Yet, in the age of globalism, Jacobinism has assumed more of an internationalist character. Ryn states:

The old Jacobins assumed that their principles were for all but as they faced pressing and specific obstacles near to home and were culturally focused on France and Europe they did not, for the most part, think globally. The new Jacobins do. They put stress on the international implications of their principles. The new Jacobinism is indistinguishable from democratism, the belief that democracy is the ultimate form of government and should be installed in all societies of the world. The new Jacobinism is the main ideological and political force behind present efforts to turn democracy into a worldwide moral crusade. (21)

It is interesting to note that the new Jacobinism makes democracy synonymous with freedom. However, pure-strain democracies historically fall prey to demagoguery and mobocracy. Typically, such a sociopolitical climate precedes the emergence of some sort of dictatorship. Ryn expands on the subject of democracy and its neo-Jacobin context:

To demonstrate the ideological and practical import of neo-Jacobin democratism it is necessary to make a distinction between two different forms of popular government. . . Although both notions of government are referred to as "democracy," they are not different versions of the same kind of government. They imply radically different understandings of human nature and society and have radically different institutional entailments. They are ultimately incompatible. The one may be called constitutional or representative democracy, the other plebiscitary or majoritarian democracy. The former is compatible with the old Western view of man's moral predicament. The latter flows from the kind of ideas that Rousseau and the Jacobins advocated. Constitutional democracy means popular rule under self-imposed restraints and representative, decentralized, institutions. Its aim is not to enact the popular wishes of the moment but to articulate what in American constitutional parlance is called the "deliberate sense" of the people. Plebiscitary democracy aspires to rule according to the popular majority of the moment. To ensure the speediest possible implementation of their wishes it seeks the removal of representative, decentralized and decentralizing practices and structures that limit the power of the numerical majority. Plebiscitary democrats recognize that, especially in large modern societies, the people need government officials, "representatives," to serve them, but the proper role of these officials is not to exercise independent judgment in determining the public interest. They should be agents of the people. (50)

According to Ryn, constitutional democracy acknowledges the moral complexities of the human condition:

Constitutional democracy assumes a human nature divided between higher and lower potentialities and sees a need to guard against merely self-serving, imprudent, and even tyrannical impulses in the individual and the people as a whole. According to John Adams, who was a key figure in shaping early American political thinking, the people can be as tyrannical as any king. The Framers of the American Constitution gave no power to the people as a national entity and placed various restraints on the ability of majorities to get their way. So also did the Framers try to protect against the abuse of power by government officials by instituting checks and balances. In a constitutional democracy the people and their representatives adopt restraints on power to arm themselves in advance against their own moments of weakness and shortsightedness. In awareness of the flaws of human nature, they do not wish to be governed according to their own impulse of the moment. Sound government requires that the popular opinions of the moment be carefully scrutinized, sometimes that they be resisted by responsible leaders. The real, more considered, and enduring will of the people emerges over time through the interplay between popular opinion, as expressed in elections and public debate, and the informed, independent judgment of popular representatives. (50-51)

Ryn correctly observes the Founding Father's derision for plebiscitary democracy:

> The American Framers set up a government that, by the standards of their own time, had a strong bias in favor of popular consent. They can even be said to have created a special kind of democracy. To understand the Framers' conception of good government it is necessary, however, to remember that they had a very low opinion of what they called "democracy" or "pure democracy." They associated it with demagoguery, rabble-rousing, opportunism, ignorance, and general irresponsibility. One of their chief aims was to protect against such possible manifestations of popular government. While envisioning broad popular participation in politics, they sought to shield most of those charged with making decisions from the momentary popular will. Except in the case of the House of Representatives voters could affect the decisions of the national institutions of government only indirectly, and even the members of the House could, for the two years of their term of office act independently of their constituencies, should they see the need to do so. The Framers never contemplated universal suffrage in the modern democratic sense, expecting only male property-owners to participate in politics and voting. Fearful of the passions of popular majorities, they sought to create structures conducive to responsible articulation of the long-term interests of the people as a whole. (51)

It comes as little surprise that an ideologue of the old Jacobin tradition also religiously espoused democracy: Karl Marx. Marx recognized democracy as the catalyst for installing a dictatorship of the proletariat, which typically disguised the rule of a hidden oligarchy. It is precisely this Marxist variety of democracy that neo-Jacobins are working to establish.

In addition, neo-Jacobins also share Marx's dichotomous logic. Ryn explains:

> The neo-Jacobin view of the world splits humanity into two camps, those wishing the good of humanity and those perversely resisting it. Here as in so many other ways the new Jacobins follow Rousseau, for whom political opinions are

either moral or immoral: the line between good and evil is not blurry; good and evil are not found on all sides. For Karl Marx, social and political existence was defined not by moral community or nationality but by class, but Marx, too, split humanity into separate camps that could not hope to co-exist or accommodate each other. His notion of the inevitable, unrelieved, and intensifying conflict between the bourgeoisie and the proletariat denied the existence of a common humanity that would make it possible for human beings to see beyond their different economic interests and through which they might come to recognize the legitimacy of claims other than their own. Marx denied, in effect, that people of different backgrounds might resonate to, and be humanized by, the same poetry, art, music, or moral example. Had Marx admitted the possibility of a shared human frame of reference that operated right across the disagreements and diversity of human beings, he would have had to accept the possibility of a muting of hostilities, of a give-and-take between interests, even of reconciliation among them.

The new Jacobinism resembles Marxism in that it does not contemplate compromise or mutual respect between those who are doing mankind's bidding and those who stand in their way. Universal principles must prevail. You are either for or against political virtue, which is to say that conflict will always be in the air. (181-82)

Thus, neo-Jacobins seek to universalize the maxims of democracy without compromise. It is just such rigid thinking that has provided a rationale for America's recent militaristic campaigns in Iraq and Afghanistan. The neo-Jacobin crusade for world democracy is little more than the latest incarnation of Trotsky's world communist revolution. However, the neo-Jacobins have yet to develop the sort of political cunning boasted by their Marxist progenitors. Ryn elaborates:

The new Jacobins have worked to expand their influence within a democratic society whose laws and norms of appropriate behavior, though increasingly shaky, still impose some limits on political action. The new Jacobins have not yet exhibited the kind of political ruthlessness that was characteristic of Marxism and French Jacobinism before it. (191)

However, the neo-Jacobins are swiftly attaining parity with their devious predecessors:

Yet their diligent and cliquish pursuit of power and their ideological fervor are reminiscent of those of the Marxists and the old Jacobins. There are grounds for suspecting that, upon gaining a firmer hold on power, the new Jacobins will gravitate in the direction of more despotic methods. They are already employing systematic demonization and ostracism of their critics and the attempted destruction not only of their reputations but also of their livelihood. As the restraints of American constitutionalism continue to deteriorate, military or other emergencies will provide neo-Jacobin leaders widening opportunities for silencing their opponents as well as for imposing general restrictions on civil liberties. The neo-Jacobin movement already exercises great influence in the United States, especially within the foreign policy establishment, and it has potentially unobstructed access to great military power, the CIA, the FBI, and the Department of Homeland Security. (191)

Of course, these facts are more than a little unsettling. With such immense power within their grasp, the neo-Jacobins could eventually transform America into a scientific dictatorship reminiscent of revolutionary France. Ryn reveals the variety of individual that constitutes the new Jacobinism:

A sign of the power of the new Jacobinism is that it is well represented across the political-intellectual spectrum. It is common among liberals and socialists, many of whom consciously trace their own ideological lineage to the French Revolution. Paradoxically, in the United States the new Jacobinism also finds expression among people called "conservatives" or "neoconservatives." This is a curious fact considering that modern, self-conscious conservativism originated in opposition to the ideas of the French Revolution. The person commonly regarded as the father of modern conservativism, the British statesman and thinker Edmund Burke (1729-97) focused his scorching critique on the French Revolution precisely on Jacobin thinking. (21-22)

Indeed, it is paradoxical that neoconservativism purports to be an outgrowth of traditional conservativism. It is equally paradoxical that most neoconservative Jacobins claim to be Christians. Neither traditional conservativism nor Christianity is reconcilable with the Jacobin principles espoused by the neoconservative. Appellations aside, the neoconservative is actually just another variety of Marxist and, therefore, qualifies as a sociopolitical Darwinian.

While they peddle empty "Christian" rhetoric, neoconservatives still view their variety of technocratic government as the desired outcome of humanity's alleged political evolution. The neoconservative believes that this political evolution can be consciously guided through the alchemy of geopolitics. The geopolitical schema of neoconservativism automatically lends itself to the promulgation of violent conflict, which is central to the evolutionary process. In the end, neoconservativism relocates the Darwinian "struggle for survival" within a sociopolitical context. Even if the neoconservative believes himself or herself to be sincere about their "Christian" faith, their Jacobin thinking unavoidably predisposes him or her to sociopolitical Darwinism.

As Ryn previously stated, the new Jacobins are growing closer to consolidating and controlling the machinations of power. Just how close are they to this objective? Seymour Hersh may have already answered this question:

> [O]ne of the things that you could say is, the amazing thing is we are been taken over basically by a cult, eight or nine neo-conservatives have somehow grabbed the government. (No pagination)

The title of James Billington's seminal book, *Fire in the Minds of Men*, was a literary allusion to Fyodor Dostoevsky. In his novel *The Possessed*, Dostoevsky uses the same phrase to describe the revolutionary faith that was emerging in 19th century Russia. During his second inaugural address, George W. Bush invoked the exact same phrase:

> By our efforts, we have lit a fire as well—a *fire in the minds of men*. It warms those who feel its power, it burns those who fight its progress, and one day this untamed fire of freedom will reach the darkest corners of our world. (No pagination)

Has America been designated the new headquarters of the elite's next 'scientific dictatorship?' One thing is certain. Although she is no longer worshipped in the cathedral of Notre Dame, the Goddess of Reason has never relinquished her crown.

The Rise of Modern Scientific Dictatorships

Darwinism shares the Hegelian framework with two other belief systems. In *The Secret Cult of the Order*, Antony Sutton states: "Both Marx and Hitler have their philosophical roots in Hegel" (118). It is here that one arrives at the Hegelian nexus where Darwin, Marx, and Hitler intersect. Recall that Nietzsche-ism, Darwinism and Marxism were all mentioned together in the *Protocols of the Wise Men of Sion.* This was no accident. Nazism (a variant of fascism) sprung from Nietzsche-ism (Carr, *Pawns in the Game*, xiv). Communism sprung from Marxism. Both communism and Nazism (a variant of fascism) were based upon Hegelian principles. Moreover, both were "scientific dictatorships" legitimized by the "science" of Darwinism. Ian Taylor elaborates:

> However, Fascism or Marxism, right wing or left-- all these are only ideological roads that lead to Aldous Huxley's brave new world [i.e. scientific dictatorship], while the foundation for each of these roads is Darwin's theory of evolution. Fascism is aligned with biological determinism and tends to emphasize the unequal struggle by which those inherently fittest shall rule. Marxism stresses social progress by stages of revolution, while at the same time it paradoxically emphasizes peace and equality. There should be no illusions; Hitler borrowed from Marx. The result is that both Fascism and Marxism finish at the same destiny--totalitarian rule by the elite. (Taylor 411)

The interest of both Hitler and Marx in Darwinian evolution is a matter of history. The interest of both Hitler and Marx in Darwinian evolution is a matter of history. In his comprehensive book, *American Socialists and Evolutionary Thought 1870-1920*, author Mark Pittenger describes Marx's elation over the publication of Darwin's theory and its subsequent adoption by socialist movements:

> In December of 1859, shortly after the publication of *The Origin of Species*, Friedrich Engels wrote to Marx: "Darwin, whom I am just reading, is splendid." Marx responded: "Although it is developed in the crude English style, this is the book which contains the basis in natural history for our view." Over the ensuing decades, the theorists of scientific

socialism would often praise Darwin for having convincingly historicized nature, naturalized humankind, and discredited all metaphysical and teleological world-views (15).

In the late 1860s, Marx was reported to have declared: "Nothing gives me greater pleasure than to have my name linked onto Darwin's. His wonderful work makes my own absolutely impregnable. Darwin may not know it, but he belongs to the Social Revolution" (Pittenger 17). While he was living in London, Karl Marx attended lectures on evolutionary theory delivered by T.H. Huxley (Taylor 381). Recognizing the odd synchronicity between the communist concept of class war and the Darwinian principle of natural selection, Marx sent Darwin a copy of *Das Kapital* in 1873 (Taylor 381). Within this work, Marx called Darwin's theory "epoch-making" (Pittenger 17). Enamored of evolution, Marx asked Darwin the permission to dedicate his next volume to him six year later. Troubled by the fact that it would upset certain members of his family to have the name of Darwin associated with an atheistic polemic, Charles politely declined the offer (Taylor 381).

Numerous authors have established firm connections between Darwinism and Hitler's Nazism. Evolutionary theory underpinned the very philosophy of the Third Reich:

> One of the central planks in Nazi theory and doctrine was. . .evolutionary theory [and]. . .that all biology had evolved. . .upward, and that. . .less evolved types. . .should be actively eradicated [and]. . .that natural selection could and should be actively aided, and therefore [the Nazis] instituted political measures to eradicate. . .Jews, and. . .blacks, whom they considered as "underdeveloped" (Wilder-Smith 27).

Commenting on the Darwinian influence upon Hitler, historian Hickman writes:

> (Hitler) was a firm believer and preacher of evolution. Whatever the deeper, profound, complexities of his psychosis, it is certain that [the Darwinian notion of perpetual struggle was significant because]. . .his book, *Mein Kampf*, clearly set forth a number of evolutionary ideas, particularly those emphasizing struggle, survival of the fittest and the extermination of the weak to produce a better society. (51-52)

The title for Hitler's own manifesto, *Mein Kampf* (translated: *My Struggle*), was inspired by the Darwinian concept of the struggle for survival. In an analysis of *Mein Kampf*, contemporary author Werner Maser reveals that Darwin was the crucible for Hitler's "notions of biology, worship, force, and struggle, and of his rejection of moral causality in history" (Taylor 409). In fact, in *Evolution and Ethics*, Darwinian Arthur Keith candidly stated: "The German Fuhrer as I have consistently maintained, is an evolutionist; he has consciously sought to make the practice of Germany conform to the theory of evolution" (230).

Returning to the Hegelian nexus that binds Darwinism, Marxism, and Nazism, both the fascist and communist "scientific dictatorships" represented political enactments of the dialectical framework resident in evolutionary theory. Marx was greatly influenced by Hegel (Taylor 381). The concept of class struggle, which paralleled Darwinian natural selection, resulted from Marx's redirection of the Hegelian dialectic towards the socioeconomic realm. The proletariat (thesis) comes into conflict with the bourgeois (antithesis), resulting in a classless Utopia (synthesis). Marx, however, rejected the concept of a world spirit and relocated the revolution's causal source within the proletariat itself.

The same Hegelian framework was evident within Hitler's genocidal Final Solution. The German people (thesis) came into conflict with the Jew (antithesis) in hopes of creating the Aryan (synthesis). In both the case of communism and Nazism, the results were enormous bloodbaths. This is the natural consequence of Darwinian thinking and the legacy of the "scientific dictatorship."

Freemasonry played an integral role in the ascendancy of both of these "scientific dictatorships." This continuity of Masonic involvement in the formation of Darwinian-dignified oligarchies suggests that evolutionary theory was developed specifically for the purposes of creating and legitimizing Technocracies.

The Freemasonic connection to communism can be identified in the person of Adam Weishaupt, founder of what is probably the most infamous branch of Illuminism, the Bavarian Order of Illuminists. Librarian of Congress James Billington introduces this secret society:

> The Order of Illuminists was founded on May 1, 1776, by a professor of canon law at the University of Inglostadt in Bavaria, Adam Weishaupt, and four associates. The order was secret and hierarchical, modeled on the Jesuits (whose long domination of Bavarian education ended with their

abolition by the Papacy in 1773) and dedicated to Weishaupt's Rousseaian vision of leading all humanity to a new moral perfection freed from all established religious and political authority. (Billington 94)

Billington explains that Weishaupt's Illuminist activities involved Masonry:

> The Illuminists attempted to use the ferment and confusion in Freemasonry for their own ends. Weishaupt joined a Masonic lodge in Munich in 1777; and attempted to recruit "commandos" (groups of followers) from within the lodges of the Bavarian capital. (95)

Why did Weishaupt consider the Masonic lodges a good vehicle for his Illuminist conspiracy? Some researchers hold that Illuminist doctrine and the teachings of Masonry have always been similar, if not identical. Deceased former Naval Intelligence officer William Cooper contended that the Illuminist involvement in Masonry was not the result of infiltration:

> Allegations that the Freemason organizations were infiltrated by the Illuminati during Weishaupt's reign are hogwash. The Freemasons have always contained the core of Illuminati within their ranks, and that is why they freely and so willingly took in and hid the members of Weishaupt's group. You cannot really believe that the Freemasons, if they were only a simple fraternal organization, would have risked everything, including their very lives, by taking in and hiding outlaws who had been condemned by the monarchies of Europe. It is mainly Freemason authors who have perpetuated the myth that Adam Weishaupt was the founder of the Illuminati and that the Illuminati was destroyed, never to surface again. (77)

One Freemasonic source supporting Cooper's contention is Kenneth Mackenzie. In his *Royal Masonic Cyclopaedia*, Mackenzie refers to Illuminist involvement in Masonry as "an attempt to purify Masonry, then in much confusion" (133). Thirty-third degree Freemason Albert Mackey even goes as far as to call Weishaupt "a Masonic reformer" (843). It seems Freemasonry was not the unwitting victim of Illuminist infiltration and corruption. Weishaupt was merely enacting corrective measures within Masonry, thus restoring its Illuminist tradition. In this context, Weishaupt is tantamount to a Martin Luther of Illuminism.

Billington explains that even though Illuminism was suppressed, it did not cease to exist. Fleeing members and remnants of Weishaupt's sect kept Illuminism alive:

> The order was subjected to ridicule, persecution, and formal dissolution during 1785-87. Weishaupt was banished to Gotha and kept under surveillance. But the diaspora of an order that had reached a membership of perhaps two thousand five hundred at its height in the early 1780s led to a posthumous impact that was far greater throughout Europe than anything the order had been able to accomplish during its brief life as a movement of German intellectuals. (96)

There is strong evidence for the transmission of Illuminism from Weishaupt, the "Masonic reformer," to Karl Marx. This transmission was facilitated through various revolutionaries and revolutionary groups. One revolutionary influence upon Marx was Francois-Noel Babeuf. In 1795, Babeuf published a work that would lay the foundation for Marx's *Communist Manifesto.* According to James Billington, this work, entitled *Plebian Manifesto*, was "the first in the new genre of social revolutionary manifestos which would culminate in Marx's *Communist Manifesto*" (74). Was this precursor to Marx's work influenced by Illuminism? Billington presents evidence that this was the case:

> Occult--possibly Illuminist--influence is detectable in Babeuf's first clear statement of his communist objectives early in 1795--inviting a friend to "enter into the sacred mysteries of agrarianism" and accepting fidelity from a *chevalier de l'ordre des egaux.* Babeuf's subsequent first outline for his conspiracy spoke of a "circle of adherents" "advancing by degree" from *les pays limotrophes* to transform the world. Babeuf's secret, hierarchical organization resembled that of Illuminists and of Bonneville. The strange absence of references by Babeuf and the others to the man who formulated their ultimate objectives, Sylvain Marechal, could be explained by the existence of an Illuminist-type secrecy about the workings of the inner groups. The conspirators may have viewed Marechal as the "flame" at the center of the circle." As such, he would have had to be protected by the outer circle against disclosure to profane

outsiders. His mysterious designation of Paris as "Atheopolis" and himself as *l'HSD* (*l'homme sans dieu*) represented precisely the ideal of Weishaupt's inner Areopagites: man made perfect as a god-without-God. (97)

An even stronger connection between Karl Marx and Illuminism was apparent in the revolutionary group known as the League of the Just. Gary Allen explains:

Karl Marx was hired by a mysterious group who called themselves the League of Just Men to write the Communist Manifesto as demagogic boob-bait to appeal to the mob. In actual fact the Communist Manifesto was in circulation for many years before Marx's name was widely enough recognized to establish his authorship for this revolutionary handbook. All Karl Marx really did was to update and codify the very same revolutionary plans and principles set down seventy years earlier by Adam Weishaupt, the founder of the Order of Illuminati in Bavaria. And, it is widely acknowledged by serious scholars of this subject that the League of Just Men was simply an extension of the Illuminati which was forced to go deep underground after it was exposed by a raid in 1786 conducted by the Bavarian authorities. (25-26)

According to James Billington, Marx did, in fact, contact and cement an alliance with the League of the Just during the period of 1846-47 (270). The League of the Just was an outgrowth of an earlier revolutionary group known as the League of the Outlaws (184). While preparing the group's structure, Outlaw leader Theodore Schuster "borrowed directly from Buonarroti's final fantasy of a Universal Democratic Carbonari" (183). The Universal Democratic Carbonari was "the last effort to realize Buonarroti's dream of an international revolutionary organization" (176). Buonarroti's influence was also felt through Johann Hoeckerig, a key member of the tailor faction of both the Outlaws and the League of the Just (185). Hoeckerig "was a protégé and visitor of Buonarroti in his last days" (185). It is through the revolutionary Buonarroti that the Illuminist influence on the Outlaws and their successor, the League of the Just, comes into clearer focus. Billington writes:

The organizational plan that Buonarroti distilled from two decades of revolutionary experience in Geneva (and basically remained faithful to for the rest of his life) was simply lifted from the Bavarian Order of Illuminists. (93)

In fact, Buonarroti:

had been fascinated with Illuminism even before the revolution. Already in 1787, he drew ideas from Mirabeau and noted the struggle of Illuminism with Catholicism in Bavaria. (Billington 97)

Buonarroti's own revolutionary comrades recognized Illuminism as the driving force behind his revolutionary beliefs. Billington elaborates:

Gioacchino Prati, a young student from Trentino who later became one of Buonarroti's closest collaborators, traced the Illuminist connection when he contended that Buonarroti's first revolutionary organization, the Sublime Perfect Masters, "was instituted during the first French Revolution." (98)

Concerning the Illuminist influence upon Buonarroti, Billington provides this summation:

Whether or not Buonarroti was in effect propagating an Illuminist program during his revolutionary activity of the 1790s, he had clearly internalized a number of Illuminist ideas well before the massive borrowing in his revolutionary blueprint of 1810-11. He had adopted the Illuminist pretension of recovering a natural religion known only to "Illuminated" sects in the past. . .He followed Weishaupt and Bonneville in attaching special importance to the Jesuits, whom he sought both to imitate and to liquidate. His secret ideal was from the beginning, according to Prati, the egalitarian Illuminist one of breaking down all "marks of private property." (99)

Both the Outlaws and their successor, the League of the Just, had internalized Buonarroti's Illuminist ideas and were continuing with the same revolutionary tradition. It is possible that the Outlaws and the League had embraced Buonnarroti's Illuminist concepts because they were already Illuminist. This contention is supported by the organization's German origin. Billington writes:

> One decade after Buonarroti's death in 1837 and eight years after Blanqui's eclipse, the social revolutionary tradition gave birth to the Communist League. A small group of young German émigrés created this short-lived but historic organization. They took over the struggle within the German emigration "between national republicans and communist republicans," and produced a leader for the latter camp in Karl Marx. (182)

Weishaupt's Illuminist and the Communist League's parent, the Outlaws, share a common birthplace. According to researcher Ralph Epperson:

> After the Illuminati was discovered in Bavaria, Germany, its members scattered throughout Europe. The League was an "off-shoot of the Parisian Outlaws League, founded by German refugees." One can only wonder if those refugees were the scattering Illuminati. (*The Unseen Hand*, 94)

If these links are too tenuous, consider the testimony of Christian G. Rakovsky, one of the founders of Soviet Bolshevism. Before World War II, Rakovsky was a victim of Stalin's show trials (Griffin 253). During an interrogation conducted by a NKVD officer, Rakovsky admitted forthright that communism was merely a continuation of Weishaupt's Illuminism:

> You know that according to the unwritten history known only to us, the founder of the First Communist International is indicated, of course secretly, as being Weishaupt. You remember his name? He was the head of the masonry which is known by the name of the Illuminati; this name he borrowed from the second anti-Christian conspiracy of that era-gnosticism. (278)

There seems to be little doubt that Illuminism had found vessels in both the Outlaws and its tributary, the League of the Just. In this context, Marx can be seen as merely reiterating "Masonic reformer" Weishaupt's teachings with only minor modifications. *Communist*

Manifesto may just as well have been entitled *Illuminist Manifesto*. In its pages, serious students will find many Illuminated Masonic ideas and concepts stripped of Masonic vernacular. While it may turn not a few stomachs on the Left, the ideology that was supposedly free of all mysticism actually finds its origins in mysticism. Billington put it quite succinctly when he wrote:

> The revolutionary faith was shaped not so much by the critical rationalism of the French Enlightenment (as is generally believed) as by the occultism and proto-romanticism of Germany. This faith was incubated in France during the revolutionary era within a small subculture of literary intellectuals who were immersed in journalism, fascinated by secret societies, and subsequently infatuated with "ideologies" as a secular surrogate for religious belief. (3-4)

Many of the ideologies that were to act as surrogates for religious belief were formulated behind Lodge doors. Furthermore, Billington states that the revolutionary faith began its spread "when some European aristocrats transferred their lighted candles from Christian altars to Masonic lodges" (5-6). Behind the revolutionary symbol of a clenched fist hides the Masonic handshake.

Masonry also played a role in the creation of the Nazi scientific dictatorship. This role has gone largely undetected due to Nazi persecution of masons and other occultists before and during World War II. After all, the skeptical researcher contends, why would Hitler and his followers entertain the concepts and beliefs of a group that they have suppressed? Author Joseph Carr provides a possible explanation to this question:

> So why would an occultist like Adolf Hitler, or his own Thule Society, seek to suppress or destroy other occultists? Thulists tried to murder Steiner, and when the concentration camps opened after 1933, Freemasons and other occultists were imprisoned along with the Jews and common criminals. Why? At first, I believed that the response of Hitler and his Thulists to other occultics was similar to the response of the medieval church to heresy: there is no sinner quite so bad as a believer who has fallen into error! Perhaps the seemingly minor difference between the various esoteric belief systems loom very large to other occultic "true believers". But an

alternate and somewhat more viable answer was provided to me by a friend who was knowledgeable in such matters. His explanation of Hitler's behavior is that Hitler did not want the public to know of his occultic connection and was afraid that other initiates into "secret knowledge" would easily recognize and expose his involvement. Although Christians would have a little heartburn with the idea that not all occultists are outright Satanists, there are many occultists who are genuinely unaware of the demonic nature of their experiences; they believe that their knowledge is good and proper for man. That segment of the occultic world, plus those occultist ethicists who claim to derive supposedly high moral standards from occultic teachings, would have trumpeted Hitler's true nature from the housetops if they had not been suppressed. (*The Twisted Cross*, 89-90)

Carr continues:

After the Nazis came to power they perceived a need for respectability. Occultism is almost universally regarded in the west as the province of cranks and lunatics, and that image did not conform to the appearances that the Nazis wanted to project. It is also probable that the Nazis desperately wanted to conceal the occultic origins of their movement. In suppressing the record left by these groups they effectively hid the facts from view. Only a few times before World War II did anyone perceive the occultic origins of National Socialism. (*The Twisted Cross*, 91-92)

Suppression of occultists successfully concealed the occult influence evident within Nazism. One such influence was Madame Helena P. Blavatsky, the founder of Theosophy in 1875. When U.S. Army historians catalogued Hitler's personal library, they found amongst the various occult volumes a copy of Blavatsky's *The Secret Doctrine* (Carr 93). According to Carr:

The Secret Doctrine was introduced to Hitler by Dietrich Eckart during the early 1920s', and he was taught its secrets by Professor-General Karl Haushofer. (*The Twisted Cross*, 93)

Could Blavatsky's influence on Hitler be considered Masonic? Dennis Cuddy presents evidence that this may be the case:

> Madame Blavatsky founded the Theosophical Society in 1875. On November 24, 1877, she was sent her Masonic certificate, and she and Masonic leader Albert Pike had been seen walking together in Washington, D.C. In 1887 Blavatsky began publication of a journal named Lucifer, which she would later coedit with Annie Besant who wore a swastika on a necklace (on February 19, 1922, Besant's co-Masonry will form an alliance with the Grand Orient Masonic Lodges of France). (19)

A more substantial connection between Nazism and Masonry lies in the Thule group. This occult secret society seems to have been the hidden power behind the Nazis and one of Hitler's most closely guarded secrets. Trevor Ravenscroft elaborates:

> Above all he [Hitler] kept silent about the fact that the [German Workers Party] Committee and the forty original members of the forty original members of the New German Workers' Party were all drawn from the most powerful Occult Society in Germany which was also financed by the High Command–The Thule Gessellschaft. (102)

When the New German Workers' Party transformed into the Nazi party, it took Thule personnel and beliefs with it. Those beliefs appear to be Masonic in origin. Jean-Michel Angebert points out that Thule was "but a fragment of a much more important secret society known as the Germanic Order founded in 1912" (164). The Germanic Order had "gathered together certain lodges of Prussian Freemasonry"(167). One influential member of the Germanic Order and founder of Thule was Baron Rudolf von Sebottendorff. Sebottendorff had been initiated into a Masonic society while in Egypt (Levenda 49).

This Freemasonic influence upon Nazism had to be concealed. To achieve this end, the Nazis employed what has come to be known as "the blood libel." Carl Raschke states: "Historically, the blood libel seems to have been a dodge by aristocrats practicing satanism" (*Painted Black*, 231). To divert attention away from them, occultists will accuse despised minorities of engaging in the very occult activities that they themselves practice. In this case, the despised minority was the Jews.

To accomplish this goal, the Nazis employed *The Protocols of the Elders of Zion*, which Raschke classifies as the "most notorious of the blood libel documents" (*Painted Black* 231).

Heinrich Himmler was cribbing liberally from *The Protocols* when he told his therapeutic masseur Dr. Felix Kersten that Freemasonry was controlled by a "world Jewish conspiracy" (Carr, *The Twisted Cross*, 92-93). Jews and ordinary masons were imprisoned and executed. Meanwhile, the fact that Nazism was derivative of Freemasonic occultism was successfully obfuscated.

So emerged the modern "scientific dictatorships" of communism and fascism. In the years to come, these two "scientific dictatorships" would engage in numerous skirmishes for primacy. However, they were always little more than variants of the same Masonic concept. The core commonalities of these two political movements can be demonstrated through simple linguistic analysis.

The appellation of "communism" comes from the Latin root *communis*, which means "group" living. Fascism is a derivation of the Italian word *fascio*, which is translated as "bundle" or "group." Both fascism and communism are forms of coercive group living, or more succinctly, collectivism. The only substantial difference between the two is fascism's limited observance of private property rights, which is ostensible at best given its susceptibility to rigid government regulation. In 1933, the Fuehrer candidly admitted to Hermann Rauschning that: "the whole of National Socialism is based on Marx" (Martin 239). Nazism (a variant of fascism) is derivative of Marxism. The historical conflicts between communism and fascism were merely feuds between two socialist totalitarian camps, not two dichotomously related forces.

As systems of socialism, both fascism and communism represent a continuation of Francis Bacon's technocratic tradition. In fact, technocratic concepts have pervaded the very fabric of socialist totalitarian regimes throughout history. Fischer elaborates:

In practice, the technocratic concept of the administrative state has been most influential in the socialist world of planned economies. Given their emphasis on comprehensive economic and social planning, the technocratic theory is ready-made both to guide and to legitimate the centralized bureaucratic decision-making systems that direct most socialist regimes. Easily aligned with the ideas and techniques of scientific planning, particularly those shaped by Marxist economists, technocratic concepts have played an important role in the evolution of socialist theory and practice. (25)

Communism and fascism are merely technocratic kissing cousins. Yet, many were seduced by both. Mentally spayed by Hegelian epistemology, people chose sides and participated in the fraudulent ideological melee. Ayn Rand probably provided the most eloquent summation of this dialectic:

It is obvious what the fraudulent issue of fascism versus communism accomplishes: it sets up, as opposites, two variants of the same political system; it eliminates the possibility of considering capitalism; it switches the choice of "Freedom or dictatorship?" into "Which kind of dictatorship?"—thus establishing dictatorship as an inevitable fact and offering only a choice of rulers. The choice—according to the proponents of the fraud—is: a dictatorship of the rich (fascism) or a dictatorship of the poor (communism). (180)

Indeed, the dialectic of communism against fascism leaves one with only one choice: "Which kind of dictatorship?" Whether one selects the communist variant or the fascist variant, the final choice is always a "scientific dictatorship."

Both communists and fascists have murdered millions. Professor R.J. Rummel estimates that the Nazis slaughtered roughly 20,946,000 (*Freedom, Democide, War*, no pagination). Meanwhile, the Soviet communists exterminated approximately 61,911,000, surpassing even their fascist brethen (*Freedom, Democide, War*, no pagination). Both of these Technocracies were premised upon evolutionary thought. Darwinian natural selection mirrored Hegelianism's dialectical framework, which was politically enacted by both the Nazis and Soviets. The collective atrocities of both fascists and communists throughout the twentieth century echo a single rationale: "Survival of the fittest." This is the legacy of the "scientific dictatorship."

Engineering Evolution: The Alchemy of Eugenics

A common misnomer that has been circulated by academia's anointed historians is that the alchemists of antiquity were attempting to transform lead into gold. In truth, this was a fiction promulgated by the alchemists themselves to conceal their ultimate objective . . .the transformation of man into a god. Among one of the various occult organizations that aspired to complete this alchemical mission was Freemasonry. Providing a summation of Masonry's supreme goal, Masonic scholar W.L. Wilmshurst writes:

> This--the *evolution of man into superman*--was always the purpose of the ancient Mysteries, and the real purpose of modern Masonry is, not the social and charitable purposes to which so much attention is paid, but the expediting of the spiritual evolution of those who aspire to perfect their own nature and transform it into a more god-like quality. And this is a definite science, a royal art, which it is possible for each of us to put into practice; whilst to join the Craft for any other purpose than to study and pursue this science is to misunderstand its meaning. (47)

According to this alchemical mandate, humanity is a gradually developing deity requiring scientific assistance in its evolution. In *Mystic Masonry*, 32nd degree Mason J.D. Buck reiterates this theme of man as a progressively apotheosizing organism: "Humanity, 'in-toto', then, is the only Personal God" (136). Of course, the concept of evolution would later be disseminated on the popular level as Darwinism and become the veritable cornerstone of contemporary science.

Sir Francis Galton could be considered an early evolutionary alchemist. His own cousin's theory of evolution was one of his chief inspirations. In *Memories of My Life*, Galton wrote:

> The publication in 1859 of the *Origin of Species* by Charles Darwin made a marked epoch in my own mental development, as it did in that of human thought generally. Its effect was to demolish a multitude of dogmatic barriers by a single stroke, and to arouse a spirit of rebellion against all ancient authorities whose positive and unauthenticated statements were contradicted by modern science. (287)

Viewing evolutionary theory in conjunction with the alchemical mandate for man's consciously engineered apotheosis, one inevitably recognizes a belief system that exhibits all of the characteristics of a religion. This revelation is most clearly illustrated by Galton's statements in *Inquiries into Human Faculty and its Development*:

The chief result of these Inquiries has been to elicit the religious significance of the doctrine of evolution. It suggests an alteration in our mental attitude, and imposes a new moral duty. The new mental attitude is one of a greater sense of moral freedom, responsibility, and opportunity; the new duty which is supposed to be exercised concurrently with, and not in opposition to the old ones upon which the social fabric depends, is an endeavor to further evolution, especially that of the human race. (337)

That Galton recognized the "religious significance of evolution" is no accident. Throughout the years, this Masonically inspired religion of emergent deities has resurfaced under various appellations. Wagar enumerates its numerous manifestations:

Nineteenth- and early twentieth-century thought teems with time-bound emergent deities. Scores of thinkers preached some sort of faith in what is potential in time, in place of the traditional Christian and mystical faith in a power outside of time. Hegel's *Weltgeist*, Comte's *Humanite*, Spencer's organismic humanity inevitably improving itself by the laws of evolution, Nietzsche's doctrine of superhumanity, the conception of a finite God given currency by J.S. Mill, Hastings Rashdall, and William James, the vitalism of Bergson and Shaw, the emergent evolutionism of Samuel Alexander and Lloyd Morgan, the theories of divine immanence in the liberal movement in Protestant theology, and du Nouy's telefinalism--all are exhibits in evidence of the influence chiefly of evolutionary thinking, both before and after Darwin, in Western intellectual history. The faith of progress itself--especially the idea of progress as built into the evolutionary scheme of things-- is in every way the psychological equivalent of religion. (106 -07)

This emergent deity, Man (spelled with a capitalized M to connote his purported divinity), would be fully enthroned through the efforts of alchemists themselves. Galton would reintroduce the concept of alchemy under the appellation of "eugenics," a term derived from Greek for "well born." The basic precepts of eugenics were delineated in Galton's *Hereditary Genius*, a racist polemic advocating a system of selective breeding for the purposes of providing "more suitable races or strains of blood a better chance of prevailing over the less suitable" (24).

According to Galton, society should be eugenically regimented. The framework of such a society would be a caste system where status was assigned according to genetic superiority. In an article in the January 1873 edition of *Fraser's Magazine*, Galton stated:

> I do not see why any insolence of caste should prevent the gifted class, when they had the power, from treating their [lower caste] compatriots with all kindness, as long as they maintained celibacy. But if these continued to procreate children, inferior in moral, intellectual, and physical qualities, it is easy to believe that the time may come when such persons would be considered as enemies to the State, and to have forfeited all claims to kindness. (Qutd. in Chase 100)

Galton hoped that such societal regimentation would promote "eugenically sound" breeding amongst the citizenry. Summarizing Galton's objectives, Allan Chase explains: "What Galton was talking about here was the power to breed people as we breed pigs" (101). Of course, as George Orwell opined in *Animal Farm*, some pigs are more equal than others. Galton's cousin and racialist progenitor, Charles Darwin, shared this contention. In *The Descent of Man*, he writes:

> With savages, the weak in body or mind are soon eliminated; and those that survive commonly exhibit a vigorous state of health. We civilised men, on the other hand, do our utmost to check the process of elimination; we build asylums for the imbecile, the maimed, and the sick; we institute poor-laws; and our medical men exert their utmost skill to save the life of every one to the last moment. There is reason to believe that vaccination has preserved thousands, who from a weak constitution would formerly have succumbed to small-pox. *Thus the weak members of civilised societies propagate their kind. No one who has attended to the breeding of domestic animals will doubt that this must be highly injurious to the race of man.* It is surprising how soon a want of care, or care wrongly directed, leads to the degeneration of a domestic race; *but excepting in the case of man himself, hardly any one is so ignorant as to allow his worst animals to breed.* (No pagination; emphasis added)

According to Darwin, the pigs of higher stock were the Anglo-

Saxons. This becomes evident in Darwinian Josiah Strong's manifesto, *America's Destiny*. Quoting Darwin, Strong wrote:

> "At the present day," says Mr. Darwin, "civilized nations are everywhere supplanting barbarous nations, excepting, where the climate opposes a deadly barrier; and they succeed mainly, though exclusively, through their arts, which are the products of the intellect." He continues: "Whether the extinction of inferior races before the advancing Anglo-Saxon seems to the reader sad or otherwise, it certainly appears probable…Is there room for reasonable doubt that this race, unless devitalized by alcohol and tobacco, is destined to dispossess many weaker races, assimilate others, and mold the remainder, until, in a very true and important sense, it has Anglo-Saxonized mankind?" (165-80)

Of course, it comes as no real surprise that such thinking underpinned the racialist policies of Nazi Germany, which was a scientific dictatorship edified by Darwinian evolution. It comes as even less a surprise that Leonard Darwin, son of Charles, was vice-president of both the 1912 and 1921 International Eugenics Congresses. The first of these two meetings was the outgrowth of a 1911 gathering of the International Society for Racial Hygiene, a predominantly German organization. That Germany would see the full enactment of eugenical policies is hardly a coincidence.

Although the Nazis' eugenical Holocaust of WWII constituted an enormous public relations disaster for proponents of eugenics, the movement would later resurface under the banner of population control and radical environmentalism. Researchers Tarpley and Chaitkin document this transmogrification:

> The population control or zero population growth movement, which grew rapidly in the late 1960s thanks to free media exposure and foundation grants for a stream of pseudoscientific propaganda about the alleged "population bomb" and the limits to growth," was a continuation of the old prewar, protofascist eugenics movement, which had been forced to go into temporary eclipse when the world recoiled in horror at the atrocities committed by the Nazis in the name of eugenics. By mid-1960s, the same old crackpot eugenicists had resurrected themselves as the population-control and

environmentalist movement. Planned Parenthood was a perfect example of the transmogrification. Now, instead of demanding the sterilization of the inferior races, the newly packaged eugenicists talked about the population bomb, giving the poor "equal access" to birth control, and "freedom of choice." (203)

Indeed, Planned Parenthood successfully carried the banner of eugenics into the post-WWII era. Planned Parenthood was founded by Margaret Sanger, a virulently racist woman who touted the slogan: "Birth Control: to create a race of thoroughbreds." Her manifesto, entitled *The Pivot of Civilization*, thoroughly delineates the mission of Planned Parenthood and its allied organizations in the eugenics movement. In this treatise, which featured an introduction written by Freemason and Fabian socialist H.G. Wells, Sanger reveals the true motives underpinning the promotion of birth control:

> Birth Control, which has been criticized as negative and destructive, is really the greatest and most truly eugenic method, and its adoption as part of the program of Eugenics would immediately give a concrete and realistic power to that science. . .as the most constructive and necessary of the means to racial health. (*The Pivot of Civilization*, 189)

Sanger believed that society's tolerance of "morons," "human weeds," and the "feeble-minded' was encouraging dysgenics. To remedy this purported genetic threat, Sanger unabashedly promoted the implementation of authoritarian measures:

> The emergency problem of segregation and sterilization must be faced immediately. Every feeble-minded girl or woman of the hereditary type, especially of the moron class, should be segregated during the reproductive period. . .we prefer the policy of immediate sterilization, of making sure that parenthood is absolutely prohibited to the feeble-minded. (*The Pivot of Civilization*, 101-02)

Understand, these are the words of a so-called "proponent of reproductive rights." Moreover, Sanger desired to see the establishment of a gulag system within America for the internment of the "feeble-minded." In an issue of *Birth Control Review*, she wrote:

To apply a stern and rigid policy of sterilization and segregation to that grade of population whose progeny is already tainted. . .to apportion farm lands and homesteads for these segregated persons where they would be taught to work under competent instructors for the period of their entire lives. . . ("Plan of Peace,"107-08)

Although Sanger's gulag system was not formally enacted in the United States, her vision saw horrible fulfillment in Nazi Germany. It comes as little surprise that Planned Parenthood's board of directors included Nazi supporters such as Dr. Lothrop Stoddard, author of a racist tract entitled *The Rising Tide of Color Against White Supremacy*. In fact, *Birth Control Review* acted as a conduit for the dissemination of Nazi propaganda in America. In April of 1933, Dr. Ernst Rudin, Hitler's director of genetic sterilization and a founder of the Nazi Society for Racial Hygiene, published an article in *Birth Control Review*. Entitled "Eugenic Sterilization: An Urgent Need," the article presented the following appeal:

The danger to the community of the unsegregated feeble-minded woman is more evident. Most dangerous are the middle and high grades living at large who, despite the fact that their defect is not easily recognizable, should nevertheless be prevented from procreation. . . In my view we should act without delay ("Eugenic Sterilization: An Urgent Need," 102-04)

Of course, in Rudin's native country, the "feeble-minded" did not remain "unsegregrated" for very long. The same year that Sanger's publication printed Rudin's article, Ernst collaborated with Heinrich Himmler on Germany's 1933 sterilization law. This genocidal edict stipulated the sterilization of all Jews and "colored" German children. Eventually, the "undesirables" were collected, segregated, and systematically murdered. The final result of the Nazi eugenics program was the Holocaust, which claimed six million lives.

Yet, how many people would have been segregated for orderly disposal according to Sanger's vision? Upon examination of army statistics, Sanger concluded that:

. . .nearly half-47.3 per cent-of the population had the mentality of twelve-year-old children or less-in other words that they are morons.(*The Pivot of Civilization*, 263)

Sanger expressed dismal hopes for a vast segment of the population, declaring that: "only 13,500,000 will ever show superior intelligence" (*The Pivot of Civilization*, 264). Thus, only a meager 13.5% of the population would be permitted to procreate. The rest would be segregated for orderly disposal. Evidently, Sanger's holocaust would have even dwarfed Hitler's Final Solution.

In typical Darwinian fashion, Sanger showed little mercy towards the weak. In fact, Margaret expressed a distinct aversion towards the poor. Chapter Five of her book is entitled "The Cruelty of Charity." Reiterating Malthus' proposal to "disclaim the right of the poor to support," she wrote:

Organized charity itself is. . .the surest sign that our civilization has bred, is breeding and is perpetuating constantly increasing numbers of defectives, delinquents and dependents.(Sanger, *The Pivot of Civilization*, 108)

Sanger particularly loathed:

. . .a special type of philanthropy or benevolence,. . .which strikes me as being more insidiously injurious than any other. . .to supply gratis medical and nursing facilities to slum mothers (*The Pivot of Civilization*, 114)

According to Margaret, such an investment of time, effort, resources, and love represented the height of futility:

. . .we are paying for and even submitting to the dictates of an ever increasing, unceasingly spawning class of human beings who never should have been born at all. . .(*The Pivot of Civilization*, 187)

Planned Parenthood retains an active role in the scientific dictatorship's project of eugenical regimentation today. Despite revelations of Nazi atrocities constituted public relations disaster for the organization, Planned Parenthood survived and continues to tangibly enact Sanger's vision. In many ways, Sanger's vision synchronized with the Weltanschauung of the power elite. The underlying reason for this synchronicity is elementary. Simply stated, Sanger and other Malthusian ideologues are instrumental in class war. E. Michael Jones explains:

> As Scrooge had shown in *A Christmas Carol*, the Malthusian Ideology had always been used by the plutocrats as a way of diverting employees who wanted higher wages into thinking about ways to "decrease the surplus population." Birth control had always been the Malthusian answer to the worker clamoring for higher wages, and in 1930 Margaret Sanger, sensing a new window of opportunity, tried to use the Depression as a way of promoting birth control. The Rockefellers and the ethnic interests they represented funded Sanger to do just this: promote birth control as a solution to poverty of the Depression and divert the working classes thereby from asking for higher wages. (282-83)

This synchronicity was demonstrated by the wealthy Rockefeller dynasty's financial and political support of Sanger. In 1934, Sanger and Msgr. John A. Ryan "testified before congress on a bill that would make it legal to distribute contraceptives" (283). In a way, their testimonies represented the latest incarnation of the debate between Malthus and William Godwin (283). Obviously, Sanger carried the torch for Malthus. Exploiting the fear generated by stock-market crash of the 30s, Sanger argued that the Depression was directly attributable to overpopulation. To reinforce this contention, Sanger read a letter purporting to be authored by a poor mother before Congress:

> "My husband," Sanger began citing a letter from an anonymous woman impoverished by the Depression, "has been gone for more than 2 weeks looking for work, and I don't know where he is. I am almost barefoot and have only 2 badly worn dresses. . .and my 15 yr. old girl has been in the hosp. since Jan. So, Mrs. Sanger, if my poor miserable letter that comes from bitterness and want can help other wives and mothers to have less babies and more common sense and comfort then for God's sake use it." (283)

Needless to say, the author's anonymity certainly raises suspicion. It is possible that the letter was a fiction concocted to suit Sanger's radical agenda. Its hyperbolic and melodramatic language certainly reinforce this contention. Moreover, the letter's conclusion is fraught with fallacious reasoning. Jones elaborates:

Touching as we may find this letter, it is not self-evident that this woman's poverty came from the number of children she had, nor is it self-evident that if Margaret Sanger sent her birth control that her husband would get a job, or if he did get a job that it would pay him a decent wage, one whereby he could support his family. In fact, as the Malthusian ideology developed, it became, more often than not, an excuse not to pay a decent wage to the worker, since any increase in his well-being would only urge him to procreate more fervently, thereby once more outstripping the resources available to him. (283)

Meanwhile, Ryan reiterated the position of Godwin:

Malthus's "law," Ryan argued, was both an ideology, tailored to the interests of the ruling class of England, and a self-fulfilling prophecy. It became the lens through which economic injustice was rationalized in both England and America. Transmuted into a "law" of the sort proposed by Newtonian physics, the unequal distribution of wealth was transformed from a case of injustice into an inescapable scientific fact, and therefore a justification for maintaining an intolerably unjust status quo. Malthus may not have intended it as such, but his ideas quickly took on a life of their own and were adopted by the wealthy classes of both England and America as the rationale for their essentially unjust business practices. Malthus was, in fact, eventually to repudiate his belief that human populations would inevitably follow the growth trajectory of animal populations, but by then his ideas had taken on a life of their own, primarily because of their benefit to those who wanted to maintain the status quo. (282)

On June 13, 1934, Sanger's bill passed (285). However, it was recalled by Nevada Senator Pat McCarran and was eventually terminated (285). Yet, Sanger's religion became John D. Rockefeller III's crusade:

On March 1, 1934, at the height of the Ryan/Sanger debate in Congress, John D. Rockefeller III, scion of the Rockefeller family, wrote to his father urging him, in spite of shutting down the Bureau of Social Hygiene, to continue his support of both the American Birth Control League, which was to get $10,000, and Sanger's National Association for Federal Legislation, which was to get $1,000. He also announced that he had "one further statement in regard to my interest in birth control. I have come pretty definitely to the conclusion that it is the field in which I will be interested, for the present at least to concentrate my own giving, as I feel that it is so fundamental and underlying." (285)

The Rockefeller family's support of Sanger is most ironic. In April 1914, shortly after John D. Rockefeller, Jr. sanctioned the infamous Ludlow Massacre, Sanger demanded his assassination (142). In one of her articles, Sanger admonished readers to "Remember Ludlow" (142). This mandate prompted a warrant for Sanger's arrest and she was forced to flee to England (142). Later, she would claim that the indictments that precipitated her exile only mentioned "sending birth-control material through the mails, not the call to murder the scion of the Rockefeller family" (142).

This irony would only be compounded by the overwhelming hypocrisy exhibited by Sanger later into her life. Jones reveals that "by the time she was famous enough to be interviewed by the press, she was also the beneficiary of Rockefeller money as well" (142). Jones continues:

Within a period of ten years, Sanger went from a position calling for Rockefeller's murder, to having him as one of her major benefactors. Of all the benefactors she had during her lifetime, only one gave her more money than John D. Rockefeller, Jr., and that man was her second husband Noah Slee, who devoted his entire fortune to Sanger's cause. The philosopher's stone that worked this alchemy was contraception, and the alliance between the left and the wealthy which contraception forged has proved enduring indeed. They say that politics makes strange bedfellows, and if so sexual politics makes bedfellows even stranger. But the alliance between John D. Rockefeller, Jr., and the woman who once urged Americans to rise up and assassinate his is stranger than most. Yet after a little thought not so strange at all. In fact, the alliance is with us still because the mutual needs it resolved. (142)

Indeed, the alliance remains firmly intact. Sanger's Planned Parenthood continues to work in concert with the power elite towards mutually beneficial ends. In fact, so-called "conservative, pro-life, pro-family, Christian" President George Bush Sr. pledged his whole-hearted support to the group. Researchers Tarpley and Chaitkin explain:

> Although Planned Parenthood was forced, during the fascist era and immediately thereafter, to tone down Sanger's racist rhetoric from "race betterment" to "family planning" for the benefit of the poor and racial minorities, the organization's basic goal of curbing the population growth rate among "undesirables" never really changed. Bush publicly asserted that he agreed "1,000 percent" with Planned Parenthood. (195)

It comes as little surprise that, during his congressional career, Bush Sr. would supply infamous population control advocate Paul Ehrlich with an audience. In *The Population Bomb*, Ehrlich argued that rampant environmental degradation and the depletion of vital resources were the result of overpopulation. According to Ehrlich, there was not a single aspect of life on earth that was not impacted by humanity's population growth. He declares:

> It is fair to say that the environment of every organism, human and nonhuman, on the face of the Earth has been influenced by the population explosion of Homo sapiens. (26)

Throughout the course of the text, Ehrlich presented several eschatological predictions, all of which were thoroughly refuted by the arrival of the projected dates. Yet, Ehrlich managed to arouse considerable concerns, thus creating a pretext for the proposal of eugenical "solutions" to the alleged problem. Predictably, these "solutions" included abortion. In response to anti-abortion criticism, Ehrlich writes:

> Biologists must promote understanding of the facts of reproductive biology which relate to matters of abortion and contraception. They must do more than simply reiterate the facts of population dynamics. They must point out the biological absurdity of equating a zygote (the cell created

by joining of sperm and egg) or fetus (unborn child) with a human being. As Professor Garrett Hardin of the University of California pointed out, that is like confusing a set of blueprints with a building. People are people because of the interaction of genetic information (stored in a chemical language) with an environment. Clearly, the most "humanizing" element of that environment is the cultural element, to which the child is not exposed until after birth. When conception is prevented or fetus destroyed, the *potential* for another human being is lost, but that is all. That potential is lost *regardless* of the reason that conception does not occur—there is no biological difference if the egg is not fertilized because of timing or because of mechanical or other interference. (138)

Clearly, such an argument is fraught with fallacies. It is premised upon collectivistic assumptions. Ehrlich argues that the individual only derives his/her value from the collective, which he euphemistically calls "culture." Since unborn babies have not been exposed to cultural interchange yet, they do not qualify as human beings. Ehrlich is merely reiterating the collectivist mandate for the subordination of the individual to the collective.

Ehrlich also declares that a zygote cannot be "humanized" without the presence of the "cultural element." In other words, to be human, a person must be exposed to other people. To be sure, there is a social component in the formative processes of human identity. However, privation of this component does not automatically preclude one from being considered "human." Yet, despite the inherent fallacies of Ehrlich's pro-abortion arguments, his overpopulation thesis still received serious credence from Bush Sr. and his GOP Task Force on Earth Resources and Population.

Ehrlich's wife is a member of the Club of Rome, another globalist machination. One of the principal founders of the Club was Aurelio Peccei, an Italian Freemason who once remarked to Secretary of State Alexander Haig that he felt like Adam Weishaupt reincarnated (Coleman 15).

In 1972, the Club of Rome's research team at MIT published *The Limits to Growth*, which presented contentions paralleling those of Ehrlich. The only difference was the MIT team's projected year for the arrival of an impending environmental holocaust: 2000 A.D. This fraudulent eschatological claim declined as 1999 swiftly welcomed the millennium. While the Club flaunted an advanced computer-based system by which it arrived at its dismal conclusions, respected economist Gunnar Myrdal was anything but impressed:

> "The use of mathematical equations and a huge computer, which registers the alternatives of abstractly conceived policies by a 'world simulation model,' may impress the innocent general public but has little, if any, scientific validity. That this 'sort of model is actually a new tool for mankind' is unfortunately not true. It represents quasilearnedness of a type that we have, for a long time, had too much of . . ." (Simon and Kahn 34-35)

Myrdal had little reason to be impressed. Peccei later confessed that the Club's "new tool" had been preprogrammed to deliver the desired conclusion (*Executive Intelligence Review Special Report* 16). The motive for this deception, Peccei contends, is purely an altruistic one. Apparently, the "noble lie" provided necessary "shock treatment" to compel nations to adopt measures of population control (16). In a critique of *The Limits to Growth*, Christopher Freeman characterized the MIT group as a collective "Malthus with a computer" (5). Freeman's characterization proves itself accurate when read in conjunction with the core contention of *The Limits to Growth*: "Entirely new approaches are required to redirect society toward goals of equilibrium rather than growth" (Meadows 196-97).

As its title suggests, *The Limits to Growth* is replete with recommendations for technological apartheid and the restriction of infrastructural development. A Marxist program of wealth redistribution was also part of the report's recommendations, as is evidenced by the MIT group's proposal "to organize more equitable distribution of wealth and income worldwide" (Meadows 196-97). This was certainly not a new idea. Years before the report was published, other Malthusian ideologues were establishing similar mandates. In *The Impact of Science on Society*, Fabian socialist Bertrand Russell suggested a program for the "equitable distribution" of global wealth and resources:

> A scientific world society cannot be stable unless there is world government . . . It will be necessary to find ways of preventing an increase in world population. If this is to be done otherwise than by wars, pestilences and famines, it will demand a powerful international authority. This authority should deal out the world's food to the various nations in proportion to

their population at the time of the establishments of the authority. If any nation subsequently increased its population, it should not on that account receive any more food. The motive for not increasing population would therefore be very compelling. (111)

This could be the very end towards which globalist financial institutions like the IMF and World Bank are working (see Chapter One: The IMF and World Bank for further explication). This contention is reinforced by other globalist machinations that predate these two institutions and are devoted to eugenics. One such organization is United Nations Educational, Scientific and Cultural Organization (UNESCO). Julian Huxley, brother of Aldous, was the first director general of UNESCO and penned the organization's manifesto in 1947. Entitled *UNESCO: Its Purpose and Its Philosophy*, this document presents the following mission statement:

Thus even though it is quite true that any radical eugenic policy will be for many years politically and psychologically impossible, it will be important for Unesco to see that the eugenic problem is examined with the greatest care, and that the public mind is informed of the issues at stake so that much that now is unthinkable may at least become thinkable. (21)

As the unthinkable becomes thinkable, the fictional becomes factual and *Brave New World* becomes a reality. Aldous Huxley's "scientific dictatorship" may not be confined to the pages of classic literature for much longer. The Peak Oil movement graphically illustrates the extent to which the "unthinkable has become thinkable." Motivated by the eschatological assertion that oil production has peaked and global supply is perpetually declining, Peak Oil advocates portend a future of increasing energy shortages. Invariably, Peak Oil proponents attribute this inescapable period of scarcity to humanity's alleged overpopulation of Earth. The accusatory finger that edifies this Malthusian attribution becomes evident in a report from the elitist Council on Foreign Relations (CFR). Entitled "Strategic Energy Policy Challenges for the 21st Century," this report states: "the American people continue to demand plentiful and cheap energy without sacrifice or inconvenience" (No pagination). While the report singles out the American people, there can be no doubt that the power elite blames common people worldwide for the problem.

An article for the Association for the Study of Peak Oil (ASPO) written by William Stanton illustrates how Peak Oil is being used as a pretext for eugenics and population reduction. Stanton begins this article making it clear that depletion of fossil fuel makes population reduction inevitable. Stanton states:

> Recent articles in the ASPO Newsletter have agreed that the explosion of world population from about 0.6 billion in 1750 to 6.4 billion today was initiated and sustained by the shift from renewable energy to fossil fuel energy in the Industrial Revolution. There is agreement that the progressive exhaustion of fossil fuel reserves will reverse the process, though there is uncertainty as to what a sustainable global population would be. (No pagination)

The Third World has always been a primary target of the eugenics movement. Apparently, that fact will not change with Peak Oil. Stanton elaborates:

> In Third World nations, without oil, that can neither buy food nor grow it in adequate quantity without mechanised agriculture, a Darwinian struggle for shrinking resources of all kinds will be in full swing. Tribe against tribe, religion against religion, family against family, the imperative to survive will be driving strong groups to take what they want from weak ones. The concept of human rights will be irrelevant: "How can the weak have rights to food, when there is not enough even for the strong"? (No pagination)

Stanton even goes as far as to suggest that the West will employ nuclear weapons against the population to prevent further oil depletion. He writes:

> It may well be that, in the West, the same argument will affect the thinking of militarily powerful nations. "If billions must die, and we have the technology to ensure that they are others, not us, why should we hold back"? Instantaneous nuclear elimination of population centres might even be considered merciful, compared to starvation and massacres prolonged over decades. Eventually, probably before 2150, world population will have fallen to a level that renewable energy,

mainly biomass, can sustain. It is likely to be similar to the population before the Industrial Revolution. (No pagination)

According to Stanton, the end of the fossil fuel era will also usher in a mindset that no longer holds human life as sacred. Stanton prepares his audience for a time when man will become an expendable organism:

> Probably the greatest obstacle to the scenario with the best chance of success (in my opinion) is the Western world's unintelligent devotion to political correctness, human rights and the sanctity of human life. In the Darwinian world that preceded and will follow the fossil fuel era, these concepts were and will be meaningless. Survival in a Darwinian resource-poor world depends on the ruthless elimination of rivals, not the acquisition of moral kudos by cherishing them when they are weak. (No pagination)

What about those whose memories stretch back to the atrocities carried out during the nineteenth and twentieth centuries in the name of eugenics. Will the warning be heeded? Stanton makes it clear that their voices will be ignored:

> So the population reduction scenario with the best chance of success has to be Darwinian in all its aspects, with none of the sentimentality that shrouded the second half of the 20th Century in a dense fog of political correctness. . . (No pagination)

Stanton continues with his persecution of the anti-eugenics crowd:

> To those sentimentalists who cannot understand the need to reduce UK population from 60 million to about 2 million over 150 years, and who are outraged at the proposed replacement of human rights by cold logic, I would say "You have had your day, in which your woolly thinking has messed up not just the Western world but the whole planet, which could, if Homo sapiens had been truly intelligent, have supported a small population enjoying a wonderful quality of life almost for ever. You have thrown away that opportunity." (No pagination)

Stanton finishes up his prescriptions in true Nazi fashion:

The scenario is: Immigration is banned. Unauthorised arrives are treated as criminals. Every woman is entitled to raise one healthy child. No religious or cultural exceptions can be made, but entitlements can be traded. Abortion or infanticide is compulsory if the fetus or baby proves to be handicapped (Darwinian selection weeds out the unfit). When, through old age, accident or disease, an individual becomes more of a burden than a benefit to society, his or her life is humanely ended. Voluntary euthanasia is legal and made easy. Imprisonment is rare, replaced by corporal punishment for lesser offences and painless capital punishment for greater. (No pagination)

Before the people of the world adopt Stanton's recommendations and find themselves on a trajectory already visited by Nazi Germany and Communist Russia, it is important to determine whether the Peak Oil movement's contentions are true. There are two sides in the Peak Oil controversy. The first side holds that the world has been bled dry and man is about to become a post-industrial Barney Rubble hunting pterodactyl burgers with spear in hand. The other side holds that it is time to break out the bathing suits because the world is actually swimming in crude. As is usually the case, the truth lies somewhere in the middle.

Those oil fields controlled by the power elite do seem to be in decline. However, the world's supply in general has not peaked. Instead, it has been suppressed and remains largely untapped. News analyst and researcher Joel Skousen elaborates:

Yes, Peak Oil is coming—not so much because the supply has really peaked, but because the manipulated supply is peaking. The U.S. is withholding vast Arctic and offshore resources in order to keep an ace in the hole for the coming war. . . There has been a conspiracy to restrict refining capacity and buy out the little guys to cap supply--and that isn't likely to change in our lifetime. We're stuck with these powerful controlling forces, and will be--thanks to a dumbed-down electorate-- until it's too late to do anything about it. (Quoted in Monteith, no pagination)

A statement in a 1996 Texaco internal memo supports this contention:

> As observed over the last few years and as projected well into the future, the most critical factor facing the refining industry on the West Coast is the surplus refining capacity, and the surplus gasoline production capacity. (The same situation exists for the entire U.S. refining industry.) Supply significantly exceeds demand year-round. This results in very poor refinery margins, and very poor refinery financial results. Significant events need to occur to assist in reducing supplies and/or increasing the demand for gasoline. (No pagination)

The message in this memo is fairly obvious: "Abundance equals lower prices and low profits. Scarcity equals higher prices and greater profits. Crowd out any and all competition and don't be too generous with the supply." This may explain why there hasn't been a new refinery constructed in America since 1975. It may also explain why many independent refineries have been bought by one of the remaining four oil cartels and then mysteriously shut down.

Evidence suggests that certain factions of the elite have manufactured oil crises in the past. Many reputable people believe that the oil crisis of the 1970s was contrived. One such individual is Sheikh Ahmed Zaki Yamani, the OPEC minister for Saudi Arabia at the time of the seventies oil scare. The former OPEC minister has stated:

> "I am 100 per cent sure that the Americans were behind the increase in the price of oil. The oil companies were in real trouble at that time, they had borrowed a lot of money and they needed a high oil price to save them." (Morgan and Islam, no pagination)

Sheikh Ahmed Zaki Yamani was persuaded to believe the oil scare was a scam by a conversation he had with the Shah of Iran. The former OPEC minister recounts the incident:

> He says he was convinced of this by the attitude of the Shah of Iran, who in one crucial day in 1974 moved from the Saudi view, that a hike would be dangerous to Opec because it would alienate the US, to advocating higher prices. "King Faisal sent me to the Shah of Iran, who said: 'Why are you against

the increase in the price of oil? That is what they want? Ask
Henry Kissinger - he is the one who wants a higher price.'"
(No pagination)

Two powerful groups orchestrated the seventies oil emergency.
The first were the oil companies. Former Justice Department lawyer
John Loftus and investigative reporter Mark Aarons explain:

The sudden shortage of gasoline was a propaganda ploy by the
oil industry, an opportunistic bid to frighten the American
people into allowing it to increase their profits by removing
domestic price caps. (336)

Loftus and Aarons go on to lay out the oil industry's motives in
three parts:

(1) lift the laws holding down the price on domestic American
oil; (2) provide a panic to excuse raising world oil prices; and
(3) appease the Arabs, by arming the Saudis instead of the
Israelis. (339)

The second group involved was anti-Jewish elements in the
Central Intelligence Agency who were connected to Big Oil. This
faction hoped to deceive President Carter "into arms deals that
marked the beginning of the end of Israel's military supremacy over its
hostile Arab neighbors" (331). Their role in the seventies oil scam was a
manifestation of the "level battlefield doctrine" (LBD) subscribed to by
anti-Jewish individuals among the elite and the intelligence community.
Joel Bainerman provides an accurate definition of LBD:

The LBD became and some say still is today the cornerstone
of U.S. policy towards Israel. The doctrine is based on a Saudi
Arabian notion that the problem in the Middle East is not
with the Arabs, but in Israel's reckless use of its military
superiority. The Administration must ensure that the Arabs
are put on parity with Israel militarily so that Israel will be
pressured into concessions which will lead to a comprehensive
Middle East peace. (184-85)

Like the 2003 invasion of Iraq, the idea of oil scarcity in the seventies was promoted through intelligence fraud. Loftus and Aarons elaborate:

> The CIA was used to produce phony oil data to show that the world's two greatest oil producers, the Soviet Union and Saudi Arabia, were running out of oil. (332)

Using the phony oil shortage as a foundation, the intelligence arm of the conspiracy began to paint a bleak picture meant to scare President Carter into fulfilling the conspirators' wishes. This picture included a devastating war for resources with the Soviets. Becoming the victor in such a conflict called for arming the Arabs, even if Israel would be vulnerable as a result. Loftus and Aarons explain:

> The oil men in the intelligence services promoted the fear that the Soviets would have no other option than to move down into the Middle East, invade Saudi Arabia, Kuwait, and Iran, and seize U.S. oil for themselves. If Carter didn't move soon, there would be no hope of withstanding a Soviet invasion of the Middle East. The most powerful army in the area was Israel's. But the Jews could hardly be expected to go to war to save the Arab's oil for the West's often anti-Semitic oil companies. The time had come for the president to make a clear-cut decision: Either bow to the Jews' resolute opposition to U.S. arms sales to the Arabs and risk losing Middle Eastern oil to the Soviets, or arm the Arabs, thereby ending Israel's military supremacy—a sacrifice that would have to be made in the American national interest. (339)

According to Loftus and Aarons, the oil scare produced the desired results:

> The oil fraud of the 1970s succeeded in getting the U.S. president to change his policy toward Israel and the Arabs completely and directly promoted the massive defense expenditures of the 1980s, all based on the false premise that the Soviets needed oil and were planning to invade the Middle East. (334)

While the President had been fooled by the oil scam, one of Loftus' and Aarons' intelligence sources made it clear that the exact opposite was, in fact, the case:

> "Don't you get it?" asked one of our sources. "The gas shortage during the Carter administration was as phony as the CIA's prediction about the Soviet oil shortage. The god damn Middle East was swimming in oil during the Carter administration, but less and less of it was shipped to America. For chrissakes, there was so much oil in South America that they had to shut down the refineries in the Caribbean to keep it away from the U.S." (353)

Given the success of the seventies oil scam, it is quite possible that the current crisis is merely a recycling of the same play book. It is interesting to note that both the past and current crises have a common connection: the Bush dynasty. As this book is being published, George W. Bush occupies the White House. His father, George H. W. Bush, may have played a pivotal role in the seventies oil crisis. Loftus and Aarons go into the details of Bush's involvement:

> It should be recalled that George Bush was the director of the CIA at the time the oil scam was put in place in 1976. There is some evidence to suggest that it was Bush himself who passed the fake oil estimates to Carter. In the immediate aftermath of Carter's win, Bush traveled to Plains, Georgia, to brief the incoming president. According to Bush's own autobiography, halfway through a five-hour session with Carter "one of my deputies, Dan Murphy, began outlining long-range national security problems facing the country. He mentioned a particular problem, due to come to a head around 1985...." Now, it may be just a coincidence, but the "CIA report, prepared in late 1976" that influenced Carter's energy policy also predicted that 1985 would be the crunch year for Soviet oil production. Perhaps Bush was referring to some other matter "due to come to a head around 1985." But it seems like a very improbable coincidence. (334-35)

George W. Bush can hardly be described as a master conspirator. His dimwitted public displays make it painfully clear that he is not the brightest crayon in the box. However, it is interesting that both father and son have been involved in oil crises. This seems to suggest that may be some of the same forces that were behind George H.W. Bush are now behind George W. Bush. This would include those forces wishing to precipitate a panic surrounding the black gold that humanity is so dependent upon.

In light of scientific discoveries, technological advancements, and previously overlooked logistical considerations, finitude in regards to oil becomes virtually meaningless. Julian Simon explains:

Energy is particularly important because it is the "master resource;" energy is the key constraint on the availability of all other resources. Even so, our energy supply is non-finite, and oil is an important example. (1) The oil potential of a particular well may well be measured, and hence is limited (though it is interesting and relevant that as we develop new ways of extracting hard-to-get oil, the economic capacity of a well increases). But the number of wells that will eventually produce oil, and in what quantities, is not known or measurable at present and probably never will be, and hence is not meaningfully finite. (2) Even if we make the unrealistic assumption that the number of potential wells in the earth might be surveyed completely and that we could arrive at a reasonable estimate of the oil that might be obtained with present technology (or even with technology that will be developed in the next 100 years), we still would have to reckon the future possibilities of shale oil and tar sands —- a difficult task. (3) But let us assume that we could reckon the oil potential of shale and tar sands. We would then have to reckon the conversion of coal to oil. That, too, might be done; yet we still could not consider the resulting quantity to be "finite" and "limited." (4) Then there is the oil that we might produce not from fossils but from new crops —-palm oil, soybean oil, and so on. Clearly, there is no meaningful limit to this source except the sun's energy. The notion of finiteness does not make sense here, either. (5) If we allow for the substitution of nuclear and solar power for oil, since what we really want are the services of oil, not necessarily oil itself, the notion of a limit makes even less sense. (6) Of course the sun may eventually run down. But even if our sun were not as vast as it is, there may well be other suns elsewhere. (49)

In reality, world oil supplies may never be fully depleted. Deceased astrophysicist Thomas Gold and several Soviet geologists held that oil is created by tectonic, non-biological processes deep within the earth's crust and mantle ("Abiogenic petroleum origin," no pagination). There is a body of evidence to support this contention. Meteors, comets, and moons in the Solar System have been found to contain methane (no pagination). There have been natural gas explosions, flames and eruptions during the eruption of some volcanoes and earthquakes (no pagination). Mantle methane and hydrocarbons are apparently what cause deep focus and intraplate earthquakes (no pagination). All of this may suggest that decomposing fossils may have little to do with oils origins. Biological debris found in hydrocarbon fuels may merely be from bacteria feeding on the oil ("Thomas Gold," no pagination).

However, the power elite could never be accused of being rational. Those elites not involved in perpetrating the Peak Oil myth are becoming converts and true believers. To them, depletion of the blueblood-controlled oil supplies translate to world oil supplies being depleted. Of course, the elites' favorite prescription for the scarcity problems is population control. William Stanton's prescription will inevitably become very attractive to the bluebloods.

Ironically, depopulation as a means of averting a Peak Oil disaster may lead to an even greater catastrophe. Due in large measure to various elite depopulation campaigns, the world population is now imploding as opposed to exploding. David Francis describes the situation:

> For decades, much has been written about the world's exploding population. But 60 countries, about a third of all nations, have fertility rates today below 2.1 children per woman, the number necessary to maintain a stable population. Half of those nations have levels of 1.5 or less. In Armenia, Italy, South Korea, and Japan, average fertility levels are now close to one child per woman.
> Barring unforeseen change, at least 43 of these nations will have smaller populations in 2050 than they do today. (No pagination)

Governments once involved in the campaign of depopulation are now concerned about the impact that this population implosion will have upon their economic and political stability. Economic vitality and future funding of pension programs and healthcare will be seriously affected by the baby dearth (Francis, no pagination). Anxious governments are now implementing measures that will encourage procreation. Francis lists some of those measures:

> Starting this year, France's government has been awarding mothers of each new baby 800 euros, almost $1,000.
> In Italy, the government is giving mothers of a second child 1,000 euros.
> South Korea has expanded tax breaks for families with young children and is increasing support for day-care centers for working women.
> Last year parliament members in Singapore called on the government to do more to keep Cupid and the stork busy.
> Japanese prefectures have been organizing hiking trips and cruises for single people - dating programs to halt the baby bust. (No pagination)

If elites subscribing to Peak Oil eschatology begin a new depopulation campaign, no amount of government pro-natal measures will be effective in preventing complications stemming from a population implosion. The crusade to preserve hydrocarbon energy may contribute to bringing humanity to the verge of extinction.

What the Peak Oil movement and its other allied Malthusian organizations overlook is the centrality of population density to the welfare of nations. Examining the reasons for the prosperity of America's past, Julian Simon observes:

> How and why did total output and productivity per worker and per acre increase so fast? Supply increased so fast because of agricultural knowledge gained from research and development induced by the increased ability of farmers to get their produce to market on improved transportation systems. (71)

Malthusians argue that technological development and infrastructural growth cause demographic expansion. Of course, Malthusians view such demographic expansion as potentially dangerous. Therefore, they promote technological apartheid, compulsory sterilization, and abortion as the only solutions. However, this contention is premised on some heavily distorted thinking.

In truth, technological development and infrastructural growth are not the causes of demographic expansion. To assert otherwise is to presuppose that technological advancements and infrastructural progress are almost magical phenomena that occur independent of people. People are the driving force of invention and innovation. Population control proponents are placing the cart before the horse, so-to-speak.

Instead, it is demographic expansion that causes technological development and infrastructural growth. More people mean more potential innovators and inventors. More innovators and inventors mean more technological advancements and infrastructural progress. More technological advancements and infrastructural progress means a greater abundance of resources and food. Simon concludes that "in the long run additional people actually *cause* food to be less scarce and less expensive, and cause consumption to increase" (69; emphasis in original).

Given the correspondence between population density and increasing food supplies, terms like "carrying capacity" and "sustainability" are virtually meaningless. Julian Simon and Herman Kahn reiterate in *The Resourceful Earth*:

> Environmental, resource, and population stresses are diminishing, and with the passage of time will have less influence than now upon the quality of human life on our planet. These stresses have in the past always caused many people to suffer from lack of food, shelter, health, and jobs, but the trend is toward less rather than more of such suffering. Especially important and noteworthy is the dramatic trend toward longer and healthier life throughout all the world. Because of increasing throughout the decades and centuries and millennia to such an extent that the term "carrying capacity" has by now no useful meaning. (34-35)

As for scarcity and hunger, the true source of these ills is production output, which is being significantly retarded globally. Simon explains:

> We know for sure that the world can produce vastly more food than it now does, even (or especially) in such places as India and Bangladesh. If low-production countries were to produce even at the present level of agriculture in Japan and Taiwan, with present technology and without moving toward the much higher yields found under experimental conditions, world food production would increase dramatically and would more than feed any foreseeable population. Of course such an

increase in output would impose costs in the short run, but it could reduce costs as well as improve the food supply in the long run. (67)

Along with increasing output, nations could make better use of the scientific advancements in food production. Simon recapitulates:

In addition to already proven methods of raising output, there are many promising scientific discoveries still being developed. These include such innovations as orbiting giant mirrors that could reflect sunlight onto the night side of the earth and thereby increase growing time, increase harvest time, and prevent crop freezes; and meat substitutes made of soybeans that produce the nutrition and enjoyment of meat with much less resource input. (67)

In light of these neglected methods of food production, one is left with some very disturbing questions. Are the output levels of food production being intentionally retarded? IMF and World Bank bailout conditions, which impose restrictions upon production levels in recipient nations, seem to answer in the affirmative (see <u>Chapter One: IMF and the World Bank</u> for further explication). Are technologies that could potentially raise food production output being suppressed? Malthusian policies, which seem to be status quo in governments both domestic and abroad, stipulate forms of technological apartheid for those segments of the population exhibiting undesirable demographic trends... namely growth. In every instance, there remains one invariant: authoritarianism. As national governments become more and more anti-democratic in character, more and more people starve. Francis Moore Lappe of the Institute for Food and Development Policy comments:

"If the cause of hunger is neither scarcity of food, nor scarcity of land, we've come to see that it's a scarcity of democracy. That may sound rather contrived, because in the West we tend to think of democracy as a political concept and not as an economic concept. But democracy is really a principle of accountability; in other words, those making the decisions must be accountable to those who are affected by them. Once we understand hunger as a scarcity of democracy, what we are saying is that from the village level to the level of international commerce, fewer and fewer people are making decisions, and more and more anti-democratic structures are being entrenched. This is the cause of hunger." (Qutd. in Keith, *Casebook on Alternative Three* 91)

Ironically, the very anti-democratic structures that Malthusians are trying to entrench within the governments of the world constitute the ultimate cause of hunger and starvation. Scarcity does not increase according to population density, but according to design. As scientific dictatorships continue to rise and gradually migrate toward coalescence into a technocratic world government, growing numbers of people shall go hungry.

The paradoxical argument that the woes of humanity stem from too many people is no longer a viable excuse. Population control theories are fraught with demographic inconsistencies. For instance, Malthusians insist that China and India exhibit extreme population densities. It is within these two countries that many of the most authoritarian population control measures have been implemented. However, in actuality, these two nations have population densities that more closely parallel the United Kingdom and Pennsylvania (Kasun 50).

In light of demographic revelations like these, the true motives for population control come into question. No doubt, several Malthusian theoreticians are guided by the delusion that they are serving a humanitarian agenda. Meanwhile, other proponents are just plain sociopaths and misanthropes. In both cases, irrationality and mental instability are to blame. However, the motive underpinning the power elite's advocacy of population control is far more sinister. A series of memoranda produced by the National Security Council may provide a microcosm of the ruling class' Malthusian rationale. Entitled *Population and National Security: A Review of U.S. National Security Policy 1970-1988*, this collection of NSC documents examined the potential corollaries of diminishing demographic expansion in America and rising population growth in the Third World. The memoranda observe that:

"[t]he United States and its Western allies are declining as a percentage of world population. Whereas 6 percent of the world's people resided in the United States in 1950, the U.S. accounted for only 5 percent of the world's people in 1988, and its population is expected to be no more than 4 percent of the world total by the year 2010." (Qutd in Jones 532)

Evidently, this demographic decline held significant ramifications for both NATO and the Warsaw Pact. The memoranda state:

> "Declining fertility rates will make it increasingly difficult for the United States and its North Atlantic Treaty Organization (NATO) allies and the Soviet Union and its Warsaw Pact allies alike to maintain military forces at current levels." (Qutd. In Jones 532)

Thus, the war making capabilities of the United States and the former Soviet Union would inevitably weaken. The memoranda proceed to examine the waning dominance of the Cold War military powers that accompanied the preceding demographic decline:

> "That demographic fact, in effect, defused the East/West confrontation and may have contributed to its ultimate resolution following the collapse of the Soviet empire. However, unlike the former Soviet Union and other Warsaw Pact nations, the nations of the southern hemisphere have birth rates much higher than those of the West, leading to a North/South confrontation that will succeed the Cold War as the major arena of military concern for people like the folks at the National Security Council. The same analysis that saw a demographic stand-off in East/West relations projected that 'exceptionally high fertility rates' in the developing world 'could lead to expanded military establishments in affected countries as a productive alternative to unemployment,' and that developing nations 'may have a built-in momentum to capitalize on unused manpower for purposes of both internal and external security.'" (Qutd. in Jones 532-33)

Within the context of these NSC findings, E. Michael Jones offers an interesting assessment of the State Department's "foreign aid" policies:

> Decline in fertility and birth rate, in other words, means a decline in national power and military might. If this is true for the United States military, it is true for other countries as well. So U.S. "aid" in helping other countries lower their birth rates is really an attempt by the United States to weaken

them militarily, as the NSC document and other recently declassified documents make perfectly clear. (533)

Like the nations of the world, the supranational elite also wage wars. Theirs, however, are wars of class and their chief opponents are all those who do not occupy their particular layer of socioeconomic strata. Yet, the oligarchs have a significant strategic disadvantage. During the 20th century, it became demographically clear that the wealthy dynasties of the ruling class were having smaller families (535).

One region where this demographic decline has become evident is the British Commonwealth, the birthplace of the Cecil Rhodes Round Table groups and other elitist machinations. By 1944, Britain's fertility rate had declined significantly, prompting King George VI to form the Royal Commission on Population. The Commission discovered that the demographic implosion advanced "fastest among the higher occupational categories" (qutd. in Jones 536). As a result, a significant fertility differential between aristocrats and commoners was gradually making itself apparent:

> "Of the social groups, those with the highest incomes, and among individual parents within each social group, the better educated and the more intelligent, have smaller families on the average than others. We are not in a position to evaluate the expert evidence submitted to us to the effect that there is inherent in this differential birth rate a tendency towards lowering the average level of intelligence of the nation, but there is here an issue of the first important which needs to be thoroughly studied." (Qutd. in Jones 536)

Of course, the ruling class had always practiced selective breeding to preserve their genetically insular bloodlines. Thus, it comes as little surprise that the fertility differential that now beset the oligarchs was attributable to a tradition of "deliberate family limitation" (qutd. in Jones 535). Jones eloquently concludes:

> The ideology of population control is, simply a combination of fact#1: people produce economic wealth and military power, and fact #2: the affluent have smaller families. The English upper classes converted to Darwinism at the same time that they stopped having large families. As a result, they began to be concerned about something they referred to as "differential

fertility," which meant that while the "best people" (i.e., people of their class) limited the size of their families, the rest of the world, especially the pullulating races of the Southern Hemisphere, did not. As good Darwinians they realized that the population with the higher fertility rate would eventually replace the population with lower fertility rate. Out of that fearful realization the idea of population control was born. (536)

The mentor of Cecil Rhodes, John Ruskin, vividly illustrates the oligarchs' fear concerning "differential fertility." He voiced such fears during a speech that inspired Rhodes to initiate a campaign to establish an Anglo-Saxon global government. Carroll Quigley writes:

Ruskin spoke to the Oxford undergraduates as members of the privileged ruling class. He told them that they were possessors of a magnificent tradition of education, beauty, rule of law, freedom, decency, and self-discipline, but that this tradition could not be saved, and did not deserve to be saved, unless it could be extended to the lower classes in England itself and to the non-English masses throughout the world. If this precious tradition were not extended to these two great majorities, the minority of upper class Englishmen would ultimately be submerged by these majorities and the tradition lost. To prevent this, the tradition must be extended to the masses and to the Empire. (130)

Population control is, in essence, class war. The elite are not concerned with carrying capacity. They are concerned with the capacity of their control. They have good reason to be. In their war on humanity, they are hopelessly outnumbered.

The IMF and World Bank

Machinations for the eugenical regimentation of society are already in place. Through several political, social, and financial institutions, the ruling elite have attempted to impose the sanctions of natural selection upon the rest of humanity. Simultaneously, globalization has fostered the rise of a global social Darwinism. Two of the principal enabling mechanisms for this global social Darwinism have been the IMF and World Bank.

If technology and infrastructure prevent natural selection from being applied to the common man, then the "Country Assistance Strategy" of the IMF and World Bank can be used to remedy the "problem." One country that can attest to this fact is Ecuador. In 1983, Ecuador's elite had accrued a tremendous debt to foreign banks (Palast 46-47). The IMF forced Ecuador's government to take over the private debt, leading to the country borrowing $1.5 billion (46-47). The door had been opened not for financial recovery, but for eugenical regimentation. Investigative journalist Greg Palast provides the details:

> For Ecuador to pat back this loan, the IMF dictated price hikes in electricity and other necessities. And when that didn't drain off enough cash, yet another "Assistance Plan" required the state to eliminate 120,000 workers.
> Furthermore, while trying to pay down the mountain of IMF obligations, Ecuador foolishly "liberalized" its tiny financial market, cutting local banks loose from government controls and letting private debt and interest rates explode. Who pushed Ecuador into this nutty romp with free market banking? Hint: the initials are I-M-F-which made liberalization of the nation's banking sector a condition of another berserker Assistance Plan. (47)

Victimization of Ecuador continued into the 21st century. Using the IMF's own documents, Palast elaborates:

> So I thumbed through my purloined IMF "Strategy for Ecuador" looking for a chapter on connecting Ecuador's schools to the world wide web. Instead, I found a secret schedule. Ecuador's government was *ordered* to raise the price of cooking gas by 80 per cent by November 1, 2000, it says. Also, the government had to eliminate 26,000 jobs and cut real wages for the remaining workers by 50 per cent in four steps in a timetable specified by the IMF. By July 2000, Ecuador would grant British Petroleum's ARCO unit rights to build and own an oil pipeline over the Andes.
> That was for starters. In all, the IMF's 167 detailed loan conditions looked less like an "Assistance Plan" and more like a blueprint for a financial *coup d'etat*. (46)

Tanzania is yet another country that has had the misfortune of receiving the IMF's "assistance." Greg Palast tells the story:

The IMF and its sidekick, the World Bank, have lent a sticky helping hand to scores of nations. Take Tanzania. Today, in that African state, 1.3 million people are getting ready to die of AIDS. The IMF and World Bank have come to the rescue with a brilliant neoliberal solution: require Tanzania to charge for hospital appointments, previously free. Since the Bank imposed this requirement, the number of patients treated in Dar Es Salaam's three big public hospitals has dropped by 53 per cent. The Bank's cure must be working.
The IMF/World Bank also ordered Tanzania to charge fees for school attendance, then expressed surprise that school enrolment dropped from 80 per cent to 66 percent. (47)

Ecuador and Tanzania are only two victims of the IMF/World Bank syndicate. Greg Palast sums up the situation:

From 1980 to today, life under structural assistance has got brutish and shorter. Since 1985, in 15 African nations the total number of illiterate people has risen and life expectancy fallen... In the former Soviet states, where IMF and World Bank shock plans hold sway, life expectancy has fallen off a cliff-adding 1.4 million a year to the death rate in Russia alone. Tough luck, Russia! (48)

Countries that do not need the IMF's assistance can be pushed into a condition of dependency. Such was the case with Jamaica. In 1972, Jamaica democratically elected a new government made up of the People's National Party and headed up by Prime Minister Michael Manley (Phillips 475-76). Manley became a problem for the elite, namely those behind four U.S. aluminum companies: Kaiser, Reynolds, Alcoa, and Revere (477). Jamaica was one of the world's largest exporters of bauxite, the mineral that is processed into aluminum (477). All four American companies and the Canadian company Alcan had dominance over the Jamaican Bauxite industry (477). Manley's government had begun to tip the power balance by negotiating for 51 percent controlling interest in Jamaica's bauxite industry (477).

The transnationalists who ran the companies used their monopoly over Jamaica's bauxite to make the country conform to their vision of sociopolitical Darwinism. Manley's government had chosen a different route for Jamaica. Unlike the transnationalists' Darwinian model, the Manley government's approach did not seek to consolidate wealth and power for elites. James Phillips describes the crusade to remove Manley's government:

> In 1975, a destabilization campaign-reminiscent of the one aimed at Allende's Chile-was launched against Jamaica. Jamaica was closed out of the international public and private lending market. Total U.S. assistance dropped from $13.2 million in 1974 to $2.2 million in 1976, and further AID assistance was embargoed. "Even though many in the State Department objected to what one mid-level official called an 'effort to cut Jamaica off at the knees,' observes one writer, the embargo on new lending remained in effect until the Carter Administration revoked it in early 1977." Jamaica's credit rating with the U.S. Export Import bank dropped from a top to a bottom category. During 1976 the Jamaican Government was unable to secure a single private bank loan. Jamaican capitalists went "on strike" along with foreign capital, closing down the factories, cutting back production, and laying off workers. Emigrating wealthy Jamaicans smuggled an estimated $200 million from the island. (478)

This campaign also involved covert operations and dirty tricks. Phillips elaborates:

> Meanwhile, toward the end of 1975 and during the first half of 1976, a concerted plan (later exposed as "Operation Werewolf") was put into effect by agencies within the U.S.-apparently including the CIA-acting in concert with forces within the JLP, the major opposition party (misappropriately called the Jamaican Labour Party). In January 1976, a series of increasingly violent events began. People by the dozens were burned out of their homes (whole areas were torched) while paid gunmen shot or threatened those trying to escape. Most of the victims were PNP supporters. Food supplies were poisoned, and rumor of poisoning of water supplies were rife. The police and defense Force uncovered guns, explosives, and hundreds of rounds of ammunition of a kind never seen before in Jamaica. The violence escalated until Manley, exercising his

constitutional powers, declared a state of national emergency in June, hinting strongly that outside forces were at work. The foreign Western press, especially in the U.S., picked up on the theme of violence in Jamaica, often misrepresenting or distorting the actual facts of events. Later, Manley, his ministers, and the leaders of several other Caribbean nations (including Guyana and Barbados) charged that a plot to "destabilize" the Jamaican government was under way. (478-79)

Attempts by Manley and other Caribbean leaders to thwart the destabilization campaign came too late. The panic project had already radically altered Jamaica's economic landscape. James Phillips paints the picture:

The campaign of fear and violence during the first half of 1976 had several important consequences for the Jamaican economy. It slowed production as fear and disturbance spread among the working population. It frightened any remaining potential foreign investors and financial leaders. It had a devastating effect upon the tourist industry, Jamaica's third most important source of revenue. In effect, the campaign of violence greatly aggravated the problems of the Jamaican economy, already in trouble because of the increase in oil prices (Jamaica imports all of its oil), the activities of the bauxite companies, and U.S. controlled or dominated public and private lending agencies. (479)

In other words, Jamaica had been made a prime target for the IMF's loan shark techniques. In 1977, Manley began to negotiate with the IMF for assistance (481). The IMF granted a $74 million dollar standby loan to be paid back in four installments (481). However, the money came at a price. James Phillips explains:

As conditions for the standby arrangement, the IMF demanded 40 percent devaluation of the Jamaican dollar, a wage freeze, and cutbacks in government spending on social programs. (481)

Needless to say, Jamaicans suffered as these measures were implemented. Because wages were frozen, the people had no way of dealing with jumps in prices (481). This brought an inevitable decline in the standard of living (481).

Observers on the left usually identify greed as being the sole motivation behind this particular form of exploitation. The reality is that there are multiple motives for depreciating living conditions in the Third World and developing nations. The Darwinian agenda becomes evident when establishing motive. The power elite view improvements in living standards as an impediment to natural selection. Removing those improvements permits the sanction of natural selection to be imposed. In this sense, the IMF and World Bank serve a eugenical purpose. By selectively dismantling vital infrastructure among certain segments of the population, these institutions leave the so-called "dysgenics" susceptible to the elements. Subsequently, the "inferiors" are naturally expunged by disease, natural disasters, and exposure.

As long as sociopolitical Darwinians are shaping the world system, there is no hope of dismantling the IMF/World Bank system of usury. According to Malachi Martin, sociopolitical evolution is being ushered in on "a real and living and evolving tripod that will carry us on its three legs into the globalist community of the near future" (316). The first leg of this tripod is international trade (316). The second leg of the tripod is an international system of payment that keeps the first leg from collapsing (316). Martin explains that the IMF is part of the second leg:

> As the first leg of the tripod, international trade, got its modern footing with the help of GATT, so the second leg was set on the right path by means of an international agency established in the same year, 1947.
>
> Because the basic agreements making this new monetary agency possible were signed in Bretton Woods, New Hampshire, they are often referred to collectively as the Bretton Woods Agreement(s). The agency itself, however, was named for its function: the International Monetary Fund, or IMF. (318)

Certainly, no one wishes to do anything that would collapse the system. Victimizing and criminal though it may be, the fact is that growing interdependence has locked the better part of humanity into it. That being said, major reform and overhaul is needed to stop the cycle of oppression people have suffered at the hands of the IMF and World Bank. Any true reform would remove these two inherently corrupt organizations from the picture altogether.

National Security Study Memorandum 200

Eugenical regimentation of the world's population is also a chief concern of the oligarchs of America's Establishment. National Security Study Memorandum (NSSM) 200 illustrates this with brutal candor. The National Security Council (NSC) under the Henry Kissinger's guidance put this document together in 1974. The thesis was quite simple: population growth in lesser-developed countries constitutes a threat to national security. NSSM 200 named target countries:

> In order to assist the development of major countries and to maximize progress toward population stability, primary emphasis would be placed on the largest and fastest growing developing countries where the imbalance between growing numbers and development potential most seriously risks instability, unrest, and international tensions. These countries are: India, Bangladesh, Pakistan, Nigeria, Mexico, Indonesia, Brazil, The Philippines, Thailand, Egypt, Turkey, Ethiopia, and Colombia. Out of a total 73.3 million worldwide average increase in population from 1970-75 these countries contributed 34.3 million or 47%. (No pagination)

Interestingly, all of the countries named as threats are non-white nations. This seems to suggest a Darwinian theme running throughout the document. It should be remembered that Darwin referred to non-Caucasians as "anthropomorphic apes" (178). According to the Darwinian Weltanschauung, the extermination of the "anthropomorphic ape" represents an evolutionary step forward. Of course, this means that an increase in population amongst the "anthropomorphic apes" would constitute an evolutionary step backwards.

How would Kissinger and the NSC address this "problem"? The prescriptions put forward in NSSM 200 are genocidal in nature. One measure is reminiscent of Stalin's forced famine in the Ukraine from 1932 to 1933, which resulted in the death of seven million people. This measure suggests cutting food supplies to lesser-developed nations. Such an action would not only starve out many "anthropomorphic apes," it would force target nations to comply with birth control policies. This means that target nations would impose eugenical regimentation upon themselves in hopes of receiving food. An action taken in the name of survival would actually lead to extinction. The report states:

There is also some established precedent for taking account of family planning performance in appraisal of assistance requirements by AID [U.S. Agency for International Development] and consultative groups. Since population growth is a major determinant of increases in food demand, allocation of scarce PL 480 resources should take account of what steps a country is taking in population control as well as food production. In these sensitive relations, however, it is important in style as well as substance to avoid the appearance of coercion. (No pagination)

The report goes on to ask: "Would food be considered an instrument of national power? . . . Is the U.S. prepared to accept food rationing to help people who can't/won't control their population growth" (No pagination)?

NSSM 200 was reaffirmed as the cornerstone of the United States' population policy on November 26, 1975 when Brent Scowcroft signed National Security Decision Memorandum 314 (NSDM 314) (Jones 527). This document endorsed the policy recommendations presented in NSSM 200 (527). NSSM 200's reaffirmation was clearly at odds with world opinion. Just a year later, opposition towards population proposals like NSSM 200 arose at a United Nations-sponsored population conference in Bucharest. According to author E. Michael Jones:

There the Holy See along with Communist and Third World countries, led by Algeria, denounced the United States for practicing what they called "contraceptive imperialism." (526)

Although NSDM 314 was declassified in the late '80s, it is still very much in force today (527). As long as Western elites are dedicated to the erection of a global scientific dictatorship, "contraceptive imperialism" will remain the order of the day.

Pax Britannia: An Anglophile Scientific Dictatorship

In the British Isles, the very birthplace of Darwinism, another scientific dictatorship emerged. As was the case with communism and fascism, Freemasonry and its occult doctrine of evolution were integral to this emergence. In 1870, John Ruskin "hit Oxford like an earthquake," proselytizing students in the imperialistic gospel of the British Empire

(Quigley 130). In *Tragedy and Hope*, Dr. Carroll Quigley provides a brief summation of this gospel:

> Ruskin spoke to the Oxford undergraduates as members of the privileged ruling class. He told them that they were possessors of a magnificent tradition of education, beauty, rule of law, freedom, decency, and self-discipline, but that this tradition could not be saved, and did not deserve to be saved, unless it could be extended to the lower classes in England itself and to the non-English masses throughout the world. If this precious tradition were not extended to these two great majorities, the minority of upper class Englishmen would ultimately be submerged by these majorities and the tradition lost. To prevent this, the tradition must be extended to the masses and to the Empire. (130)

Among one of the undergraduates who wholeheartedly embraced this message was Cecil Rhodes, who would keep his longhand copy of Ruskin's inaugural lecture for thirty years (Quigley 130-31). However, while this message comprised the nucleus of Rhodes' Weltanschauung, there were two other significant belief systems that would shape Cecil's vision: Freemasonry and Darwinism. Having already established the Masonic origins of Darwinism, it comes as little surprise that the two would find an intersection within the man of Cecil Rhodes. Indeed, Freemasonry and Darwinism are natural correlatives. The two seem to be inextricably linked. Where one goes, the other seems to invariably follow.

In June 1877, Rhodes became a life member of Freemasonry's Apollo Chapter at Oxford University (Rotberg 90). There have been questions of whether or not Rhodes regarded his membership in the Lodge as a "serious venture" (Rotberg 90). Author Robert Rotberg elaborates:

> At a banquet marking his induction, the story goes, he became angry at some criticism and, not untypically, shocked the assembled brethren of the Order by babbling away about the mystic cult secrets of the 33rd Degree Rite into which he had been admitted. (90)

Indeed, Rhodes candidly admitted his derision for a group devoted "to what at times appear the most ridiculous and absurd rites without an object and without an end" (qutd. in Rotberg 90). However, despite disagreements with the organization, exposure to the Lodge "presumably helped shape Rhodes' 'Confession of Faith,' the later wills, and the protean thinking which led ultimately to the scholarships" (91). Rhodes' "Confession of Faith" articulated his vision for a British world government or, as it has been called in more Anglophilic language, a *Pax Britannia*.

Predictably, the Freemasonic influence on Rhodes was accompanied by its natural correlative: Darwinism. The primary transmitter of Darwinian thought to Rhodes was William Windwood Reade, author of *The Martyrdom of Man*. Rotberg explains:

> William Windwood Reade, the then-obscure British Darwinian, influenced Rhodes' search for understanding. An unsuccessful novelist, Reade visited West Africa twice in the 1860s, the second time while Rhodes was in Natal, and published *The Martyrdom of Man* in 1872. Begun as an attempt to revise England's accepted and critical view of the contribution of Africans to human civilization, *The Martyrdom* became a universal history of mankind, with long sections on Rhodes' favorite mysteries: ancient Egypt, Rome, Carthage, Arab Islam, and early Christianity. *The Martyrdom* consisted of the kind of late nineteenth-century pseudo-science that appealed to Rhodes. It was larded with philosophically impressive arguments about the true "meaning" of man based on the post-Hegelian as well as neo-Darwinian notion that man's suffering on earth (his martyrdom) was essential (and quasi-divinely inspired) in the achievement of progress. Man was perfectable, but only by toil. He could not be saved, nor would his rewards be heavenly, for Reade was a pre-Tillichean Gnostic who believed in God's existence but, at the same time, not in deism and certainly not in the accessibility of an anthropomorphic Christian God. The rewards of man were in continuing and improving the human race. "To develop to the utmost our genius and our love, that is the only true religion," wrote Reade. Reade was Rhodes' Ayn Rand or Antoine Saint Exupery. Or perhaps his Jules Verne, too, for Reade prophesied a locomotive force more powerful than steam, the manufacture of flesh and flour chemically, travel through space, and the discovery by science of a destructive force which would be so horrible as to end all wars. Rhodes read Reade only shortly after its publication and later said

that it was a "creepy book." He also said, mysteriously, that it had "made me what I am." (99-100)

The impact of Reade's work on Rhodes was unmistakable. This impact validates the appropriation of evolutionary thought as a foundation for the racist and genocidal doctrine of *Pax Britannia*. Raging Anglophile that he was, Rhodes was already predisposed to Darwinian thought. His oppressive colonial system was premised upon the evolutionary primacy of the Anglo-Saxon. Darwin himself was an advocate of Britain's "colonial warfare," a contention that he developed during his voyage on the *Beagle*:

> At every outpost the *Beagle* crew had witnessed the destruction: the Tasmanians were all but exterminated, the aborigines were dying from European diseases, General Rosa's policy was deliberate genocide. But Darwin believed that colonial warfare was necessary "to make the destroyers vary" and adapt to the new terrain. (Desmond and Moore 266)

Inspired by Freemasonry and the theistic Darwinism of Reade, Rhodes began the construction of his Anglophile "scientific dictatorship." Rhodes established his scientific dictatorship in South Africa, where he monopolized the diamond fields through DeBeers Consolidated Mines (Quigley 130-31). Instrumental in the formation of this diamond cartel were Lord Rothschild and Alfred Beit, who provided Rhodes with financial support (Quigley 130-31). Yet, the borders of Rhodes' African empire did not end there. Cecil also:

> rose to be prime minister of the Cape Colony (1890-1896), contributed money to political parties, controlled parliamentary seats both in England and South Africa, and sought to win a strip of British territory across Africa from the Cape of Good Hope to Egypt and to join these two extremes together with a telegraph line and ultimately with a Cape-to-Cairo Railway.(Quigley 130-31)

This colonial expansion received the absolute sanction of Darwin. Darwin's theory of evolution was hardly based upon "objective scientific observations." Instead, it was based on a sociopolitical bias favoring the British oligarchy:

The "stronger [are] always extirpating the weaker," and the British were beating the lot. This imperial expansion ended the isolation of the indigenous races, and thwarted their development in other ways. As whites spread out from the Cape, the black tribes were pushed together in the interior, blending races and ending their species-making isolation. Had this not happened, Darwin speculated, in "10,000 years [the] Negro [would] probably [have become] a distinct species." (Desmond and Moore 267)

In this sense, Rhodes' scientific dictatorship was also a project in selective breeding. His colonial system was eugenically regimented to "extirpate the weak," specifically the "Negro." Meanwhile, fettered by colonial slavery and cloistered in abject poverty, native Africans lived under the yolk of this Anglophile "scientific dictatorship." Of course, according to the Darwinian doctrine of *Pax Britannia*, this was the natural order of things. After all, in the evolutionary ladder, the Negro occupied a lower rung than did the Caucasian.

The United Nations: A Global Scientific Dictatorship
Although the concept of a "scientific dictatorship" was outwardly expressed through the socialist totalitarian systems of Nazi Germany and the Soviet Union, the concept had not yet reached the global scope of Huxley's *Brave New World*. Yet, there was another embryonic scientific dictatorship waiting to be birthed. On October 24, 1945, shortly after the fall of the Nazi scientific dictatorship, another one called the United Nations was created.

The United Nations finds its proximate origins with the architects of *Pax Britannia*, an Anglophile variant of the scientific dictatorship concept. Recall Cecil Rhodes' "Confession of Faith," which articulated his vision for a British world government. This vision was inspired by John Ruskin, a professor at Oxford University. However, Cecil Rhodes not the only adherent of Ruskin's imperialistic message. Evidently, others had taken to heart the Anglophilic gospel of Ruskin and, eventually, became associated with Rhodes. Together, this network would establish a secret society devoted to the cause of British expansionism. Carroll Quigley elaborates:

Among Ruskin's most devoted disciples at Oxford were a group of intimate friends including Arnold Toynbee, Alfred (later Lord) Milner, Arthur Glazebrook, George (later Sir

George) Parkin, Philip Lyttelton Gell, and Henry (later Sir Henry) Birchenough. These were so moved by Ruskin that they devoted the rest of their lives to carrying out his ideas. A similar group of Cambridge men including Reginald Baliol Brett (Lord Esher), Sir John B. Seeley, Albert (Lord) Grey, and Edmund Garrett were also aroused by Ruskin's message and devoted their lives to the extension of the British Empire and uplift of England's urban masses as two parts of one project which they called "extension of the English-speaking idea." They were remarkably successful in these aims because of England's most sensational journalist William Stead (1849 - 1912), an ardent social reformer and imperialist, brought them into association with Rhodes. This association was formally established on February 5, 1891, when Rhodes and Stead organized a secret society of which Rhodes had been dreaming for sixteen years. In this secret society Rhodes was to be leader; Stead, Brett (Lord Esher), and Milner were to form an executive committee; Arthur (lord) Balfour, (Sir) Harry Johnston, Lord Rothschild, Albert (Lord) Grey, and others were listed as potential members of a "Circle of Initiates;" while there was to be an outer circle known as the "Association of Helpers" (later organized by Milner as the Round Table organization). Brett was invited to join this organization the same day and Milner a couple of weeks later, on his return from Egypt. Both accepted with enthusiasm. Thus the central part of the secret society was established by March 1891. It continued to function as a formal group, although the outer circle was, apparently, not organized until 1909-1913. This group was able to get access to Rhodes' money after his death in 1902 and also to funds of loyal Rhodes supporters like Alfred Beit (1853-1906) and Sir Abe Bailey (1864-1940). With this backing they sought to extend and execute the ideals that Rhodes had obtained from Ruskin and Stead. Milner was the chief Rhodes Trustee and Parkin was Organizing Secretary of the Rhodes Trust after 1902, while Gell and Birchenough, as well as others with similar ideas, became officials of the British South Africa Company. They were joined in their efforts by other Ruskinite friends of Stead's like Lord Grey, Lord Esher, and Flora Shaw (later Lady Lugard). In 1890, by a stratagem too elaborate to describe here, Miss Shaw became Head of the Colonial Department of the Times while still remaining on the payroll of Stead's Pall Mall Gazette. In this past she played a major role in the next ten years in carrying into execution the imperial schemes of Cecil Rhodes, to whom Stead had introduced her in 1889. (131-32)

When Rhodes died, the continuation of his imperialistic vision fell upon the shoulders of chief Rhodes Trustee Alfred Milner. Under Milner's coordination, the Rhodes network would establish a stateside surrogate organization that would be instrumental in the formation of the United Nations. Quigley continues:

As governor-general and high commissioner of South Africa in the period 1897-1905, Milner recruited a group of young men chiefly from Oxford and from Toynbee Hall, to assist him in organizing his administration. Through his influence these men were able to win influential posts in government and international finance and become the dominant influence in British imperial and foreign affairs up to 1939. Under Milner in South Africa they were known as Milner's Kindergarten until 1910. In 1909-1913 they organized semisecret groups, known as Round Table Groups, in the chief dependencies and the United States... In 1919 they founded the Royal Institute of International Affairs (Chatham House) for which the chief financial supporters were Sir Abe Bailey and the Astor Family (owners of The Times). Similar Institutes of International Affairs were established in the chief British dominions and in the United States (where it is known as the Council on Foreign Relations) in the period of 1919-1927. (132-33)

The Council on Foreign Relations (CFR) was the chief organizational conduit for the importation of the Anglophile scientific dictatorship into the United States. With the machinations of Rhodes' *Pax Britannia* successfully relocated, the Rhodes tradition would continue in the United States. The CFR would create the United Nations to act as a vehicle for realizing Rhodes' vision globally. Shoup and Minter describe the CFR's involvement in the United Nations' birth:

The planning of the United Nations can be traced to the secret steering committee established by Secretary Hull January 1943. This informal Agenda Group, as it was later called, was composed of Hull, Davis, Taylor, Bowman, Pasvolsky, and, until he left the government in August 1943, Welles. All of them, with the exception of Hull, were members of the Council on Foreign Relations. They saw Hull regularly to plan, select, and guide the labors of the Department's Advisory Committee. It was, in effect, the coordinating agency for all the State Department postwar planning...

In late 1943, the Agenda Group began to draft the U.S. proposal for a United Nations organization to maintain international peace and security. The position eventually taken at the Dumbarton Oaks Conference was prepared during the seven-month period from December 1943 to July 1944. Once the group had produced a draft for the United Nations and Hull had approved it, the Secretary requested three distinguished lawyers to rule on its constitutionality. Myron C. Taylor, now on the Council's board of directors, was Hull's intermediary to Charles Evans Hughes, retired chief justice of the Supreme Court, John W. Davis, Democratic presidential candidate in 1924, and Nathan L. Miller, former Republican governor of New York. Hughes and Davis were both Council members and John W. Davis had served as president of the Council from 1921 to 1933 and a director since 1921. The three approved the plan, and on 15 June 1944, Hull, Stettinius, Davis, Bowman, and Pasvolsky discussed the draft with President Roosevelt. The chief executive gave his consent and issued a statement to the American people that very afternoon.

Although the Charter of the United Nations underwent some modification in negotiations with other nations at the Dumbarton Oaks and San Francisco conferences during 1944 and 1945, one historian concluded that "the substance of the provisions finally written into the Charter in many cases reflected conclusions reached at much earlier stages by the United States government." The Department of State was clearly in charge of these propositions within the U.S. government, and the role of the Council on Foreign Relations within the Department of State was, in turn, very great indeed. The Council's power was unrivaled. (149-150)

It comes as little surprise that the U.S delegation to the UN's founding San Francisco Conference was replete with people who had been or would later become members of the CFR. Among them were:

Theodore C. Achilles
James W. Angell
Hamilton Fish Armstrong
Charles E. Bohlen
Isaiah Bowman
Ralph Bunche
John M. Cabot
Mitchell B. Carroll
Andrew W. Cordier
John S. Dickey
John Foster Dulles
James Clement Dunn
Clyde Eagleton
Clark M. Eichelberger
Muir S. Fairchild
Thomas K. Finletter
Artemus Gates
Arthur J. Hepburn
Julius C. Holmes
Philip C. Jessup
Joseph E. Johnson
R. Keith Kane
Foy D. Kohler
John E. Lockwood
Archibald MacLeish
John J. McCloy
Cord Meyer, Jr.
Edward G. Miller, Jr.
Hugh Moore
Leo Pasvolsky
Dewitt C. Poole
William L. Ransom
Nelson A. Rockefeller
James T. Shotwell
Harold E. Stassen
Edward R. Stettinius, Jr.
Adlai E. Stevenson
Arthur Sweetser
James Swihart
Llewellyn E. Thompson
Herman B. Wells

Francis Wilcox
Charles W. Yost (Lee 243)

Peopled by the adherents of Rhodes' vision for an Anglophile scientific dictatorship, the UN pursued the same goals on a global scale. The UN was designed to preserve Rhodes' system of oppressive colonialism, a cold fact candidly voiced by former UN Secretary-General Boutros Boutros-Ghali: "Even the charter of the UN was based on maintaining colonialism, through the system of trusteeship" (no pagination).

However, because of the UN's admission of other elite interest groups, its Anglophilic vision did not remain completely unadulterated. *Pax Britannia* would eventually become *Pax Universalis*. Still, as the product of globalist architects within the CFR, the UN retains the globalist agenda of sociopolitical Darwinism. Claire Chambers reiterates: "Since its inception, the U.N. has advanced a world-wide program of population control, scientific human breeding, and Darwinism." (3)

The United Nations' sociopolitical Darwinian agenda was clearly demonstrated in the Balkans at the time of the NATO air operation against Yugoslavia. Fortress America founder Julie Makimaa elaborates:

On April 8, 1999, the UN Population Fund (UNFPA) announced that it was sending 350,000 "Emergency Reproduction Health Kits" to Albania to be distributed among Kosovo Albanian refugees. Joseph Meaney of Human Life International, who inspected health care facilities in northern Albania during the air war and refugee crisis, recounts that these kits included "condoms, birth control pills, 'emergency contraception' or 'morning after' pills [that is, chemical abortifacients], intrauterine devices (IUDs), and manual vacuum aspirators," which are used for early term abortions. Although the packages were originally labeled "Pregnancy Termination" kits, the name was changed to "reduce the risk of offending sensitivities and possibly make [them] more acceptable," in the words of a UN document on "refugee situations" issued in 1995. (35)

Under the pretext of "humanitarianism," the United Nations also conducted a sterilization campaign against unsuspecting women in the Balkans. Julie Makimaa continues:

> Dr. Enza Ferrara, who works in a hospital in Scutari, Albania, testifies that the UN's anti-natal campaign in that nation began in earnest in 1995. "She saw that women were being surgically sterilized without their knowledge or consent after delivering by C-section in the hospital," recalls Meaney, who interviewed Dr. Ferrara at length. "Time and again it fell upon her to inform women to consult her that a tubal ligation was the cause of their infection or inability to have children." When Dr. Ferrara protested to Albanian government officials, she received a candid response from a representative of Albania's Ministry of Health: "We have accepted international aid on condition of reducing births." To carry out a similar mission among Kosovo Albanians, UNFPA helped establish an office of the British birth control organization Marie Stopes International in Pristina. (35)

These activities bear eerie resemblance to the eugenics programs of Nazi Germany. While the United Nations has been presented to the public as a benevolent organization and "humanity's last best hope for peace," there can be no question that it is merely the Power Elite's attempt to erect a global scientific dictatorship.

Cross-examining Darrow

Traditionally, Darwinians have held the Scopes "monkey" trial aloft as a victory. Certainly, this historically significant event edified the evolutionist movement and guaranteed the popularization of Darwinism. Yet, while historians attribute this victory to the polemical resilience of Darwinian proponents and irrefutable scientific evidence for evolutionary theory, the truth is that it was a consciously engineered media circus. Clarence Darrow, the pro-Darwinian attorney in the Scopes "monkey" trial, was a member of the Society for Psychic Research (Chaitkin 452-54). Author Anton Chaitkin reveals the mission of the Society:

> This club, using newly developed techniques of psychology and organization, would serve as an advance detachment in the war of the old European feudalists against the American system. Around its central core members, the British Foreign

Office would construct a deadly transatlantic political machine to demoralize and divert the United States from its commitment to global technological development, and to close the American West to further settlement. (452)

The Malthusian character of the Society should be fairly apparent. Its opposition to America's technological growth echoes Malthus' mandate for the limiting of infrastructural development among the "undesirable" segments of the population. The Society was but one more machination birthed by the sociopolitical Darwinians of the British elite. In fact, the president of the Society in 1893 was Arthur Balfour, who was also involved in the formation of an elitist network devoted to realizing the globalist aspirations of Cecil Rhodes (460). As such, the Society occupied the same category as the East India Company and the Round Table groups. All were guided by the common theme of creating a scientifically managed feudal system governed by the Darwinian precept, "Survival of the fittest."

However, unlike its kindred organizations, the Society had cultivated a much more complex and pervasive methodology for subjugating populations: psychological warfare. Chaitkin explains:

> The practical theme of the Society's work was psychological experimentation, on two levels. First, to study the extent of power over men's minds that could be achieved with hypnosis and hysteria-inducing trances. Second, to try to break down the subject-victim's faith in rationality, in the lawfulness of nature, and in the coherence of his own mind. (454)

Clarence Darrow was certainly proficient in the dismantling of "subject-victim's faith in rationality, in the lawfulness of nature, and in the coherence of his own mind." This fact is evident in his performance during the Scopes "monkey" trial. Christian philosopher Ravi Zacharias provides an excellent analysis of Darrow's questionable polemical approach during the 1925 debate:

> If today one were to analyze the questioning by Clarence Darrow of William Jennings Bryan, it would be readily seen that Darrow's answers to an equally adept challenge would have been at least as unconvincing. His whole scheme was to persuade Bryan to take the stand in defense of the miraculous and then destroy him. Bryan thought he was up to it, and

for him, it was the equivalent of getting O.J. Simpson to try
on the glove. The supernatural elements of the Scripture as
caricatured by Darrow did not fit the "scientific" framework,
and Bryan looked bedraggled and defeated. (169)

Darrow's approach was purely scientistic in character. His
questions presupposed the falsity of the supernatural because it
was disproportionate with the "yardstick" of science. However, the
"yardstick" of science was clearly not applicable to the supernatural,
which circumvents the narrow parameters of naturalistic interpretation.
Again, the epistemological imperialism of scientism becomes painfully
evident.

In addition to the epistemological selectivity of Darrow's questions,
his cross-examination of Bryan was actually a thinly camouflaged
assault on the particulars of the Scriptures. Such an approach ignores
the overall theme of the Christian Weltanschauung, which requires
examination in order to understand the context and connotative
meaning of the specific Biblical passages referenced by Darrow. In the
absence of a macrocosmic analysis, the particular instances cited by
Darrow could be effectively distorted and bowdlerized so as to appear
unreliable. Zacharias explains:

But was that really the way to determine whether the Bible could be
trusted as a document on origins? Herein is the fallacy. Can particulars of
a world-view be defended without first defending the world-view itself?
It defies logic that something so methodologically tendentious could
be taken as compelling proof. Any brilliant lawyer can tell you that in
most trials, when only selected facts are permitted into the courtroom,
any adept wordsmith can construct a farce. The added component of
the media only compounds the sham. (169-70)

Zacharias provides an example:

Think of this. One of the questions for which Mr. Darrow
demanded an answer of William Jennings Bryan was where
Cain got his wife. That could be a fair question if it were
permitted that the Bible could first be defended in its intent
and content, and if the assertion were also made that it
contained every detail of how human reproduction began.
But none of that was even given possibility. (170)

This technique was the sort of polemical trickery promulgated by the Society for Psychic Research. Darrow's sophistry working in conjunction with the epistemological selectivity of scientism effectively destroyed Bryan's "faith in rationality, in the lawfulness of nature, and in the coherence of his own mind." Faith in a miraculous Creator, which qualifies as a supra-rational belief, was portrayed as superstitious irrationalism.

In addition to discrediting his opposition through force of rhetoric, Darrow also drew the attention of the media with cheap publicity stunts. Exploiting the petty theological differences endemic to Christian denominations, Darrow employed a clever strategy of divide and conquer. Zacharias elaborates:

> On the third day of the trial, the judge asked a minister present to open in prayer. The controversy engendered was almost a circus in itself. But in spite of Clarence Darrow's strong objection, the judge allowed the prayer to proceed. Darrow's team of attorneys then rounded up a group of ministers to sign a petition objecting to the prayer on the grounds that their particular theological persuasion was not represented in it. That objection was denied by the judge. Finally, they submitted another petition signed by two Unitarian ministers, one Congregationalist minister, and one rabbi. It stated that they believed that God had shown Himself as much in the wonders of the world as He had in the written Word, and hence, a prayer that did not reflect that was abhorrent to them.
>
> One can only shake one's head in disbelief. How ironic that "the wonders of the world" were placed on equal footing with God's spoken Word, while all along the very case being argued was whether these wonders required natural or supernatural explanation. You see, the real issue was not the explicability of the material world. The real issue was whether God had spoken through language as well as through nature. (171)

Darrow was truly the Johnny Cochrane of his day. Public opinion had been manipulated. Minds were swaying towards a more secular Weltanschauung. It was vintage psychological warfare. The word "psychological warfare" is derivative of the German word *Weltanschauungskrieg*, a term cribbed from the conceptual lexicon of the Nazis in 1941 (Simpson 24). The word literally means "worldview warfare" (24). Darrow's polemical deception was one of the opening salvos launched in what would become an ongoing war between the Christian Weltanschauung and the Darwinian Weltanschauung.

Zacharias raises an interesting question: What if Darrow was placed on the stand and faced the same style of interrogation? Would the adept agent of psychological warfare be able to sufficiently answer? Entertaining this hypothetical scenario, Zacharias begins his cross-examination:

> How did human sexuality and marriage emerge in the evolutionary scheme of things? I would like to have asked Mr. Darrow to explain how the "Big Bang" came to confer on sexuality the enormous combination of intimacy, pleasure, consummation, conception, gestation, nurture, and supererogatory expressions of care and love. All this came from the explosion of a singularity? In no other discipline would so much information density be swallowed up under the nomenclature of chance. In case Mr. Darrow was not forthcoming with an answer, I could help him even with the most modern research.
> William Hamilton of Oxford has offered one theory (this is serious, by the way): "Sex is for combating parasites." You see, in warm and rich climates where microscopic parasites threaten the stable health of their hosts, the hosts mess up the attacking power of these foes through sex and procreation. That is the reason sex came to be: to stay ahead of the game! My! How different prescriptions look today to ancient cures. Imagine what the late-night comedians could do with this material. The laughter could be even more hilarious than the derision afforded to Bryan. (17)

Another interesting question for Mr. Darrow would have been, "What is your motive?" Exactly what did Darrow have to gain from orchestrating such a judicial travesty? Perhaps Darrow had already answered this question when he said: "Chloroform unfit children. Show them the same mercy that is shown beasts that are no longer fit to live" (qutd. in Dowbiggin 26). Like the many other architects of the scientific dictatorship, Darrow was committed to creating a Darwinian world where the so-called "unfit" would be eradicated.

This leads one to ask, "Just who constitutes the unfit?" The answer lies with John Scopes and his classroom curriculum. Scopes, who violated Tennessee's ban on the teaching of evolution, taught from a book entitled *A Civic Biology Presented in Problems*. Authored by George William Hunter, the book presented the following racialist contention:

> At the present time there exist upon the earth five races or varieties of man, each very different from the other in instincts, social customs, and, to an extent, in structure. These are the Ethiopian or negro type, originating in Africa; the Malay or brown race, from the islands of the Pacific; The American Indian; the Mongolian or yellow race, including the natives of China, Japan, and the Eskimos; and finally, the highest type of all, the caucasians, represented by the civilized white inhabitants of Europe and America. (196)

With this supremacy doctrine effectively enshrined in academic institutions, the new class distinction of race attained a semblance of scientific credibility. This mentality would contribute to the power elite's efforts to erect a scientific dictatorship in the West. This is one of the many important details that the film *Inherit the Wind* intentionally omitted. Seduced by the propaganda pouring from the movie screen, audiences cheered for Clarence Darrow. Little did they know that the very man they canonized would have advocated their extermination if any of them proved to be "genetically inferior." Yet, how could they have known? The Scopes "monkey" trial was now history. It was too late for anyone to cross-examine Darrow.

Traversing the Moral Rubicon
In *Morals and Dogma*, Albert Pike wrote:

> . . .no human being can with certainty say. . .what is truth, or that he is surely in possession of it, so every one should feel that it is quite possible that another equally honest and sincere with himself, and yet holding the contrary opinion, may himself be in possession of the truth. . . (160)

Evident in this statement is the overall relativistic Weltanschauung of Freemasonry. This Weltanschauung is the dominant paradigm among all correlative elitist groups as well. As adherents to relativism, the ruling class rejects absolute truths and moral certainties. Over the years, this Weltanschauung has been vigorously promulgated by the elite and, thus, has become the dominant paradigm of society. The mantra of "Do what thou wilt" is continually reiterated by academia, the media, and pop culture. With each successive generation, humanity continues its inexorable drift towards amorality. Of course, this drift serves the interests of the ruling class. The further away humanity drifts from morality, the closer it drifts towards enslavement. C.S. Lewis reiterated this contention in *Christian Reflections*:

> The very idea of freedom presupposes some objective moral law which overarches rulers and ruled alike. Subjectivism about values is eternally incompatible with democracy. We and our rulers are of one kind only so long as we are subject to one law. But if there is no Law of Nature, the ethos of any society is the creation of its rulers, educators and conditioners; and every creator stands above and outside his own creation. (81)

Amorality facilitates the dialectic of freedom followed by Draconian control (Jones 15). With the enshrinement of moral relativism, society invariably assumes a progressively more anarchistic trajectory. Impulses are entertained and excesses are indulged. Meanwhile, objective moral law is increasingly disregarding. Eventually, individual liberties are subordinated to hedonist appetites. Fleeting pleasures are ravenously sought, even at the expense of others. The ensuing chaos provides a pretext for the imposition of authoritarian policies to restore order. Of course, there is always a self-appointed elite that establishes and benefits from such systems.

Paradoxical, the power elite is equally as amoral as those they would fetter in the name of the law. Ruling class thought is permeated with relativistic notions. It is just such relativism that allows the oligarchs to believe that they can act as the arbiters of the dominant societal ethos. It comes as little surprise that Oxford Professor Carroll Quigley, a self-avowed elitist and apologist for the ruling class, rebuked the lower classes for their rejection of "complex relativisms" (Quigley 980). Of course, Quigley's endorsement of "complex relativisms" was irreconcilable with his endorsement of an absolutist world oligarchy. After all, one cannot lay claim to an absolute right to rule if there are no absolutes at all.

In addition to promoting amorality, relativism encourages the embracing of irrationality. The problem with relativism is a systemic one, a dilemma intrinsic to the view itself. Relativism is predicated upon the contention that there are no absolutes. Yet, if there are no absolutes, then one cannot absolutely declare that there are no absolutes. In fact, declarative statements cannot exist because they are statements of fact. Facts are absolutes and, according to relativism, do not exist. Immediately, the position implodes, crushed by its own intrinsic irrationality. Relativism is a self-refuting philosophical position.

However, Darwinism cosmetically obfuscated the irrationality that blemished relativistic Weltanschauungs. By undermining the foundations of Christianity with so-called "scientific proof," Darwinism banished moral absolutes and edified the unstable premises upon which relativism tottered. In *The Outlines of History*, H.G. Wells writes:

> If all animals and man evolved, then there were no first parents, no paradise, no fall. And if there had been no fall, then the entire historic fabric of Christianity, the story of the first sin, and the reason for the atonement collapses like a house of cards. (616)

With Christianity's "house of cards" effectively toppled, relativistic ideas could be actively promulgated with less resistance. Such ideas were certainly nothing new and had been promoted before by ideologues like Hume, Bacon, Rousseau, Descartes, Kant, and Weishaupt (Jasper *Global Tyranny. . .Step by Step* 262). Yet, Darwinism was different. Cribbed from Freemasonic doctrine and promoted through the British Royal Society, Darwin's theory of evolution promised to "scientifically" legitimize relativistic Weltanschauungs. This included, of course, the relativistic outlook of the Royal Society's would-be sculptors of a new societal ethos. The nature of this emergent ethos becomes apparent one considers the technocratic proclivities of the Royal Society's early Masonic founders.

In light of these observations, it becomes clear that Darwinism was an epistemological weapon developed for the technocratic restructuring of society. Jane H. Ingraham elaborates:

> But Darwin's role was to dignify[relativistic] these ideas with "scientific" backing and to make them accessible to the average man in terms he could understand. His shattering "explanation" of the evolution of man from the lower animals through means excluding the supernatural delivered the *coup de grace* to man's idea of himself as a created being in a world of fixed truth. Confronted with the "scientific proof" of his own animal origin and nature, Western man, set free at last

from God, began the long trek through scientific rationalism, environmental determinism, cultural conditioning, perfectibility of human nature, behaviorism, and secular humanism to today's inverted morality and totalitarian man. (Qutd. In Jasper *Global Tyranny. . .Step by Step* 262)

As the "objective moral law which overarches rulers and ruled alike" continued to disappear with the belief in a transcendent God, human society began to witness the rise of "totalitarian man." Of course, the rise of relativism also saw the rise of mass irrationality. This mass irrationality, which is the natural corollary of relativistic thought, is especially prevalent in orthodox academia. This irrationality was most vividly illustrated during a discussion between Christian philosopher Ravi Zacharias and a group of students at Oxford University. Zacharias relates the details of this shocking discourse:

I asked a group of skeptics if I took a baby and sliced it to pieces before them, would I have done anything wrong? At my question, there was silence, and then the lead voice in the group said, "I would not like it, but no, I could not say you have done anything wrong." My! What an aesthete. He would not like it. My! What irrationality--he could not brand it wrong. (115)

What irrationality indeed! It is especially ironic that the very same school of skepticism that repeatedly asks the question, "How can there be a good God when there is so much evil in the world." How can one reject the existence of God on such grounds when one rejects moral absolutes in the same breath? Such thinking has been commensurate with the rise of scientific dictatorships during the 20th century.

On November 29, 1994, Stone Phillips conducted an interview infamous serial killer Jeffrey Dahmer. During the interview, Dahmer made a rather revealing confession:

"If a person doesn't think there is a God to be accountable to, then-—then what's the point of trying to modify your behavior to keep it within acceptable ranges? That's how I thought anyway. I always believed the theory of evolution as truth, that we all just came from the slime. When we, when we died, you know, that was it, there is nothing. . ." (No pagination)

If this is how far humanity has traveled beyond the moral Rubicon, then the next step that the evolutionary Weltanschauung will take man is frightening indeed. Time and time again, history has demonstrated the consequences of relativistic thought. Perhaps the best historical example can be found in the Scriptures. Presented with a sinless man who was the obvious target of a malevolent conspiracy, Pilate merely responded, "What is truth?" (John 18:38). In the book *Life of Christ*, Fulton J. Sheen offers an eloquent summation of this response and its ramifications:

> Then he [Pilate] turned his back on truth—better not on it, but on Him Who is Truth. It remained to be seen that tolerance of truth and error in a stroke of broadmindedness leads to intolerance and persecution; "What is truth?" when sneered, is followed up with the second sneer, "What is justice?" Broadmindedness, when it means indifference to right and wrong, eventually ends in a hatred of what is right. He who was so tolerant of error as to deny an Absolute Truth was the one who would crucify Truth. It was the religious judge who challenged Him, "I adjure thee;" but the secular judge asked, "What is truth?" He who was in the robe of the high priest called upon God the things that are God's; he who was in the Roman toga just professed a skepticism and doubt. (364)

Pilate's question was a rhetorical one, inferring that truth did not exist. Meanwhile, the Truth stood right before him, enveloped in a profound silence. Still, it was easier for Pilate to resort to the frivolity of pragmatism and utilitarianism. Despite the clear absence of evidence to convict this guiltless man of any crime, judicial protocol was circumvented and He was crucified. Of course, the Truth did not remain buried for very long.

Over two thousand years later, it would appear as though man has come no further. As moral absolutes are jettisoned in favor or relativism, technocratic social engineers continue to shape a totalitarian ethos. In *Brave New World Revisited*, Aldous Huxley wrote:

> . . .a new Social Ethic is replacing our traditional ethical system. . .the system in which the individual is primary.
> the social whole has greater worth and significance than its individual parts. . .that the rights of the collectivity take precedence over. . .the Rights of Man. (23)

Nietzsche's world "beyond good and evil" is more closely akin to Skinner's world, which is "beyond freedom and dignity." As the moral Rubicon is traversed, so is the line separating freedom from slavery. It is a scientific dictatorship, dignified by Darwinism and built on the ashes of morality.

Sci-fi "Predictive Programming"

Aldous Huxley first presented the scientific dictatorship to the public imagination in his book *Brave New World*. In *Dope, Inc.*, associates of political dissident Lyndon LaRouche claim that Huxley's book was actually a "mass appeal" organizing document written "on behalf of one-world order" (*Dope, Inc.* 538). The book also claims the United States is the only place where Huxley's "science fiction classic" is taught as an allegorical condemnation of fascism (*Dope, Inc.* 538). If this is true, then the scientific dictatorship presented within the pages of his 1932 novel *Brave New World* is a thinly disguised *roman a' clef*—a novel that thinly veils real people or events— awaiting tangible enactment.

Such is often the case with "science fiction" literature. According to researcher Michael Hoffman, this literary genre is instrumental in the indoctrination of the masses into the doctrines of the elite:

> Traditionally, "science fiction" has appeared to most people as an adolescent genre, the province of time-wasting fantasies. This has been the great strength of this genre as a vehicle for the inculcation of the ideology favored by the Cryptocracy. As J.H. Towsen points out in *Clowns*, only when people think they are not buying something can the real sales pitch begin. While it is true that with the success of NASA's Gemini space program and the Apollo moon flights more serious attention and respectability was accorded "science fiction," nonetheless in its formative seeding time, from the late 19th century through the 1950s, the predictive program known as "science fiction" had the advantage of being derided as the solitary vice of misfit juveniles and marginal adults. (205)

Thus, "science fiction" is a means of conditioning the masses to accept future visions that the elite wish to tangibly enact. This process of gradual and subtle inculcation is dubbed "predictive programming." Hoffman elaborates: "Predictive programming works by means of the propagation of the illusion of an infallibly accurate vision of how the world is going to look in the future" (205). Also dubbed "sci-fi inevitabilism" by Hoffman, predictive programming is analogous to a virus that infects its hosts with the false belief that it is:

Useless to resist central, establishment control.
Or it posits a counter-cultural alternative to such control which is actually a counterfeit, covertly emanating from the establishment itself.
That the blackening (pollution) of earth is as unavoidable as entropy.
That extinction ('evolution") of the species is inevitable.
That the reinhabitation of the earth by the "old gods" (Genesis 6:4), is our stellar scientific destiny. (8)

Memes (contagious ideas) are instilled through the circulation of "mass appeal" documents under the guise of "science fiction" literature. Once subsumed on a psychocognitive level, these memes become self-fulfilling prophecies, embraced by the masses and outwardly approximated through the efforts of the elite.

In addition to spreading virulent strains of thought, sci-fi has also been instrumental in the promulgation of Darwinism. For instance, the sci-fi literature of Freemason H.G. Wells would play an important role in promulgating the concept of evolution. J.P. Vernier reveals Wells' religious adherence to the concept of evolution and its inspiration on him as an author of science fiction:

The impact of the theory of evolution on his [Wells'] mind is well known: it was the first felt when he attended the Lectures of T.H. Huxley, at South Kensington, in 1884 and 1885, and, ten years later, evolution was to provide him with the fundamental theme of his "scientific romances" and of many of his short stories. ("Evolution as a Literary Theme in H.G. Wells's Science Fiction," 70)

J.P. Vernier elaborates on the role of sci-fi literature, particularly Wells' "scientific romances," in promulgating evolutionary thought:

Science fiction is admittedly almost impossible to define; readers all think they know what it is and yet no definition will cover all its various aspects. However, I would suggest that evolution, as presented by Wells, that is a kind of mutation resulting in the confrontation of man with different species, is one of the main themes of modern science fiction. ("Evolution as a Literary Theme in H.G. Wells's Science Fiction," 85)

In *Orthodoxy and the Religion of the Future*, Bishop Serphim Rose expands on the role of sci-fi in the promulgation of evolutionary thought:

The center of the science fiction universe (in place of the absent God) is man—not usually man as he is now, but man as he will "become" in the future, in accordance with the modern mythology of evolution. (73)

Reiterating Vernier's contention that the sci-fi notion of evolution is "a kind of mutation resulting in the confrontation of man with different species," Rose observes:

Although the heroes of science fiction stories are usually recognizable humans, the story interest often centers about their encounters with various kinds of "supermen" from "highly-evolved" races of the future (or sometimes, the past), or from distant galaxies. The idea of the possibility of "highly-evolved" intelligent life on other planets has become so much a part of the contemporary mentality that even respectable scientific (and semi-scientific) speculations assume it as a matter of course. Thus, one popular series of books (Erich von Daniken, Chariots of the Gods?, Gods From Outer Space) finds supposed evidence of the presence of "extraterrestrial" beings or "gods" in ancient history, who are supposedly responsible for the sudden appearance of intelligence in man, difficult to account for by the usual evolutionary theory. (73)

According to Rose, science fiction's traditional depiction of religion suggests that the future will inherit a nebulous and indefinable spirituality:

Religion, in the traditional sense, is absent, or else present in a very incidental or artificial way. The literary form itself is obviously a product of the "post-Christian" age (evident already in the stories of Poe and Shelley). The science fiction universe is a totally secular one, although often with "mystical" overtones of an occult or Eastern kind. "God," if mentioned at all, is a vague and impersonal power, not a personal being (for example, the "Force" of Star Wars, a cosmic energy that has its evil as well as good side). The increasing fascination of contemporary man with science fiction themes is a direct reflection of the loss of traditional religious values. (73)

Expanding on the "mystical" themes of sci-fi, researcher Carl Raschke asserts that the literary genre invariably extends itself into the realm of the occult:

> The snug relationship between occult fantasy and the actual practice of the occult is well established in history. Writers such as H.P. Lovecraft and Edgar Rice Burroughs, progenitor of the Tarzan and Jane tales, were practicing occultists. (*Painted Black* 303)

Raschke explains that sci-fi presents a future that has rediscovered the occult traditions of its past:

> Increasingly, science fiction with its vistas of the technological future intertwines with the neopagan and the medieval. The synthesis was first achieved with polished artistry in Lucas' Star Wars trilogy. (*Painted Black* 398)

Eloquently summarizing the close correlation between science fiction and occultism, Raschke states: "Science fiction, 'science fantasy,' pure fantasy, and the world of esoteric thought and activity have all been intimately connected historically." (*Painted Black* 303)

Clearly, such ideas are fantastic to say the least. Yet, contemporary scientists have given them serious credence:

> Serious scientists in the Soviet Union speculate that the destruction of Sodom and Gomorrah was due to a nuclear explosion, that "extraterrestrial" beings visited earth centuries ago, that Jesus Christ may have been a "cosmonaut," and that today "we may be on the threshold of a 'second coming' of intelligent beings from outer space." Equally serious scientists in the West think the existence of "extraterrestrial intelligences" likely enough that for at least 18 years they have been trying to establish contact with them by means of radio

telescopes, and currently there are at least six searches being conducted by astronomers around the world for intelligent radio signals from space. (Rose 73-74)

According to Rose, the sci-fi genre's influence upon science could, in turn, provoke a shift in religious thinking:

Contemporary Protestant and Roman Catholic "theologians"--who have become accustomed to follow wherever "science" seems to be leading – speculate in turn in the new realm of "exotheology" (the "theology of outer space") concerning what nature the "extraterrestrial" races might have (see Time magazine, April 24, 1978). It can hardly be denied that the myth behind science fiction has a powerful fascination even among many learned men of our day. (74)

In his final assessment of science fiction, Rose concludes that this ostensibly "scientific and non-religious" genre is, in truth, the "leading propagator (in a secular form) of the 'new religious consciousness'" that is gradually supplanting Christianity (77). Laced with occultism and intimations of an emergent pagan spirituality, science fiction could be facilitating a paradigm shift in religious thinking.

Such a paradigm shift could already be underway. Among one of its chief "evangelists" is William Sims Bainbridge, sociologist and member of the National Science Foundation. Bainbridge concerns himself predominantly with the development of a new world religion, which he dubs the "Church of God Galactic." Expanding on the characteristics intrinsic to such a church, Bainbridge suggests, "its most likely origins are in science fiction" ("Religions for a Galactic Civilization," no pagination).

According to Bainbridge, secularization provides the religio-cultural segue for this new religion. Examining the sociological phenomenon of secularization, Bainbridge makes an interesting observation:

Secularization does not mean a decline in the need for religion, but only a loss of power by traditional denominations. Studies of the geography of religion show that where the churches become weak, cults and occultism explode to fill the spiritual vacuum. ("Religions for a Galactic Civilization," no pagination)

Secularization has been commonly associated with atheism. Indeed, past periods of secularization have seen the decline of theistic faiths and a general rejection of traditional notions of God. No doubt, the publication of *Origin of the Species* and the subsequent widespread promotion of evolutionary thought had this effect. However, periods of secularization do not represent the obliteration of religion, but the preparation of the dominant religio-cultural milieu for the arrival of a new religion. Secularization and its correlative, atheism, only act as a catalyst for an enormous paradigm shift. This begins with the realization of a significant philosophical paradox intrinsic to atheism. Authors Ron Carlson and Ed Decker explain this intrinsic paradox:

It is philosophically impossible to be an atheist, since to be an atheist you must have infinite knowledge in order to know absolutely that there is no God. But to have infinite knowledge, you would have to be God yourself. It's hard to be God yourself and an atheist at the same time! (17)

In order to be philosophically consistent, the atheist must eventually conclude that he/she is a god. Whittaker Chambers, former member of the communist underground in America, revealed the name of this faith in one's own intrinsic divinity:

"Humanism is not new. It is, in fact, man's second oldest faith. Its promise was whispered in the first days of Creation under the Tree of the knowledge of Good and Evil: 'Ye shall be as gods.'" (Qutd. in Baker 206)

Simply stated, humanism is the religion of self-deification. Its god is Man, spelled with a capital M to denote the purported divinity intrinsic to humanity. Of course, this was also the religion of Freemasonry. In fact, humanism and Masonry have shared a long historical relationship. In *The Keys of this Blood*, deceased Vatican insider Malachi Martin examined the emergence of "a network of Humanist associations" throughout early-Renaissance Italy (518-19). These organizations represented:

a revolt against the traditional interpretation of the Bible as maintained by the ecclesiastical and civil authorities, and against the philosophical and theological underpinnings provided by the Church for civil and political life. (519)

Although these groups espoused an ostensible belief in God, their notions of a Supreme Being were largely derivative of the Kabbala:

> Not surprisingly given such an animus, these associations had their own conception of the original message of the Bible and of God's revelation. They latched onto what they considered to be an ultrasecret body of knowledge, a gnosis, which they based in part on cultic and occultist strains deriving from North Africa-notably, Egypt-and, in part, on the classical Jewish Kabbala. (519)

Recall thirty-third Degree Freemason Albert Pike's revelation that "all the Masonic associations owe to it [the Kabbala] their Secrets and their Symbols" (Pike 744). According to Martin, however, this ancient Hebraic doctrine was modified considerably by the early humanists:

> Whether out of historical ignorance or willfulness of both, Italian humanists bowdlerized the idea of *Kabbala* almost beyond recognition. They reconstructed the concept of gnosis, and transferred it to a thoroughly this-worldly plane. The special *gnosis* they sought was a secret knowledge of how to master the blind forces of nature for a sociopolitical purpose. (519-20)

Many of the semiotic artifacts comprising the early humanists's iconography and jargon were also directly related to Masonry:

> Initiates of those early humanist associations were devotees of the Great Force--the Great Architect of the Cosmos--which they represented under the form of the Sacred Tetragrammaton, YHWH, the Jewish symbol for the name of the divinity that was not to be pronounced by mortal lips. They borrowed other symbols--the Pyramid and the All-Seeing Eye--mainly from Egyptian sources. (Martin 520)

The Great Architect of the Cosmos, the All-Seeing Eye, and the Pyramid also comprise the esoteric semiology of Freemasonry. What is the explanation for all of these commonalities? According to Martin, these shared characteristics were the result of a merger between the humanists and the old Mason guilds:

In other northern climes, meanwhile, a far more important union took place, with the humanists. A union that no one could have expected. In the 1300s, during the time that the cabalist–humanist associations were beginning to find their bearings, there already existed–particularly in England, Scotland and France-medieval guilds of men who worked with ax, chisel and mallet in freestone. Freemasons by trade, and God-fearing in their religion, these were men who fitted perfectly into the hierarchic order of things on which their world rested. (521)

Evidently, there couldn't have been two organizations that were more diametrically opposed than Masonry and humanism:

No one alive in the 1300s could have predicted a merger of minds between freemason guilds and the Italian humanists. The traditional faith of the one, and the ideological hostility to both tradition and faith of the other, should have made the two groups about as likely to mix as oil and water. (Martin 522)

Nevertheless, the late 1500s would witness the amalgamation of these two groups (Martin 522). The most evident corollary of this organizational coalescence was a noticeable difference in recruiting practices:

As the number of working or "operative," freemasons diminished progressively, they were replaced by what were called Accepted Masons–gentlemen of leisure, aristocrats, even members of royal families–who lifted ax, chisel and mallet only in the ultrasecret symbolic ceremonies of the lodge, still guarded by the "Charges" and the "Mason Word." The "speculative" mason was born. The new Masonry shifted away from all allegiance to Roman ecclesiastical Christianity. (Martin 522)

Indeed, the new Masonic doctrine appeared to be one that thoroughly eschewed Christian concepts:

There was no conceptual basis by which such a belief could be reconciled with Christianity. For precluded were all such ideas as sin, Hell for punishment and Heaven for reward, and eternally perpetual Sacrifice of the Mass, saints and angels, priest and pope. (522)

The new Mason was no longer an architect of freestone. Instead, he was an architect of the technocratic Utopia mandated by Bacon's *New Atlantis*. His god was Man himself, an emergent deity sculpted by the Kabalistic golem of nature through the occult process of "becoming." Of course, this concept would later be disseminated on the popular level as Darwinism and the world would call it "evolution." Coomaraswamy recapitulates:

> Evolutionary theory came along as a "chance-sent" gift providing humanists and their ilk with the sanction of "science." If mankind accepted its postulates, who needed God and who needed the Church? It is not surprising that Masons, Marxists and Modernists did everything in their power to spread this new "devil's Gospel." (No pagination)

These humanist-Masonic concepts remain firmly embedded within the science fiction genre. In an interview with humanist David Alexander, *Star Trek* creator Gene Roddenberry commented:

> "As nearly as I can concentrate on the question today, I believe I am God; certainly you are, I think we intelligent beings on this planet are all a piece of God, are becoming God." (568)

In addition to espousing this core precept of the humanist-Masonic religion, Roddenberry's *Star Trek* presented a technocratic world government under the appellation of the "Federation." Of course, one could argue that such concepts are simply part of an innocuous fiction concocted for entertainment. According to Bainbridge, however, there is "government-encouraged research" devoted to the realization of "the Star Trek prophecies" ("Memorials"). Apparently, the demarcations between fact and fiction are becoming increasingly indiscernible.

As science fiction vigorously proselytizes the masses in the humanist-Masonic religion, the spiritual vacuum left by secularization is being filled. As Bainbridge previously stated, the immediate elements to supplant the orthodox ecclesiastical authority are "cults and occultism" ("Religions for a Galactic Civilization," no pagination). The contemporary religious counterculture movement has most vividly expressed itself through the explosion of scientistic cults in the late twentieth and early twenty-first century. Bainbridge himself has been actively involved with some of these cults, which act as working models for his Church of God Galactic.

Examining the most promising model for the Church of God Galactic, Bainbridge makes the following recommendation:

> **Today there exists one highly effective religion actually derived from science fiction, one which fits all the known sociological requirements for a successful Church of God Galactic. I refer, of course, to Scientology. ("Religions for a Galactic Civilization, no pagination")**

Indeed, Scientology meets all the prerequisites for Bainbridge's Church of God Galactic, one of which being the cult's origins with science fiction. Carl Raschke explains:

> L. Ron Hubbard, architect of the controversial religion known as Scientology, openly and consciously decided to convert his science fiction work into a working belief system upon which a "church" was set up. (*Painted Black* 303)

As a derivation of science fiction, Scientology inherited a central feature of the genre: Darwinism. In *Dianetics*, Scientologist high priest L. Ron Hubbard reveals the movement's adherence to evolutionary thought:

> It is fairly well accepted in these times that life in all forms evolved from the basic building blocks: the virus and the cell. Its only relevance to Dianetics is that such a proposition works--and actually that is all we ask of Dianetics. There is no point to writing here a vast tome on biology and evolution. We can add some chapters to those things, but Charles Darwin did his job well and the fundamental principles of evolution can be found in his and other works. *The proposition on which Dianetics was originally entered was evolution.* (69; emphasis added)

Darwinian thought is especially evident in Scientology's preoccupation with survival. In *Dianetics*, Hubbard opines: "The dynamic principle of existence is survival" (52). In this statement, one can discern echoes of the Darwinian mantra: "Survival of the fittest." Hubbard proceeds to enumerate four dynamics of survival. It is within the fourth dynamic that the astute reader will recognize Darwinism's corresponding religion of self-deification: "Dynamic four is the thrust toward *potential immortality of mankind as a species*"(53; emphasis added). Of course, immortality is a trait reserved only for gods. Again, the religious theme of man's evolutionary ascent towards apotheosis becomes evident.

Eventually, Hubbard's church of Scientology "suffered religious schisms which spawned other cults" (Bainbridge, "Religions for a Galactic Civilization," no pagination). One of the resulting sects was the Process Church of Final Judgement, a satanic cult that was the subject of a five-year ethnographic study conducted by Bainbridge ("Social Construction from Within: Satan's Process," no pagination). Enamored with the group, Bainbridge praised the Process Church as a "remarkably aesthetic and intelligent alternative to conventional religion" ("Social Construction from Within: Satan's Process," no pagination).

A deeper examination of this scientistic cult reveals that its adherents probably retained much of the Darwinian thought intrinsic to its progenitor, Scientology. One case in point is the theology of the group's founder, Robert de Grimston. Bainbridge delineates this theology:

Robert de Grimston's theology was Hegelianism in the extreme. For every thesis (Christ, Jehovah) there was an antithesis (Satan, Lucifer), and the cult aimed to achieve a final synthesis of all these dichotomies in the rebirth of GOD. Indeed, one way of explaining the failure of The Process is to note that it promised a Heaven on earth to members, yet it delivered something less. ("Social Construction from Within: Satan's Process," no pagination)

Like Processean theology, Darwinian evolution also exhibits an inherently Hegelian framework. The organism (thesis) comes into conflict with nature (antithesis) resulting in a newly enhanced species (synthesis), the culmination of the evolutionary process (Marrs, *Circle of Intrigue*, 127). H.G. Wells, a Freemason and protégé of Darwinian apologist T.H. Huxley distilled a similar dialectical framework in an allegorical form. W. Warren Wagar elaborates:

> In the symbolic prologue to *The Undying Fire*, he [Wells] even likened the opposition of essence and existence to the interplay of good and evil. God was here represented as the inscrutable creator, who created things perfect and exact, only to allow the intrusion of a marginal inexactness in things through the intervention of Satan. God corrected the marginal uniqueness by creation at a higher level, and Satan upset the equilibrium all over again. Satan's intervention permitted evolution, but the ultimate purpose of God was by implication a perfect and finished and evolved absolute unity. (104-05)

The Processeans shared Wells' notion of Satan, which portrayed the Devil as a necessary element of instability:

> For Processeans, Satan was no crude beast but an intellectual principle by which God could be unfolded into several parts, accomplishing the repaganization of religion and the remystification of the world. (Bainbridge, "Social Construction from Within: Satan's Process," no pagination)

This portrait of an ongoing dialectical conflict echoes the Masonic dictum: *Ordo Ab Chao* (Latin for Order out of Chaos). The dialectical process underpins evolution, which began with the Masonic doctrine of "becoming." The final goal of a repaganized world synchronizes very well with Freemasonic occultism. All comprise the new religious consciousness being promulgated by science fiction. This is the future that the masses are being conditioned to accept by sci-fi predictive programming.

In *Religion and the Social Order*, Bainbridge presented the following mandate:

It is time to move beyond mere observation of scientistic cults and use the knowledge we have gained of recruitment strategies, cultural innovation, and social needs to create better religions than the world currently possesses. At the very least, unobtrusive observation must be supplemented by active experimentation.

Religions are human creations. Our society quite consciously tries to improve every other kind of social institution, why not religion? Members of The Process, founded mainly by students from an architecture school, referred to the creation of their cult as religious engineering, the conscious, systematic, skilled creation of a new religion. I propose that we become religious engineers. (No pagination)

To understand what sort of faith is being sculpted by the technocratic "religious engineers," one need only look to Scientology and the Process Church. Both of these scientistic cults, awash in Darwinism and its corresponding humanist-Masonic religion of apotheosized Man, are microcosms for an emergent one-world religion.

The Family: A Case Study in Religious Engineering
Scientologists and Processeans are not the only parties involved in the "conscious, systematic, skilled creation of a new religion." Religious engineering projects have also been initiated within Christianity. Just as the Processeans re-conceptualized Satan, individuals with dubious motives are re-conceptualizing Jesus.

Jeffrey Sharlet infiltrated one such group, ominously named the Family (a shared appellation with the Charles Manson cult). In an article in *Harper's* magazine, Sharlet revealed some disturbing aspects of this group. While the group consistently invokes the name of Jesus, Christian is "a term they deride as too narrow for the world they are building in Christ's honor. . ." (53). Sharlet elaborates:

> . . .the Family reject the label "Christian." Their faith and their practice seemed closer to a perverted sort of Buddhism, their God outside "the truth," their Christ everywhere and nowhere at once, His commands phrased as questions, His will as simple to divine as one's own desires. And what the Family desired. . .was power, worldly power, with which Christ's kingdom can be built. . .(63)

It is obvious that the Family's notion of Christ's kingdom breaks drastically with the Biblical concept. The concept of an Eschaton within this ontological plane is inherently Gnostic. In fact, the Gnostic mandate to "immanentize the Eschaton" is consistently reiterated by Family members. David Coe, son of the cult's current leader, tells other members that they "are here to learn how to rule the world" (59). Such statements are redolent of Gnostic thinking. Likewise, the Christ being sculpted by the Family's religious engineers is inherently Gnostic. No longer is He the Savior and Redeemer of humanity. Instead, He is a nebulous and ambiguous icon whose features are being re-configured according to the template of sociopolitical Utopianism.

To understand the Family's conception of Christ's kingdom, one must examine its founder, Abraham Vereide. Sharlet introduces this enigmatic character:

The Family was founded in April 1935 by Abraham Vereide, a Norwegian immigrant who made his living as a traveling preacher. One night, while lying in bed fretting about socialists, Wobblies, and a Swedish Communist who, he was sure, planned to bring Seattle under the control of Moscow, Vereide received a visitation: a voice, and a light in the dark, bright and blinding. (61)

Mimicking Joseph Smith, Vereide attributed his ideas to a divine encounter. These ideas would be presented under an anticommunist label, obviously manipulating justifiable fears of the communist threat to civilization. However, Vereide's goal could not be described as Christian. Sharlet writes: "In 1944, Vereide has foreseen what he called 'the new world order'" (61). The "new world order" is a term originating with and tossed about in several elitist circles seeking to establish some form of world government. It is a catch phrase for a world system best described by Donald McAlvany:

A world government, by its highly centralized nature, would be socialistic; would be accompanied by redistribution of wealth; strict regimentation; and would incorporate severe limitations on freedom of movement, freedom of worship, private property rights, free speech, the right to publish, and other basic freedoms. (287)

For many Christians, this system would resemble the kingdom of antichrist described in the thirteenth chapter of the Book of Revelation. However, many are duped by the group's "Christian" exterior. Its ranks include the influential and powerful:

The Family is, in its own words, an "invisible" association, though its membership has always consisted mostly of public men. Senators Don Nickles (R., Okla.), Charles Grassley (R., Iowa), Pete Domenici (R., N. Mex.), John Ensign (R., Nev.), James Inhofe (R. Okla.), Bill Nelson (D., Fla.), and Conrad Burns (R., Mont.) are referred to as "members," as are Representatives Jim DeMint (R., S.C.), Frank Wolf (R., Va.), Joseph Pitts (R., Pa.), Zach Wamp (R., Tenn.), and Bart Stupak (D., Mich.). (Sharlet 54)

The Family has also formed prayer groups that have given them access to the halls of power:

Regular prayer groups have met in the Pentagon and at the Department of Defense, and the Family has traditionally fostered strong ties with businessmen in the oil and aerospace industries. (Sharlet 54)

Members of the controlled conservative movement seem to make up the Family's unsuspecting prey. It is the organization behind the congressional sponsored National Prayer Breakfast held every February in Washington D.C. (Sharlet 54). Sharlet continues:

...the breakfast is regarded by the Family as merely a tool in a larger purpose: to recruit the powerful attendees into smaller, more frequent prayer meetings, where they can "meet Jesus man to man." (54)

The Jesus encountered by the Family's new recruits is hardly the Jesus one reads about in Scripture. As the Church of the God Galactic emerges, its religious engineers continue to fashion a new Christ. This evolving New Age messiah will only grow more and more unfamiliar to the eyes of the believer. It comes as little surprise that Jesus admonished, "Many shall come in my name."

Alien Co-Evolution and the Sirius Connection

In *Morals and Dogma*, 33rd Degree Freemason Albert Pike bestows special honor upon Sirius, a heavenly body that "still glitters in our Lodges as the Blazing Star" (Pike 486). Indeed, Sirius represents some foundational axiom of the Masonic Craft. Pike explains that the star is "an emblem of the Divine Truth, given by God to the first men, and preserved amid all the vicissitudes of ages in the traditions and teachings of Masonry" (Pike 136). As Pike continues, he reveals that Sirius has also held numerous other appellations: "The Blazing Star in our Lodges, we have already said, represents Sirius, Anubis, or Mercury, Guardian and Guide of Souls" (Pike, 506).

Whatever its name, the star represents an entity of great esoteric significance to Freemasonry:

> In the old Lectures they said: "The Blazing Star or Glory in the centre refers us to that Grand Luminary the Sun, which enlightens the Earth, and by its genial influence dispenses blessings to mankind." (Pike, 506)

A little later, Pike reiterates: "the Blazing Star has been regarded as an emblem of Omniscience, or the All-Seeing Eye, which to the Ancients was the Sun" (Pike 506). Recall that, before the external characteristics of the oligarchs' control apparatus were cosmetically altered to present a "scientific dictatorship," the elite ruled through institutionalized Sun worship (Keith, *Saucers of the Illuminati*, 78-79). In reference to the "Sun," Pike provides a brief glimpse of the god of Freemasonry. Although the outward features of its theocracy have changed, the deity has remained the same and his identity is associated with the star called Sirius.

According to Pike, Sirius was responsible for imparting numerous innovations to mankind:

> He was Sirius or the Dog-Star, the friend and counselor of Osiris, and the inventor of language, grammar, astronomy, surveying, arithmetic, music, and medical science; the first maker of laws; and who taught the worship of the Gods, and the building of temples. (Pike 376)

It is interesting to note that, among his various contributions, this Freemasonic deity was responsible for the introduction of several forms of science. Does Sirius also represent the Lodge's "ostensible control over the knowable?" Is the Dog-Star a symbol of the elite's "scientific dictatorship?" Michael Hoffman further elaborates on the identity of Sirius:

> The mythical Satanic bringer of civilization to earth was supposed to be an alien from the star system Sirius, around whom the Egyptians and all subsequent Hermetic systems constructed their elaborate and obsessive religio-astronomic

observances. This star Sirius also served as an astronomic secret code, an allegory of the illusory quality and inherent "trickiness" of the material world. (Hoffman 26-27)

This Freemasonic mythology of extraterrestrial intervention in human evolution may be poised for a return. Given the impossibility of spontaneous generation, Darwinism has faced a major obstacle to its unquestioned primacy. Recognizing this obstacle, scientific materialist Francis Crick presented a theory bearing an uncanny resemblance to the Sirius myth. According to Crick, technologically advanced extraterrestrials "seeded" the earth with life billions of years ago. Whether Crick was privy to the occult doctrines of the elite or was simply following the natural course of Darwinian thought, one thing is certain, he and other proponents of similar "extraterrestrial intervention" theories are paving the way for the re-introduction of Freemasonic mysticism to mainstream science.

Consider the following account of Linda Moulton Howe. During a meeting with Richard Doty, an intelligence officer with the United States military, Howe was presented with a briefing paper regarding alien visitation. In its body, Howe read an interesting claim regarding the crumbling theory of Darwinism: "It stated that all questions and mysteries about the evolution of Homo sapiens on this planet had been answered and that project was closed." (Howe 151)

How convenient! By what means did these extraterrestrials facilitate the evolutionary process? Reiterating the basic contentions of Crick, the paper stated:

[T]hese ETs have come at various intervals in the earth's history to manipulate DNA in already existing terrestrial primates and perhaps in other life forms as well. To the best of my memory, the time intervals for this DNA manipulation specifically listed in the briefing paper were 25,000, 15,000, 5,000, and 2,500 years ago. (Howe 151)

Faced with the impossibility of spontaneous generation and the inexorable collapse of Darwinism, the elite could now be invoking an "extraterrestrial intervention" myth cribbed from their own doctrines. Given Richard Doty's military intelligence connections, this remains a very real possibility. The Freemasonic doctrine of Sirius has circulated within military intelligence groups for quite some time. According to researcher James Shelby Downard, there exists a cult of Sirius adherents at the highest levels of the CIA (Keith, *Saucers of the Illuminati*, 49). Researcher Jim Keith elaborates:

> He cites as one of their ritual locations the telescope viewing room of the Palomar Observatory in California. There, he says, the adepts of the Sirius-military intelligence cult enact rituals in the telescopically-focused light of the Dog Star, in imitation of the Egyptian priesthood, astral rays bathing the viewing chamber and the participants when the telescope is aimed Sirius-ward. (Keith, *Saucers of the Illuminati*, 49)

Keith proceeds to cite the case of military intelligence officer Michael Aquino:

> Utter madness? Tell that to Colonel Michael Aquino of U.S. military intelligence, the admitted head of the satanic Temple of Set, a deity [Set] identified in occultism with Sirius. Aquino makes no bones about the fact that he is the head of his offshoot of Anton LaVey's Church of Satan, known to draw many of its leaders from military circles. Again, we see the strange conjunction of Sirius, occultism, and military intelligence. (Keith, *Saucers of the Illuminati*, 49)

Those who comprise this "strange conjunction" could also be responsible for the perpetration of a disinformation campaign, derivative of Masonic doctrine and designed to maintain the waning dominance of Darwinism.

2001: The Evangel of Sirius

The public conscious may have already been prepared for the re-introduction of the Sirius myth. In the film *2001: A Space Odyssey*, audiences are confronted by a mysterious monolith. Like Sirius, this monolith also functions as a "bringer of civilization to earth." Michael Hoffman explains:

> *2001, A Space Odyssey*, directed by Stanley Kubrick and based on the writing of Arthur C. Clarke, is, with hindsight, a pompous, pretentious exercise. But when it debuted it sent shivers up the collective spine. It has a hallowed place in the Cryptosphere because it helped fashion what the Videodrome embodies today. At the heart of the film is the worship of the

Darwinian hypothesis of evolution and the positioning of a
mysterious monolith as the evolutionary battery or "sentinel"
that transforms the ape into the space man (hence the
"odyssey").
Clarke and Kubrick's movie, *2001*, opens with a scene of the
"Dawn of Man," supposedly intended to take the viewer back
to the origins of humanity on earth. This lengthy sequence
is vintage Darwinism, portraying our genesis as bestial and
featuring man-like apes as our ancestors. In the film, the
evolution of these hominids is raised to the next rung on the
evolutionary ladder by the sudden appearance of a mysterious
monolith. Commensurate with the new presence of this
enigmatic "sentinel," our alleged simian progenitors learn to
acquire a primitive form of technology; for the first time they
use a bone as a weapon.
This bone is then tossed into the air by one of the ape-men.
Kubrick photographs the bone in slow motion and by means
of special effects, he shows it becoming an orbiting spacecraft,
thus traversing "millions of years in evolutionary time."
The next evolutionary level occurs in "2(00)1" (21, i.e. the 21st
century). In the year 2001, the cosmic sentinel that is the
monolith reappears again, triggering an alert that man is on to
the next stage of his "glorious evolution." (11-12)

This narrative strongly reflects the paradigmatic character of
the Masonic Sirius myth. The monolith facilitates humanity's ascent
from bestial primitivism to advanced civilization. What is especially
interesting is the monolith's facilitative role in human evolution.
According to Masonic scholar Wilmshurst, the completion of human
evolution involves man "becoming a god-like being and unifying his
consciousness with the Omniscient" (94). According to 33rd degree
Mason Albert Pike, Sirius, which is represented by the "Blazing Star,"
enjoys the distinction of being "an emblem of Omniscience" (321). This
Masonic interpretation of Clarke's allegory brings an added dimension
to the monolith.

The monolith may be one more semiotic permutation of the
Philosopher's Stone, the Holy Grail, and other metaphorically related
mystical artifacts. All of these fabled relics were allegedly imbued
with transformational powers. In turn, they all seem to semiotically
gesticulate towards the Gnostic concept of *gnosis* (knowing). This
concept arises from an inversion of the traditional Genesis account. As

opposed to the original Biblical version, the Gnostic account represents a "revaluation of the Hebraic story of the first man's temptation, the desire of mere men to 'be as gods' by partaking of the tree of the 'knowledge of good and evil'" (Raschke, *The Interruption of Eternity* 26). Carl Raschke elaborates:

> In *The Hypostasis of the Archons*, an Egyptian Gnostic document, we read how the traditional story of man's disobedience toward God is reinterpreted as a universal conflict between "knowledge" (*gnosis*) and the dark "powers" (*exousia*) of the world, which bind the human soul in ignorance. The *Hypostasis* describes man as a stepchild of *Sophia* ("Wisdom") created according to the "model" of *aion*, the imperishable realm of eternity. On the other hand, it is neither God the Imperishable nor Sophia who actually is responsible in the making of man. On the contrary, the task is undertaken by the archons, the demonic powers who, because of their "weakness," entrap man in a material body and thus cut him off from his blessed origin. They place him in paradise and enjoin him against eating of the tree of knowledge. The prohibition, however, is viewed by the author of the text not as a holy command but as a malignant effort on the part of the inferior spirits to prevent Adam from having true communion with the High God, from gaining authentic *gnosis*. (*The Interruption of Eternity* 26)

According to this bowdlerization, Adam is consistently contacted by the High God in hopes of reinitiating man's quest for *gnosis* (*The Interruption of Eternity* 26). The archons intervene and create Eve to distract Adam from the pursuit of *gnosis* (*The Interruption of Eternity* 26-27). However, this Gnostic Eve is actually a "sort of 'undercover' agent for the High God, who is charged with divulging to Adam the truth that has been withheld from him" (*The Interruption of Eternity* 27). The archons manage to sabotage this covert operation by facilitating sexual intercourse between Adam and Eve, an act that Gnostics contend was designed to defile the "woman's spiritual nature" (*The Interruption of Eternity* 27). At this juncture, the *Hypostasis* reintroduces a familiar antagonist from the original Genesis account:

> But now the principle of feminine wisdom reappears in the form of the serpent, called the "Instructor," who tells the mortal pair to defy the prohibition of the archons and eat of the tree of knowledge. (*The Interruption of Eternity* 27)

The serpent successfully entices Adam and Eve to eat the forbidden fruit, but the "bodily defilement" of the woman prevents man from understanding the true motive underpinning the act (*The Interruption of Eternity* 27). Thus, humanity is fettered by the archons' "curse", suggesting that the "orthodox theological view of the violation of the command as 'sin' must be regarded anew as the mindless failure to commit the act rightly in the first place" (*The Interruption of Eternity* 27). In this revisionist context, the serpent is no longer Satan, but is an "*incognito* savior" instead (*The Interruption of Eternity* 27).

Likewise, the monolith could represent an "*incognito* savior." It seems to be analogous to primordial knowledge, *gnosis*. This Gnostic conception of knowledge "bears a close affinity to the recondite revelation given to those who took part in the Hellenistic mystery cults" (*The Interruption of Eternity* 28). While Gnosticism's origins with the Ancient Mystery cults remains a source of contention amongst scholars, its promises of liberation from humanity's material side is strongly akin to the old pagan Mystery's variety of "psychic therapy" (*The Interruption of Eternity* 28). S. Angus argues that the Mystery cults were the precursors to Gnosticism:

> The Mystery-Religions were systems of *Gnosis* akin, and forming a stage to, those movements to which the name of Gnosticism became attached. They professed to satisfy the desire for the knowledge of God which became pronounced from at least the second century B.C. and increased in intensity until the acme of syncretism in the third and fourth centuries of our era. (52)

Several commonalities bear out this contention:

> Common to the Mysteries and Gnosticism were certain ideas, such as pantheistic mysticism, magic practices, elaborate cosmogonies, and theogonies, rebirth, union with God, revelation from above, dualistic views, the importance attaching to the names and attributes of the deity, and the same aim at personal salvation. (54)

In addition to "psychic therapy" and knowledge, the Ancient Mystery religion promised the:

> opportunity to erase the curse of mortality by direct

encounter with the patron deity, or in many instances by actually undergoing an *apotheosis*, a transfiguration of human into divine. (*The Interruption of Eternity* 28)

Angus contends that the central theme of the Mystery religions was "that the march of mankind is Godward" (43). Of course, not every adherent of the Mysteries necessarily subscribed to this doctrine of apotheosis. Numerous motives compelled people to accept the religion. Angus wisely observes:

> Men entered the Mystery-cults for different purposes: there were all degrees of belief and unbelief, morality and laxity, mysticism and realism. The carnal could find in orgiastic processions and midnight revels opportunities for self-indulgence; the superstitious would approach because of the magical value attributed to the formulae and sacraments; the educated could, in the material and physical, perceive symbols of the truth dear to his heart; the ascetic would look upon initiation as a means of buffeting his body and giving freedom to the spirit; the mystic would in enthusiasm or ecstasy enjoy the beatific vision by entering into communion with God or by undergoing deification. (42)

Thus, only the most ardent mystics accepted the inner doctrine of apotheosis. For the more carnally inclined, such doctrine was seldom acknowledged. In fact, few were even aware of it. The ultimate objective of self-deification was veiled by secrecy and semiotic manipulation:

> The secrecy with which the Mysteries terminated behind the veil of the temple, compared with the publicity with which they generally commenced in the streets, is explicable from the fact that the things "done" or "said" were not the things actually to be revealed but merely symbolic means of conveying the intended truth to the minds of the votaries. (62)

Modern Freemasonry operates in a similar fashion. Higher initiates are encouraged to wage semiotic warfare against lower initiates. The connotative meanings of signs are withheld from fledgling members. If inner doctrines are made available to newcomers, they are awash in an esoteric vernacular that is not easily deciphered. Through such semiotic trickery, Masonry has successfully preserved the Mystery doctrine of apotheosis.

This theme of apotheosis is reiterated in the Masonic concept of man "becoming a god-like being and unifying his consciousness with the Omniscient," which "was always the purpose of the ancient Mysteries." As Wilmshurst makes abundantly clear, this unification of human consciousness with Omniscience represents the culmination of the evolutionary process. It is precisely this process that Arthur C. Clarke's monolith facilitates. Examined from this vantage point, the monolith takes on esoteric value. It becomes a symbol of man's purported innate divinity, the potential for apotheosis through evolution. Wolfgang Smith's characterization of Darwinism as a "Gnostic myth" could not be more appropriate.

Examining the *Hypostasis* further, it is revealed that the serpent's mission is not entirely successful. Although he has helped humanity initiate its journey towards *gnosis* by enticing the first man and woman to eat the forbidden fruit, the "bodily defilement" of the woman prevents Adam from understanding the "real purpose of the deed" (*The Interruption of Eternity* 27). Thus, both humanity and the serpent are fettered by the "curse" of the archons (*The Interruption of Eternity* 27). They all must wait for the arrival of the "perfect man," the "Gnostic adept" who is insusceptible to the deception of the archons (*The Interruption of Eternity* 27).

This messianic figure is hardly the Christ of Biblical Christianity. In fact, this Gnostic myth thoroughly inverts the roles of God and Satan. As was previously stated, the serpent, which the book of Revelation clearly identifies as the Devil, becomes an "*incognito* savior." Meanwhile, God's role as benevolent Heavenly Father is vilified:

> The God of *Genesis*, who comes to reprimand Adam and Eve after their transgression, is rudely caricatured in this tale as the "Arrogant archon" who opposes the will of the authentic heavenly father. (*The Interruption of Eternity* 27)

Of course, within this Gnostic narrative, God incarnate is equally belittled. Jesus Christ, the Word made flesh, is reduced to little more than a forerunner of the coming Gnostic adept. According to the Gnostic mythology, Jesus was but a mere "type" of this perfect man (*The Interruption of Eternity* 27). He came as a "teacher and an exemplar, to show others the path to illumination" (*The Interruption of Eternity* 27-28). The true messiah has yet to come. Equally, the serpent is only a precursor to this messiah. He only initiates man's journey towards *gnosis*. The developmental voyage must be further facilitated by the serpent's predecessor, the Gnostic Christ.

Likewise, the first monolith only initiates man's evolutionary ascent towards apotheosis. It is the first exotheological Christ, a precursor to the "true messiah." The second savior arrives in 2001, symbolized by the reappearance of the monolith. By making contact with the film's central character, the monolith facilitates the next stage of human evolution and the species' attainment of deification. Man's apotheosis is symbolized by the central character's transformation into the "star child." Viewing his own development from the cradle to the grave, the central character now resides outside finitude. His evolution has culminated with the "post-human" condition, the time prophesied in Nietzsche's *Thus Spake Zarathustra*. Man has "overcome his own humanity." In a Masonic context, he has unified "his consciousness with the Omniscient" and become a "god-like being."

This myth, which the film *2001* distills in an allegorical form, is the vision to which the sociopolitical Darwinians of globalism have devoted themselves. The crusade for a New World Order is the crusade of evolution. Now at the pinnacle of its biological evolution, the homo sapien has yet to complete its political evolution. World government represents the capstone of political evolution. Lewis Mumford reiterates in *The Transformation of Man*:

[T]he destiny of mankind, after its long preparatory period of separation and differentiation, is at last to become one. . . This unity is on the point of being politically expressed in a world government that will unite nations and regions in transactions beyond their individual capacity. . . (184)

For the sociopolitical Darwinian, the "preparatory period of separation and differentiation" only represented a brief intermission in the broader narrative of political evolution. The climax is the amalgamation of all nation-states into a single global entity. History is replete with preemptive attempts to achieve just such an end: *Pax Americana* (i.e., the neoconservative's commitment to America's unilateral hegemony), *Pax Europa* (i.e., the European Union), *Pax Britannia* (i.e., the globalist aspirations of H.G. Wells, which found some fragmentary expression in Cecil Rhodes' colonial scientific dictatorship), the Roman Empire, and, of course, Nimrod's Tower of Babel project.

However, world government is only the political expression of man's purported "unity with the Omniscient." Ultimately, the power elite believe that this unification must be extended to every mind of the human race, a goal towards which the New Age movement and Transhumanism have devoted themselves. In the sci-fi novel *Childhood's End*, Arthur C. Clarke invokes yet another extraterrestrial facilitator for both of these unifications. Dubbed the Overlords, these aliens hasten the end of sovereign nation-states and the formation of a one-world government. Commenting on how the arrival of the extraterrestrial Overlords has affected the nation-state system, a character named Stormgren remarks:

> "...it is useless to cling to the past. Even before the Overlords came to Earth, the sovereign state was dying. They have merely hastened its end: no one can save it now—and no one should try." (45)

Like the *Hypostasis* myth of Gnosticism, Clarke inverts the roles of God and Satan. This inversion is semiotically inferred with Clarke's vivid description of the Overlords' eerily familiar appearance:

> It was a tribute to the Overlords' psychology and to their careful years of preparation, that only a few people fainted. Yet there could have been fewer still, anywhere in the world, who did not feel the ancient terror brush one awful instant against their minds before reason banished it forever.
> There was no mistake. The leathery wings, the little horns, the barbed tail—all were there. The most terrible of all legends had come to life, out of the unknown past. Yet now it stood smiling, in ebon majesty, with the sunlight gleaming upon its tremendous body, and with a human child resting trustfully on either arm. (68)

Commenting on this shocking visage, Clarke writes:

> In the Middle Ages people believed in the devil and feared him. But this was the twenty-first century: could it be that, after all, there was such a thing as racial memory? (70)

Like the serpent, the Overlords are humanity's "saviors." Yet, unlike the serpent, they are no longer "*incognito.*" Clarke's narrative mirrors the paradigmatic character of the Gnostic *Hypostasis*. The Devil is no longer considered evil. He is a benevolent liberator of humanity. This is the Masonic Weltanschauung, which portrays Satan as hero unjustly maligned by Christianity. Addressing the Devil by his former title of glory, Lucifer, thirty-third degree Freemason Albert Pike writes:

LUCIFER, *the Light-bearer*! Strange and mysterious name to give to the Spirit of Darkness! Lucifer, the Son of the Morning! Is it *he* who bears the *Light*, and with its splendors intolerable blinds feeble, sensual, or selfish Souls? Doubt it not. (321)

This belief system is known as Luciferianism. It lies at the core of the power elite's religion, which is evolutionary in character. Within the Luciferian mythological framework, science assumes a new role. It is no longer an instrument used to enhance the human condition and understand creation. Instead, it becomes an instrument for the transformation of humanity and the reconfiguration of reality itself. It is the chisel used to re-sculpt the gradually apotheosizing organism of man. Satan, who is referred to by his original designation of Lucifer, becomes a symbol of reason and science.

Again, this is a sentiment expressed by Clarke in *Childhood's End*. This becomes evident during a brief soliloquy delivered by an Overlord named Karellan. Commenting on a reactionary named Wainwright who opposes the Overlords' plans, Karellan opines:

"You will find men like him [Wainwright] in all the world's religions. They know that *we represent reason and science*, and however confident they may be in their beliefs, they fear that we will overthrow their gods." (23; emphasis added)

Clarke also introduces the Overmind, a psychocognitive singularity into which all human consciousness can be condensed. Herein are the themes of a one-world government and a one-world brain. The monolith of *2001* and the Overlords of *Childhood's End* could represent variants of the same exotheological motif: the "Blazing Star," Sirius. With the popularization of Van Daniken's *Chariots of the Gods* thesis and theories like panspermia, one cannot help but wonder if Masonic exotheology is insinuating itself into the public mind. By extension, Luciferianism could be seeping in with it.

Whatever the case may be, evolution remains the dominant myth of modernity. As the modern mythmakers, the "shamans of scientism" continue to alter that myth and create new ones. Drawing from esoteric sources, the shamans are developing a fiction that is very hospitable to the sociopolitical Darwinians of the scientific dictatorship.

Heralding the Technocratic Messiah

Of course, a new world religion requires a new world messiah. There is even a messianic legacy within Masonic mythology. Thirty-third degree Mason Albert Pike states:

> Behold the object, the end, the result, of the great speculation and logomachies of antiquity; the ultimate annihilation of evil, and restoration of Man to his first estate, by a Redeemer, a Masayah, a Christos, the incarnate Word, Reason, or Power of Diety. (274)

The astute reader will immediately notice the capital M in "Man," denoting humanity's intrinsic divinity. Being a god was humanity's "first estate." Thus, the Masonic messiah is not the transcendent Creator incarnated as Jesus Christ. Instead, Masonry posits that the messiah is within Man himself. According to Masonic doctrine, humanity's cognizance of its innate divinity is integral to achieving apotheosis. Pike recapitulates:

> Thus self-consciousness leads us to consciousness of God, and at last to consciousness of an infinite God. That is the highest evidence of our own existence and it is the highest evidence of His. (709)

As for the early Christians who believed that Jesus was the transcendent God clothed in flesh, Pike derisively portrays them as superstitious simpletons:

> The dunces who led primitive Christianity astray, by substituting faith for science, reverie for experience, the fantastic for the reality; and the inquisitors who for so many ages waged against Magism a war of extermination, have succeeded in shrouding in darkness the ancient discoveries of the human mind; so that we now grope in the dark to find again the key of the phenomena of nature. (732)

Pike's reprimand concerning Christianity's substitution of faith for science betrays Masonry's scientistic proclivities. Earlier in human history, such scientistic belief was less powerful. However, in this post-

Masonic era where the doctrine of the elite's epistemological cartel has been fully externalized, scientism rules the day. As such, the present scientistic society demands a scientistic messiah.

Paradoxically, this occult concept of self-deification asserts that humanity's internal deity requires an external facilitator to achieve full manifestation. Again, science fiction has played an integral role in preparing the masses for such an eventuality. One of the most significant pieces of messianic sci-fi predictive programming is Steven Spielberg's *E.T.* The central theme of the film *E.T.* is most succinctly encapsulated in the familiar shot that also adorned many of the movie's publicity posters. Of course, this is the shot of the outstretched hand of the movie's human protagonist touching the glowing fingertip of an alien hand reaching downward.

The symbolic meaning embedded within this image becomes evident when compared with Michelangelo's Sistine Chapel painting. Like the thematically axial shot in *E.T.*, Michelangelo's portrait presents Adam "with a raised arm and in fingertip union with God" (377). The synchronicity between these two pictures is clearly religious.

Both appear to be premised upon the Christian theme of God communing with His own creation. The ministry of Jesus Christ, whom Christians believe to have been God incarnate, tangibly enacted this theme. Reiterating this theme, Spielberg's film features an extraterrestrial "messiah" who reproduces many of Jesus' miracles. The most significant "miracles" performed by this visitor is its own resurrection and ascension into heaven. Yet, despite these ostensible Christian elements, Spielberg's film cannot be construed as a "Christian allegory." Both instances, it should be noted, are explained in a naturalistic context. More specifically, the "resurrection" is merely the creature's exceptional immunological response to Earth's bacteria and the "ascension" evacuation via a waiting spacecraft.

Yet, Spielberg's bowdlerization of Christian theology is anything but new or innovative. *E.T.* merely continues a tradition embodied by Michelangelo's Sistine Chapel painting. The portrait departs from the traditional Christian paradigm concerning the Genesis account and humanity's relationship with its Creator. Ian Taylor explains how Michelangelo's painting deviates from the traditional Genesis account:

Unlikely as this may seem, it is, nevertheless, a remarkable fact that when painted in 1508 Michelangelo took the bold step of departing from the biblical account of the creation of man to depict what is today seen to be a theistically evolved version. Prior to this time, artists had stuck to the Genesis description of a non-living being made from the dust of the ground becoming a "living soul" by the infusion of God's breath (Genesis 2:7). Michelangelo's now famous painting of the creation of Adam shows a human form quite evidently alive with a raised arm and in fingertip union with God. The question this painting raises is that since the creature is alive, what kind of pre-Adamic being does it represent? Enterprising Jesuit teachers have seized upon this as historical vindication of the truth of theistic evolution, so that the creature depicted must then be some kind of advanced anthropoid. There can be absolute certainty that nothing could have been further from Michelangelo's mind, yet the Greek influence and tendency to rationalize revelation is represented symbolically throughout the entire painting, not in style, but by the insertion of Greek sibyls between the Old Testament prophets. (377)

Like Michelangelo's portrait, Spielberg's *E.T.* attempts to reconceptualize man's relationship with the heavenly. The film is set in the modern age of science, a time when mystical cosmology has been supplanted by human reason. This contemporary cultural milieu is one governed by scientism. In this context, the human protagonist of *E.T.* represents an Adept or, as they are called in esoteric circles, an *Illuminatus* ("illuminated one"). With his evolutionary development augmented through extraterrestrial intervention and a paradigm shift just on the horizon, Spielberg's human protagonist is the next in a long line of Avatars. The extraterrestrial visitor is an anthropomorphic representation of Prometheus, who imparts the torch of Wisdom unto man.

As is evidenced by films like *Close Encounters of the Third Kind* and *E.T.*, the relatively recent UFO phenomenon made a significant impression upon Spielberg. In fact, the UFO mystery has prompted many to reconceptualize their relationship with the heavenly realm. Timothy Good provides an example of such a shift in thinking:

Miles Copeland, former CIA organizer and intelligence officer, related an interesting story to me involving the Agency's attempt on one occasion to use fictional UFO sightings to spread disinformation. The purpose, in this case, was to "dazzle" and intoxicate" the Chinese, who had themselves on several occasions fooled the CIA into sending teams to a desert in Sinkiang Province, West China, to search for nonexistent underground "atomic energies." The exercise took place in the early 1960s, Copeland told me, and involved launching fictional UFO sighting reports from many different areas. The project was headed by Desmond Fitzgerald of the Special Affairs Staff (who made a name for himself by inventing harebrained schemes for assassinating Fidel Castro). The UFO exercise was "just to keep the Chinese off-balance and make them think we were doing things we weren't," Copeland said. "The project got the desired results, as I remember, except that it somehow got picked up by a lot of religious nuts in Iowa and Nebraska or somewhere who took it seriously enough to add an extra chapter to their version of the New Testament!" (357)

If this UFO manipulation perpetrated by the CIA was effective enough to compel certain factions to embellish and pervert the Scriptures, imagine what a deception on a larger scale could accomplish. Rose states:

Science fiction has given the images, "evolution," has produced the philosophy, and the technology of the "space age" has supplied the plausibility for such encounters. (Rose 91)

Apparently, the idea of extraterrestrials visiting earth was so powerful that it prompted many to reconsider their traditional religious notions. No doubt, the UFO phenomenon had the same effect upon Spielberg. Herein is the ultimate theme underpinning the imagery in E.T.: the redefinition of God.

The fingertip union between terrestrial anthropoid and extraterrestrial anthropoid represents the religious mandate for the creation of a new scientistic faith. Through sci-fi predictive programming, filmmakers like Spielberg could be serving as "religious engineers" in the construction of a new messianic legacy. However, this savior is anything but the Christ of Christianity.

Returning to the briefing paper Richard Doty provided for Linda Moulton Howe, one finds a claim heralding the arrival of an individual that the film E.T. has prepared the public to accept. Howe elaborates:

There was a paragraph that stated, "Two thousand years ago

extraterrestrials created a being" that was placed on this earth to teach mankind about love and non-violence. (151)

Was Doty acting on behalf of some hidden "religious engineers?" Was he a counterfeit John the Baptist, appointed to introduce the world to a technocratic Christ? Now, it is important to recall Doty's connections with military intelligence. He has worked within circles where the Freemasonic myth of Sirius is actively circulated. If such a deception is underway, sci-fi predictive programming like *E.T.* has helped cultivate the fertile soil of public imagination.

In essence, *E.T.* is the cinematic rallying call for the reengineering of religions. Pike states: "God is, as man conceives Him, the reflected image of man himself" (223). According to the Scriptures, God made man in His own image. According to the hidden "religious engineers," it is man's time to return the favor.

Shaping Things to Come

If the concept of "predictive programming" seems fantastic, consider the case of H.G. Wells's *The Shape of Things to Come*. Published in 1933, this book seems to predict the course of human history for years to come. Darwinian apologist and Round Table member T.H. Huxley mentored Wells. Given his membership in the overtly Anglophile Round Table organization, it is very possible that Huxley passed its tradition of British elitism onto Wells. This becomes evident in Wells's own words, which bear eerie resemblance to the rhetoric of John Ruskin:

> The British Empire . . . had to be the precursor of a world-state or nothing . . . It was possible for the Germans and Austrians to hold together in their Zollverein (tariff and trade bloc) because they were placed like a clenched fist in the centre of Europe. But the British Empire was like an open hand all over the world. It had no natural economic unity and it could maintain no artificial economic unity. Its essential unity must be a unity of great ideas embodied in the English speech and literature. (*Experiments in Autobiography*, 652)

There was a significant continuity of thought maintained between both Huxley and Wells. Huxley was also the grandfather of Aldous and Julian, both of which would, in turn, eventually come under the tutelage of Wells. All of these men were members of or associated with the Freemasonic Lodge and their Masonic heritage seems to make itself evident throughout their various literary works. More specifically, Wells and the Huxley brothers collectively endorsed Sir Francis Bacon's technocratic Utopia under different banners. Such endorsements are conspicuous allusions to the Masonic concept of a "scientific dictatorship."

Wells's vision of a technocratic world state can be found in *The Shape of Things to Come*, a "mass appeal" tract disguised as a science fiction novel. Given some of its uncannily precise prognostications, the book probably should have been titled "Shaping Things to Come." Among one of the book's most notable predictions is the beginning of a second global conflict. Wells correctly identifies the Treaty of Versailles as the primary catalyst for the coming bloodbath:

It was only slowly during the decade following after the war that the human intelligence began to realize that the Treaty of Versailles had not ended the war at all. It had set a truce to the bloodshed, but it had done so only to open a more subtle and ultimately more destructive phase in the traditional struggle of the sovereign states. (*The Shape of Things to Come*, no pagination)

Wells elaborates on the Treaty's inherent flaws, which would facilitate Germany's metamorphosis into a fascist dictatorship:

Included among other amiable arrangements were clauses penalizing Germany and her allies as completely as Carthage was penalized by Rome after the disaster of Zamia--penalizing her in so overwhelming a way as to make default inevitable and afford a perennial excuse for her continued abasement. It was not a settlement, it was a permanent punishment. The Germans were to become the penitent helots of the conquerors; a generation, whole generations, were to be born and die in debt, and to ensure the security of this arrangement Germany was to be effectually disarmed and kept disarmed. (No pagination)

Indeed, Wells was correct. Built into the body of the treaty was the means by which yet another global conflict could be facilitated. Ralph Epperson elaborates:

One of the planks of the Treaty called for large amounts of war reparations to be paid to the victorious nations by the German government. This plank of the Treaty alone caused more grief in the German nation than any other and precipitated three events: 1. The "hyperinflation" of the German mark between 1920 and 1923; 2. The destruction of the middle class in Germany; and 3. The bringing to power of someone who could end inflation; a dictator like Adolf Hitler. (261)

Commenting on the Treaty, British Foreign Secretary Lord Curzon prophetically stated: "This is no peace; this is only a truce for twenty years" (261).

Simultaneously, Wells examines (and scrutinizes) the League of Nations. Wells correctly characterizes Woodrow Wilson's global organization as an extremely limited form of world government:

> The pattern conceived by him [Wilson] was a naïve adaptation of the parliamentary governments of Europe and America to a wider union. His League, as it emerged from the Versailles Conference, was a typical nineteenth-century government enlarged to planetary dimensions and greatly faded in the process; it had an upper chamber, the Council, and a lower chamber, the Assembly, but, in ready deference to national susceptibilities, it had no executive powers, no certain revenues, no army, no police, and practically no authority to do anything at all. And even as a political body it was remote and ineffective; it was not in any way representative of the peoples of the earth as distinguished from the governments of the earth. Practically nothing was done to make the common people of the world feel that the League was theirs. Its delegates were appointed by the Foreign Offices of the very governments its only conceivable rôle was to supersede. They were national politicians and they were expected to go to Geneva to liquidate national politics. The League came into being at last, a solemn simulacrum to mock, cheat and dispel the first desire for unity that mankind had ever betrayed. (No pagination)

The League's Achilles' heel, according Wells, was its observance of national sovereignty: "It was a League not to end sovereignties but preserve them" (no pagination). Wells candidly confesses that global government stipulates the consolidation of immense quantities of power within an omnipotent world entity. This, Wells contends, was Wilson's greatest mistake:

Manifestly he [Wilson] wanted some sort of a world pax. But it is doubtful if at any time he realized that a world pax means a world control of all the vital common interests of mankind (no pagination).

Ultimately, Wells attributes the inevitability of warfare to the sovereign nation-state: "The existence of independent sovereign states IS war, white or red, and only an elaborate mis-education blinded the world to this elementary fact" (no pagination). Automatically, astute readers will recognize Wells's globalist propensities. Again, this may have been a natural consequence of his tutelage under Round Table member T.H. Huxley. His promotion of Britain as a potential World-State certainly echoes the Anglophilic contentions of the Round Table group.

According to Wells, the second global conflict would result from a misunderstanding at a train station in Danzig, Poland. Wells elaborates:

War came at last in 1940. The particular incident that led to actual warfare in Europe was due to a Polish commercial traveller, a Pole of Jewish origin, who was so ill advised as to have trouble with an ill-fitting dental plate during the halt of his train in Danzig. He seems to have got this plate jammed in such a fashion that he had to open his mouth wide and use both hands to struggle with it, and out of deference to his fellow passengers he turned his face to the window during these efforts at readjustment. He was a black-bearded man with a long and prominent nose, and no doubt the effect of his contortions was unpleasing. Little did he realize that his clumsy hands were to release the dogs of war from the Pyrenees to Siberia.
The primary irritant seems to have been either an orange-pip or a small fragment of walnut.
Unhappily, a young Nazi was standing on the platform outside and construed the unfortunate man's facial disarrangement into a hostile comment upon his uniform. For many of these youths were of an extreme innate sensibility. The flames of patriotic indignation shot up in his heart. He called up three

fellow guards and two policemen—-for like the Italian Fascisti these young heroes rarely acted alone — and boarded the train in a swift and exemplary mood. There was a furious altercation, rendered more difficult by the facts that the offending Pole knew little or no German and was still in effect gagged. Two fellow travellers, however, came to his help, others became involved, vociferation gave place to pushing and punching, and the Nazis, outnumbered, were put off the train.

Whereupon the young man who had started all the trouble, exasperated, heated and dishevelled, and seeing that now altogether intolerable Jew still making unsatisfactory passes with his hands and face at the window, drew a revolver and shot him dead. Other weapons flashed into action, and the miniature battle was brought to an end only by the engine-driver drawing his train out of the station. The matter was complicated politically by the fact that the exact status of the Danzig police was still in dispute and that the Nazis had no legal authority upon the Danzig platform. (No pagination)

Within the fictional narrative of *The Shape of Things to Come*, the incident at the Danzig train station exasperates international tensions. The incident swiftly escalates, triggering the mobilization and commitment of national armies to an enormous global fray. Wells describes this world war as one of the most bloody and violent episodes of human history. Does this sound familiar? In reality, the world did experience a second global conflict. History would dub it World War II.

In approximating the actual events preceding World War II, Wells displays some uncanny precision. Indeed, the actual Second World War officially began in Poland. Wells's projected date for the war's beginning is off by only a few months. The fictional incident in *The Shape of Things to Come* is somewhat similar to the border incidents that exasperated international tensions and provided the Nazis with a pretext for war. Remaining considerably close to the actual chronology of World War II, Wells places the conflict's end in 1949.

In light of these disturbing synchronicities, the viability of sci-fi "predictive programming" certainly seems stronger. The case is only strengthened when one examines Wells's affiliations. In addition to being a Fabian socialist and Freemason, Wells was also a member of the Coefficients Club. Formed by Fabian socialist Beatrice Webb, this organization assembled some of Britain's most prominent social critics and thinkers to discuss the course of the British Empire.

One of the Club's members was none other than Fabian socialist and Malthusian ideologue Bertrand Russell. According to Russell, Wells and several other members harbored an overwhelming preoccupation with war. Russell explains:

> . . .in 1902, I became a member of a small dining club called the Coefficients, got up by Sidney Webb for the purpose of considering political questions from a more or less Imperialist point of view. It was in this club that I first became acquainted with H. G. Wells, of whom I had never heard until then. His point of view was more sympathetic to me than that of any member. Most of the members, in fact, shocked me profoundly. I remember Amery's eyes gleaming with blood-lust at the thought of a war with America, in which, as he said with exultation, we should have to arm the whole adult male population. One evening Sir Edward Grey (not then in office) made a speech advocating the policy of Entente, which had not yet been adopted by the Government. I stated my objections to the policy very forcibly, and pointed out the likelyhood of its leading to war, but no one agreed with me, so I resigned from the Club. It will be seen that I began my opposition to the first war at the earliest possible moment. (*The Autobiography of Bertrand Russell*, 230)

Indeed, the Club's proclivities towards war were strong. One Club member, Leo Maxse, had promoted war with Germany in 1902. This preoccupation with war is especially evident in much of Wells's scientific romances. Moreover, a great deal of Wells's work acknowledges the alchemical role of war in man's purported evolutionary development. J.P. Vernier observes:

> . . .I would suggest that evolution, as presented by Wells, that is a kind of mutation resulting in the confrontation of man with different species, is one of the main themes of modern science fiction. ("Evolution as a Literary Theme in H.G. Wells's Science Fiction," 85)

Recall the dialectical framework intrinsic to evolutionary theory. The organism (thesis) comes into conflict with nature (antithesis) resulting in a newly enhanced species (synthesis), the culmination of the evolutionary process (Marrs 127). In the case of Wells's work, the critical mutation within humanity was facilitating by confrontation with other species (including other species of man).

Hitler's genocidal Final Solution tangibly enacted such a dialectical framework. The German people (thesis) came into conflict with the Jew (antithesis) in hopes of creating the Aryan (synthesis). The fact that Wells "predicted" this suggests that he was privy to certain plans. Such plans may have circulated within the Coefficient Clubs, among other elitist think tanks. Whatever the case might be, World War II certainly synchronized with the evolutionary designs of Wells and his other oligarchical colleagues.

The Technocratic Movement

Wells also examined the technocratic movement of the 1930s, predicting its rise as a political movement and its role in shaping the emergent world government. In *The Shape of Things to Come*, Wells calls Technocracy an "expressive and significant word" (no pagination). In *The Technocrats: Prophets of Automation*, Henry Elsner states that Technocracy became the "new word of 1932" (1). With the Depression sinking to its nadir, the technocratic Weltanschauung "exploded into public attention" and "marked the conversation of millions of Americans" (1). The epidemic scope of this ideational contagion is made evident by the prolific media exposure that it enjoyed:

> In the closing months of 1932, as the Depression deepened and national politics seemed to be drifting helplessly, speculations about Technocracy swept across the country in almost every available form. The high point was reached in January, 1933. The *New York Times* alone had no less than sixty articles on Technocracy that month. Forty-one periodical articles and seventeen books and pamphlets on Technocracy were included in the standard indexes for the beginning of 1933. (7)

In *The Shape of Things to Come*, Wells succinctly characterizes this phase of popularity as an "outbreak." Indeed, the 30s witnessed the rampant metastasis of technocratic thought. Elaborating on the ideational epidemic of Technocracy, Wells writes:

> Everywhere in that decadence, amidst that twilight of social order, engineers, industrialists and professors of physical

science were writing and talking constructive policies. They were invading politics. (No pagination)

During his examination of this emergent technocratic movement, Wells' invokes an interesting comparative model. He begins with a critique of operative Masonry:

> Students are still working out the preservation and continuation of the art and mystery of the masons into the middle ages. There was a great loss of knowledge but also a real survival. The medieval free-masons who built those flimsy but often quite beautiful Gothic cathedrals it is now such a task to conserve, carried on a tradition that had never really broken with that of the pyramid builders. But they had no sense of politics. They had a tradition of protective guild association similar to the Trade Unions of the Capitalist age, they interfered in local affairs in order to make jobs for themselves, but there is no sign that at any time they concerned themselves with the order and stability of the community as a whole. Their horizons were below that level of intelligence. (No pagination)

Of course, the operative Mason guilds of the middle ages were initially "God-fearing in their religion" (Martin 521). Evidently, Wells regards these operative guilds with derision. In particular, he chides them for their overall lack of political ambition. With minds that thought "below that level of intelligence," the operative guilds had little hope of garnering enough political capital to suit Wells.

Conversely, Wells characterizes the men who would comprise the technocratic movement as follows:

> Now the skilled and directive men of the collapsing order of the twentieth century were of an altogether livelier quality. Their training was not traditional but progressive, far more progressive than that of any other class. They were inured to fundamental changes in scope, method and material. They ceased to be acquiescent in the political and financial life about them directly they found their activities seriously impeded. (No pagination)

Evidently, Wells considers the ideologues of the emergent

technocratic movement to be superior to the old operative Mason guilds. In light of his Masonic heritage, Wells's comparative model takes on greater dimensions. The operative guilds were eventually compromised by the occult forces of Italian humanism and speculative Masonry was born (Martin 522). One cannot help but wonder if Wells's "livelier" technocrats represented a continuation of the speculative Masonic tradition.

One semiotic item that reinforces such a contention was the technocratic movements' emblem: the Monad. Making its public debut at a 1933 Chicago convention, the Monad was:

"vermillion and French gray, which is an ancient Chinese symbol signifying unity, balance, growth, and dynamic functioning for the security of life processes." (qutd. in Elsner 95)

This "yin and yang" motif is part of the evolutionary mythology comprising Masonic doctrine. It is emblematic of the hermaphroditic god Brahma, which eventually divided itself into Viraj and Vach (Weston 38). Through the sexual union of these two, all things gradually evolved (Weston 38). Reiterating this evolutionary mythos, thirty-third degree Freemason Albert Pike claims that "all the conditions of material existence were supposed to have been evolved out of the Pythagorean Monad" (675). The technocratic movement's emblem was purported to be a geometric representation of the monad.

Ironically, the technocratic movement would become somewhat unfriendly towards Masonry. For instance, Technocracy Inc., an organization directed by technocrat Howard Scott, attacked the Democratic party for its Vice Presidential nomination of Freemason Harry Truman (Elsner 164). However, this was certainly nothing new. Many times throughout history, Masonry has become the victim of its own progenies. The technocratic movement is no exception. Besides, being a Mason is not requisite for being a part of the Lodge's conspiratorial tradition. Freemason Isindag explains:

Masonry also accepts this fact: *In the outside world there are wise people who, although they are not Masons, espouse Masonic ideology.* This is because this ideology is wholly an ideology of human beings and of humanity. (*Masonluktan Esinlenmeler* 32; emphasis added)

Again, the conspiracy to establish a scientific dictatorship is largely is a conspiracy of ideas. While there are conscious agents working to tangibly enact the Masonic concept of a "scientific dictatorship," the conspiracy also depends heavily on the continuity of thought. It is not just a plot. It is also an ideational contagion.

Wells proceeds to reveal a Fabian strategy of gradual ideological assimilation: "The movement spread from workshop to workshop and from laboratory to laboratory with increasing rapidity all over the world" (*The Shape of Things to Come*). Indeed, Technocracy did spread. James Dowell Crabtree expands on the pandemic breadth of this ideational contagion:

> Technocracy has spread from America to other parts of the world. In many recent articles, the People's Republic of China's new generation of leaders have been called "technocrats," as were some leaders in the former Soviet Union. (148)

Given the Masonic involvement in the pre-WWII rise of communism, one could contend that the technocratic contagion did not spread from America to Russia or China. Instead, the transmission chronology appears to be quite the reverse, with Technocracy beginning in the East and spreading to the West. Whatever the case may be, the technocratic contagion did not remain isolated. Observing the pervasiveness of technocratic thought, H.G. Wells' commented: "Such infection went far and deep" (The Shape of Things to Come).

Interestingly, a prominent member of the technocratic movement was M. King Hubbert, the geophysicist responsible for the Hubbert Peak and the peak oil theory ("Technocratic movement," no pagination). Peak oil theory is one of many eschatological myths that neo-Malthusians employ in their crusade to reduce the world's population. Astute percipients will recognize a Malthusian theme semiotically embedded within the technocratic monad. The two segments of the monad represent "a balance between a nation's production and consumption" ("Monad (Technocracy)," no pagination). Automatically, the familiar message of limits to growth and population control become evident. In fact, many technocrats advocated eugenics. Technocrat Harold Loeb contended that a Technocracy would eventually give rise to a new breed of man. Loeb describes the technocratic variety of *Ubermensch* as follows:

> A technocracy, then, should in time produce a race of man

superior in quality to any now known on earth, a society more exciting, interesting, and variegated than has ever been possible, and a nation in which no individual should be unhappy or discontented for remediable causes. (178)

Ironically, M. King Hubbert rejected scarcity-based economics and eventually concluded that solar power could act as a viable substitute for so-called "fossil fuels" ("M. King Hubbert," no pagination). Evidently, Julian Simon was not alone in his conviction that technological advancement guaranteed the emergence of non-finite energy sources. Many technocrats acknowledged this reality, quite in spite of themselves. Again, Loeb serves as a perfect example. In *Life in a Technocracy*, he writes:

Men live by the production, distribution, and consumption of goods. Goods are produced and distributed by effort. The incentive of effort is profit. Profit depends on price. Price depends on scarcity. Therefore, the life of man, under the capitalist system, depends on the scarcity of goods. And the scarcity of goods is being progressively destroyed by the application of science to production, known in its latest phase as technology. (7)

Given Loeb's acknowledgement in technology's capacity for developing non-finite resources, one is prompted to question his continued endorsement of a eugenically regimented society. After all, a society that suffers no scarcity has no reason to implement policies of population control and selective breeding. However, as a continuation of the elitist Masonic-Baconian tradition, the technocratic movement was always predisposed to such authoritarian thinking. Perhaps the technocratic movement publicly acknowledged what the population control and eugenics movement attempted to camouflage with cheap, pseudo-humanitarian rhetoric...programs of depopulation and selective breeding are actually about control.

With the close of the 30s, it became clear to many technocrats that a completely apolitical system was infeasible. Invariably, governmental policies and ideas become politicized. This was a harsh reality of which the fundamentalist technocrat was becoming cognizant. However, the concept of a Technocracy had left a significant impression upon the political landscape. This impression is made evident by an emergent form of elite politics that has been transforming governance over the last seventy-six years. Fischer writes:

> Already there is suggestive evidence to indicate a receptivity on the part of elite decision makers to greater emphasis on technocratic politics. Many among the political elites talk of a "governance crisis" resulting from an overload of interest-group politics and speak of depoliticizing policy issues through improved planning and management. [David] Dickson identifies this as the replacement of the "democratic paradigm" with the "technocratic paradigm." Understood in these terms, a close alliance between elites and technocrats holds out potential for a mutually beneficial strategy. Not only are technocrats specialists in policy management and planning, their techniques are packaged in ideology suited to the elite conception of the political situation. (112)

Indeed, the technocratic Weltanschauung is very compatible with the "elite conception of the political situation." Examining the world envisioned by the technocratic movement, historian William E. Akin observes:

> The technocrats attempted to pull all of these strands--their faith in positivistic science; their mechanistic view of man with his essentially animal-like irrationality, his desire for security, abundance, and tranquility; the organizational imperative caused by natural inequality; and the dominance of technology--together into one functional ideal whole. It resembled Aldous Huxley's Brave New World and the managerial society that James Burnham deplored; it could lead equally to B.F. Skinner's *Walden Two*. (148)

Given Aldous Huxley's possible role as a Masonic sci-fi predictive programmer and the Masonic elements of technocratic thought, the resemblance may have been more than a coincidence. Of course, Aldous was the member of an oligarchical dynasty that was intimately involved in the creation of the British Round Table groups and the elitist Rhodes network. Huxley's dystopian portrait of the future reflected the technocratic vision of the minds surrounding the iconic author. Such a vision synchronizes comfortably with the aspirations of today's power elite.

The oligarchs' adoption of the "technocratic paradigm" has resulted in the hybridization of political elites and managerial elites. The emergent hybrid promises to reconfigure the political terrain:

> Technocratic ideology, in short, holds out the possibility of a new—-and often seductive—-form of elite politics. Grounded in technical competence of professional expertise, such a system not only shrouds critical decisions in what would appear to be the logic of technical imperatives, it also erects stringent barriers to popular participation. Only those with knowledge (or credentials) can hope to participate in deciding the sophisticated issues confronting postindustrial society. Some, in fact, have averred that a politics of expertise will give rise to a new type of elite politician: the "politician-technician." (Fischer 112)

Modern governance has been witnessing this hybridization process for quite some time. For years, national and foreign policymaking powers have been concentrated within the hands of the "politician-technicians" that populate various elite think tanks, businesses, and organizations. These include the Council on Foreign Relations, the Trilateral Commission, the Bildeberg group, and a myriad of other cults of expertise. Such technocratic cults have almost become permanent fixtures to the American political landscape. Removing them can be extremely difficult. Pat Robertson relates an account involving Jimmy Carter that clearly illustrates this point:

> I personally was thrilled to think that a born-again Christian might gain the White House, and I gave Carter a boost with some friends of mine who were active in organized labor in Pennsylvania. After Carter won the Pennsylvania primary (believed, incorrectly I might add, to be a Northern, liberal, industrial state), the Southern peanut farmer had effectively won the Democratic nomination.
> After the general election in November, I spoke to President-Elect Carter during a three-way telephone conversation along with pro-family activist Lou Sheldon of California. I suggested to Governor Carter that he, as a strong evangelical, might want to include some evangelical Christians among his appointments. He greeted the idea with enthusiasm and

agreed to receive a list if we could get it to him within two weeks.

Lou Sheldon and I worked night and day to put together a short roster of names and resumes. The finished product was outstanding. All of the candidates were Democrats. All were highly distinguished in government, business, or education. The composition of the group looked like what Jesse Jackson sometime later called the Rainbow Coalition.

We had included male and female blacks, whites, Hispanics, natives Hawaiians, and I believe, a Chinese-American. A friend of a friend even obtained a preliminary thumbs up/ thumbs down FBI screening to keep us from wasting the president-elect's time. (We had to omit one born-again Democratic governor from the South because of alleged campaign contributions by reputed Mafia figures.)

When the document was ready. I chartered a small aircraft to take Lou Sheldon to the grassy strip in Georgia we laughingly called "Plains International." Sheldon arrived at the Carter residence to find the next president barefoot and in blue jeans. They greeted each other warmly, and Sheldon proudly presented the booklet. Carter took it, read it, and began to cry.

When he got back to Virginia Beach, Sheldon said, "Jimmy was so touched by all the work that we did that tears came to his eyes." I said, "Lou, you are wrong. The reason he cried is because the appointment process is out of his hands, and he is not going to appoint any of those people." And indeed my words were true. Not one of our recommendations—-men and women who on the surface shared every principle that Jimmy Carter espoused-—was appointed to public office, or even seriously considered. (104-05)

Keeping in mind that Robertson is not a genuine opponent of the elites, his story still demonstrates how expertise cults have a monopoly over the Washington establishment. As these various groups extend their spheres of influence and control, the technocratic paradigm is rapidly supplanting the democratic paradigm. The oligarchic "democracy of experts" prophesied by Wells and Quigley is gradually becoming a reality.

The New Deal: America's Technocratic Transformation

America's gradual transformation into a Technocracy accelerated significantly under Franklin Roosevelt's New Deal. Indeed, many adherents of Technocracy believed FDR's policies for economic recovery would facilitate the nation's final metamorphosis into a New Atlantis. In the early days of the 1932 election, economist Henry A. Porter published *Roosevelt and Technocracy*. The book concerned itself with a pivotal question: "Will TECHNOCRACY be the 'New Deal?'" (45). Based on his examination of New Deal policies, Porter gravitated towards the affirmative.

In fact, Porter voiced his resounding approval of FDR and other closely aligned New Deal liberals:

> Only skillful statesmanship--the statesmanship of a Roosevelt, and sound economic principles--the principles of Technocracy, can lead us out of the valley of Chaos and Despair into which we are plunging. (71)

Technocrats in California even advocated "the granting of dictatorial powers to Franklin D. Roosevelt" (Akin 83). Evidently, Roosevelt was the most promising candidate for the proponents of an "Administrative State." There was good reason for the technocratic movement's initial support of Roosevelt. FDR was heavily influenced by the socialist founder of the Nationalist movement, Edward Bellamy. Bellamy authored *Looking Backward, 2000-1887*, another piece of sci-fi predictive programming literature that proselytized readers on behalf of global socialism. James Crabtree synopsizes the book as follows:

> Inadvertently, Edward Bellamy and his *Looking Backward, 2000-1887* tapped into the latent American faith in technology as the solution to humanity's every problem. *Looking Backward* is a science fiction novel, telling the story of a man from 1887 thrust into the future and learning of all the changes that had taken place in America during the intervening century.
> In the year 2000 all wants are met. Society has eliminated non-progressive institutions that were corrupt in the 1880s and replaced them with a government-run industrial state. In *Looking Backward* everyone could retire at the age of 45 after a life of serving in a job for which he or she was best suited. Bellamy's future was indeed Utopian.
> Bellamy characterized society as machine. Whereas in 1887

it was a machine badly managed and inefficient, in the year 2000 society was well run. His "everyman" character, Julian West, found himself in a future where the "regime of the great consolidations of capital" had been overthrown and the "concentration of management and unity of organization" had taken control. Whether he meant to or not, Bellamy had defined the managerial concept of society. What is more important, he gave many people an ideal of an industrial, or technological, state towards which to strive. (7-8)

Looking Backward would later become "officially recommended reading for members of the Technocratic movement" (Crabtree 8). In fact, the Nationalist movement that sprouted from Bellamy's sci-fi predictive programming could be considered "a precursor to the Technocrats, and its members 'pre-Technocrats'" (Crabtree 8). Henry Elsner delineates the various commonalities shared by the technocratic movement and Bellamy's Nationalist movement:

> Despite differences in detail, a number of the basic principles of organization are remarkably similar in Bellamy's and in technocratic. (1) The organization of all industries into a few large-scale, publicly owned units, administered by technical experts who are selected from within the ranks of the units concerned. (2) A bureaucratic rather than an industrial-democratic organization of the workplace. (3) Equal, independent income issued to all members of society as a right of citizenship. (4) Income distribution through a nonmonetary accounting system wherein the registration of items purchased serves as an automatic means of estimating future production requirements. (5) The elimination of a political government, i.e., officials other than those at the heads of the productive, distributive, and professional units, and the abolition of political parties. (221)

Yet, Elsner correctly identifies various differences between Bellamyism and Technocracy (223). These dissimilarities suggest that Bellamyism acted as a conceptual segue, "a transition between an older, essentially pre-industrial 'utopian' societal socialism, and technocracy" (223). Indeed, Technocracy did appear to be on the horizon and the New Deal was its harbinger. Shortly after being sworn into office, Roosevelt outlined his plan for economic and social recovery in the book *Looking*

Forward (Crabtree 105-06). The title itself seemed to be an allusion to Bellamyism:

> By picking "Looking Forward" for the title, Roosevelt almost certainly thought to follow in Bellamy's footsteps and produce his own version of *Looking Backward* that would serve as a model for future society. (Crabtree 106)

In *Looking Forward*, Roosevelt wrote: "A greater efficiency [in government] than we have heretofore seen is urgent" (71). Within this statement, one immediately discerns the technocratic preoccupation with governmental efficiency. This preoccupation is probably attributable to the mutual doctrinal foundation of both Technocracy and New Deal liberalism: Progressivism. In particular, the ideational strand of Bellamyism is evident. Roosevelt may have owed his technocratic propensities to a darker heritage. The President was a 32nd Degree Freemason (Carrico 52). Thus, Bellamy's techno-socialism may have struck a responsive chord with Roosevelt.

The precursory technocratic concepts of Bellamy found some fragmentary expression through the policies of the New Deal. One manifestation of Bellamyism was the Social Security Act, which was inspired by a retired physician named Dr. Francis Townsend (Crabtree 105). Townsend's concepts were cribbed from Bellamy's *Looking Backward* (Crabtree 104). Among one of his theoretical policies was a federal program that would have allocated $200 a month to unemployed citizens over the age of sixty (Crabtree 104). According to Townsend's conjectural program, the recipients of this financial assistance would have thirty days within which they would be required to spend the $200 (Crabtree 104).

While the Townsend Bill did not enjoy passage by Congress, it did inspire the Social Security Act that was successfully signed into law later (Crabtree 105). Diffuse in its transmission, technocratic thought remained at the root of this New Deal machination. Crabtree explains:

> In this way, one might say that the Technocrats did indeed have an indirect influence on the New Deal, by way of one of the contributors to their doctrine (Bellamy) to an activist who espoused their ideas of guaranteed income (Townsend) and finally into law. (105)

In addition to this program of "guaranteed income," FDR's administration would introduce a plethora of government agencies. These would included the Civilian Conservation Corps (CCC), the Public Works Administration (PWA), the Agricultural Adjustment Administration (AAA), and, most notably, the National Recovery Administration (NRA) (Crabtree 106). What was initially a small federal government was soon transformed into a massive bureaucracy. In some ways, this new "[c]entralized and functional" monolith was similar to the continental administration promoted by Technocracy Inc. (Crabtree 106). Moreover, its impact upon America's Federal government would leave a "permanent mark" (Crabtree 106). Big government was born and technocracy's rise in the West had begun in earnest.

This ascendancy was semiotically intimated through the placement of the Great Seal on the U.S. one-dollar bill in 1935. The proposal for this substantial modification to American currency was made by 32nd degree Freemason Henry Wallace, who initially suggested that the Great Seal be placed on a coin. Researcher Michael Howard explains Wallace's rationale:

Wallace's reasons for wanting to introduce the Great Seal onto the American currency were based on his belief that America was reaching a turning point in her history and that great spiritual changes were imminent. He believed that the 1930s represented a time when a great spiritual awakening was going to take place which would precede the creation of the one-world state. (95)

Wallace's Masonic brother, President Roosevelt, decided to have the Great Seal placed on the dollar instead. This task was left to another Freemason, Secretary of the Treasury Henry Morgenthau, Jr. (Spenser 23-24). Morgenthau would later become the presidential candidate of the American Communist Party (Spenser 23-24). Like Wallace, Morgenthau also saw America on the brink of an enormous spiritual transformation. Quoting the Kabbalah, he prognosticated "the descent of America into the depths of purifying fires" (Spenser 23-24).

In a figurative sense, America had plunged into fires. With socialist machinations firmly embedded in the United States government, the Freedom Documents "burned." In the years to come, the size of government would continue to increase and individual rights would gradually vanish. It is appropriate at this juncture to recall the words of Hoffman:

The doctrine of man playing god reaches its nadir in the philosophy of scientism which makes possible the complete mental, spiritual and physical enslavement of mankind through technologies such as satellite and computer surveillance; a state of affairs symbolized by the "All Seeing Eye" above the unfinished pyramid on the U.S. one dollar bill. (50)

The Great Seal's public appearance signified the beginning of America's migration towards a scientific dictatorship. The truncated pyramid mounted by the "All Seeing Eye" represents the blueprint according to which society is being re-sculpted. It is the standard schematic for authoritarian governments. Ian Taylor explains:

It does not require great insight to see that power in human society takes the form of a pyramid, in which the mind-set of the general bulk of the structure largely reflects that of the mind at the top. Indeed, contrary to the common impression, modern governments are set up this way, with the apex of the pyramid often a mere figurehead representing the unseen wielders of power immediately beneath it. (33)

FDR's New Deal played a significant role in tangibly enacting this schematic. Lurking just below the apex of the pyramid were Freemasons, the "unseen wielders of power."

The Rise of the Military Industrial Complex
Interestingly enough, World War II, which was predicted by Wells in *The Shape of Things to Come*, accelerated the transformation of America into a Technocracy. Fischer elaborates:

It wasn't until World War II that the federal government succeeded in effectively pulling together its administrative machinery. In fact, it took the exigencies of a national war effort to catalyze the process that Roosevelt had begun earlier in the 1930s. World War II was truly a benchmark in the evolution of technocratic planning in the United States. (91-92)

This technocratic reconfiguration of American government is illustrated by the prolific public planning programs, which were birthed by the nation's massive war effort. Fischer states:

During this period, the country was geared to a full-scale program of public planning. The major thrust of its efforts was administered by the Office of War Mobilization, the War Production Board, the Office of Price Administration (which developed elaborate economic input-output techniques to regulate supply and demand), the Office of Scientific Research and Development, and the Office of Strategic Services. The war thus brought a vast expansion of economic intervention, administrative machinery for the coordination of national military and production efforts, and elaborate institutions for data gathering and forecasting. All were important influences on a technocratic restructuring of American society that began to emerge after the war. (92)

Of course, World War II was followed by the dialectical manipulation of the Cold War. Thanks to the deliberate mismanagement of American diplomacy shortly after the Second World War, the Soviet Union was able to expand its dominance across a sizable portion of Europe. This expansion provided the Western elite with a pretext for maintaining domestic militarism. Within this climate of perpetual fear, the technocratic "military-industrial complex" was born. Fischer writes:

The technological revolution initiated during the war, and later intensified by the "cold war" of the 1950s, was a primary impetus behind this technocratic restructuring. Scientific and technological innovation had always been a basic engine of industrialization, but after the war the relationship reached a new plateau, generally described as the rise of "Big Science." Evolving under the wing of the Pentagon and the corporate defense industry, Big Science gave birth in the postwar years to the "military-industrial complex." Increasingly, as President Eisenhower explained, this military-industrial connection was becoming the foundation of the U.S. economy. (92)

Eisenhower was acutely aware of the dangerous volumes of power being gradually consolidated by the military-industrial complex. In his farewell address, he admonished:

Yet, in holding scientific research and discovery in respect, as we should, we must also be alert to the equal and opposite danger that public policy could itself become the captive of a scientific-technological elite. (No pagination)

With the rise of Technocracy in the United States, this admonition seems prophetic. "Big Science" is now "big business," which has been married to the military establishment. World War II, the Cold War, and the current War on Terror have strengthened this union. Astride this unholy alliance, a "scientific-technological elite" now wields substantial control over "public policy." The draconian Patriot Act and the pervasive panopticism of the Department of Homeland Security woefully illustrate this fact. The primacy of the military-industrial complex is but a symptom of the scientific dictatorship's ascendance in the West.

One President worked to reverse the technocratic restructuring of the U.S. government: John F. Kennedy. Kennedy attempted to dismantle the military-industrial complex, as is evidenced by his efforts to remove U.S. forces from Vietnam. Investigative journalist Jim Marrs elaborates:

On October 11, 1963 Kennedy approved National Security Action Memorandum 263 which approved a possible disengagement in Vietnam by the end of 1965 and even ordered a quiet withdrawal of some military personnel by the end of that year. (130)

These and other similar overtures, in all likelihood, cost the President his life. More than a few people doubt the Warren Commission's conclusion that Kennedy was a victim of a lone assassin. Skeptics even included President Richard M. Nixon. On tapes of the first six months of his second term in office, Nixon said that the Warren Commission was "the greatest hoax that has ever been perpetuated" (Kevin Anderson, no pagination). While Nixon did not elaborate, one need only examine the Zapruder film to understand why the Warren Commission would be considered a "hoax." On the film Kennedy's head flies backwards from what could only be a bullet impact from the front. This flies straight in the face of the Commission's contention that Kennedy was shot by Oswald up and behind to the right in the book depository.

It may never be discovered who the actual triggerman (or men) were in the JFK assassination. However, when it comes to who planned the murder, certain elites have been implicated. In his book *Defrauding*

America, former federal inspector Rodney Stich presents the testimony of one of his informants in the intelligence community, Trenton Parker. Parker was a CIA operative tied to Pegasus, a CIA counter-intelligence unit. According to Parker, Pegasus had recorded conversations between elites who were conspiring to have Kennedy assassinated. These elites included Rockefeller, Allen Dulles, George Bush, and J. Edgar Hoover (316-317).

The tapes were turned over to Congressman Larry McDonald, a member of the Joint Armed Services Committee. When McDonald made it known that he was going to expose crime within both the government and CIA after he returned from the Far East, he was in all likelihood referring to these tapes. However, the congressman never received the opportunity to reveal any elite criminality. His plane, Korean Airlines flight 007, inexplicable flew into Russian airspace and was shot down (615). Kennedy's true murderers remained obfuscated. Meanwhile, Camelot crumbled in the shadow of the technocratic military industrial complex.

From Technocratic to Technetronic

Since the 1970s, the next developmental stage of Technocracy "has been both theorized and hailed under the banner of 'postindustrialism'" (101). Examining this shift in technocratic thinking, Fischer states: "contemporary technocratic theories are now theories of postindustrial society" (101). Yet, some technocratic ideologues regard "postindustrialism" and "postindustrial society" as potentially misunderstood or derisive characterizations. One such ideologue is Zbigniew Brzezinski, former national security advisor to President Carter and co-founder of the Trilateral Commission. Eschewing the "postindustrial" portraits of Technocracy, Brzezinski fancies the euphemism of "technetronic" society (101).

Brzezinski's "technetronic" model is no less elitist or anti-democratic than its theoretical progenitors. According to Brzezinski, this new stage in Technocracy's evolution will witness the ascendance of a "scientific/technical elite" that would seize control of the "essential flow of information and production" (Fischer 103). This epistemological cartel would subsequently direct its consolidations of knowledge toward the scientific subjugation of the masses. Fischer elaborates:

Increasingly, scientific knowledge will be used directly to plan almost every aspect of economic and social life. In the process, Brzezinski avers, class conflict will assume new forms and modes: Knowledge and culture will replace material needs in the struggle between the scientific/technical elite and the masses of people who will have to be integrated into and subordinated in the postindustrial system. (103)

Although accurate, Fischer's synopsis of Brzezinski's vision is stated in somewhat euphemistic terms. Yet, Brzezinski's own portrait is far more authoritarian in character. In *Between Two Ages: America's Role in the Technetronic Era*, Brzezinski more vividly describes the "gradual appearance of a more controlled and directed society" (252). With painful candor, the former national security advisor proceeds to paint his technocratic picture for the future:

Such a society would be dominated by an elite whose claim to political power would rest on allegedly superior scientific know-how. Unhindered by the restraints of traditional liberal values, this elite would not hesitate to achieve its political ends by the latest modern techniques for influencing public behavior and keeping society under close surveillance and control. (252)

There could hardly be a more succinct description of a scientific dictatorship. This is the objective to which Brzezinski is committed. In order to expedite this campaign for the technocratic restructuring of the world, Brzezinski has collaborated with other sociopolitical Darwinians in the establishment of certain globalist organizations. In 1973, Brzezinski and Chase Manhattan chairman David Rockefeller founded the Trilateral Commission (Sklar 1-2). The organization promotes "trilateralism," a doctrine of world order that is "rooted in a long tradition of elite ideology and corporate planning" (4). Trilateralism stipulates the following:

(1) the people, governments, and economics of all nations must serve the needs of multinational banks and corporations; (2) control over economic resources spells power in modern politics (of course, good citizens are supposed to believe as they are taught; namely, that political equality exists in Western democracies whatever the degree of economic inequality); and (3) the leaders of capitalist democracies—systems where economic control and profit, and thus political power, rest

with the few—must resist movement toward a truly popular democracy. (4)

Thus, trilateralism could be characterized as an "attempt by ruling elites to manage both dependence and democracy-—at home and abroad" (4). The Trilateral Commission itself retains the technocratic character of its co-founder, Brzezinski. Trilateralists believe that it is "'technocratic policy-oriented intellectuals' who should be cloned on a global scale" (40). Trilateralists hope that the global expansion of the technocratic paradigm will result in a world where the policy professional and the Transnationalist mutually benefit. Holly Sklar explains:

Trilateralism promises a pseudotechnocracy, where technocrats would serve the owner/managers of the global corporations. Profit would remain the force behind efficiency although it would be cloaked in the rhetoric of increasing productivity and enhancing welfare. (21)

Trilateralism represents a marriage between the technocratic affinity for efficiency and the Transnationalist hunger for profit. While it cannot promise a pure Technocracy as envisioned by Bacon and Saint-Simon, it can establish a global system where the scientists and engineers wield substantial quantities of power. Moreover, the system would not be apolitical. Instead, the alliance between technocrats and businessmen would be joined by "like-minded government officials" in a growing constellation of networks designed to "carry out national and international policy" (21). However, all three—-the technocrat, the Transnationalist, and the government official (predominantly of the Internationalist ilk)—-constitute the interests of sociopolitical Darwinism.

In turn, these sociopolitical Darwinians will comprise a "world information grid" designed to:

broaden the scope of educational-scientific and economic technological cooperation among the most advanced industrial nations that are becoming post-industrial and are in some regards moving into the post-national age. (Brzezinski, *Between Two Ages: America's Role in the Technetronic Era*, 299)

Although this model is missing some elements, it still exhibits many of the hallmarks of Bacon's *New Atlantis* and Saint-Simon's "totally scientistic society." In effect, trilateralism represents an attempt to appease all of the competing varieties of sociopolitical Darwinism. The final outcome remains the same: a scientific dictatorship.

Neoconservativism: The Cult of Techno-socialism

Of course, Brzezinski is one of the chief architects of the geostrategy that was implemented by the Bush administration immediately after September 11th, 2001. The administration of George W. Bush is peopled largely by a faction of the power elite known as the neoconservatives. Researchers of both the left and right have identified the neoconservatives as distinctly antidemocratic elements. The exposure has led to mounting opposition against the neoconservative agenda from numerous grassroots activists.

Now, several neoconservatives are launching a counterattack. The strategy is one of vilification. In an article for *National Review*, Michael Rubin characterized the neocons' opponents as anti-Semites obsessed with conspiracy theories (No pagination). Max Boot continues with the "conspiracy theory" angle, claiming that the neocons' opponents have overactive imaginations:

> A cabal of neoconservatives has hijacked the Bush administration's foreign policy and transformed the world's sole superpower into a unilateral monster. Say what? In truth, stories about the "neocon" ascendancy–and the group's insidious intent to wage preemptive wars across the globe–have been much exaggerated. And by telling such tall tales, critics have twisted the neocons' identities and thinking on U.S. foreign policy into an unrecognizable caricature. (No pagination)

Why have the neocons' retaliation been so aggressive? Do they simply wish to "set the record straight"? Are Rubin and Boot merely trying to correct several misconceptions over neoconservatism? The tone of their rhetoric and apologetics suggest another motivation: obfuscation. The neocons realize that continued exposure will eventually lead to the destruction of even the most well constructed disguise. One individual who realizes that the neocons have camouflaged their real intentions is Pulitzer Prize winning author Seymour Hersh. Hersh characterizes the neocons in the following way:

> [O]ne of the things that you could say is, the amazing thing is

we are been taken over basically by a cult, eight or nine neo-conservatives have somehow grabbed the government. (No pagination)

Cults are usually very adept at the concealment game. Many times the masquerade is so effective that a group's own members do not even realize they are part of a cult. What lies at the center of the cult of Neoconservatism?

The Neoconservative cult has always paraded around under a patriotic, pro-American, anticommunist facade. What lies behind this veneer? Frank Fischer answers this question in his book *Technocracy and the Politics of Expertise*: "...neoconservativism is at base an elitist ideology aimed at promoting a new group of conservative technocrats." (172)

To promote their own variety of Technocracy, neoconservatives present themselves as the antithesis to left-wing "policy professionals." However, the conflict between these two is superficial at best. As is the case with all good Hegelian dialectics, the neoconservative antithesis is not dichotomously related to its alleged technocratic opposition. Fischer elaborates:

Neoconservatives regularly argue that knowledge elites are a threat to democracy. But if this is their primary concern, their solution is scarcely designed to remedy the problem. Indeed, by challenging the Democratic party's use of policy expertise with a counterintelligentsia, they implicitly accept—and approve of—the evolving technocratic terrain. Developing a conservative cadre of policy analysts cannot be interpreted as a measure designed to return power to the people. (171)

Fischer correctly argues that Neoconservativism's advocacy of a so-called "conservative cadre of policy analysts" precludes citizen participation:

Neoconservatives doubtless maintain that their policy advisers speak for different political values: Rather than the welfare state and bureaucratic paternalism, conservative experts advocate democracy and free market individualism. Such an argument, however, fails to address the critical issue. As a system of decision making geared toward expert knowledge, technocracy—liberal or conservative—necessarily blocks meaningful participation for the average citizen. Ultimately only those who can interpret the complex technical languages that increasingly frame economic and social issues have access to the play of power Democratic rhetoric aside, those who nurture a conservative intelligentsia in reality only help to extend an elite system of policy-making. (171-72)

Whether under the superfluous appellations of conservative or liberal, "policy professionals" still constitute what Wells referred to as a "democracy of experts." Neoconservativism's promotion of its own "policy professionals" betrays the ideology's technocratic propensities. Rhetoric concerning "democracy" and "free market individualism" amounts to little more than pageantry. Neoconservativism is but the latest incarnation of the technocratic movement and represents another stage in the sociopolitical Darwinism's metastasis. Always looming on the horizon is a scientific dictatorship.

Neoconservativism's technocratic pedigree is also graphically illustrated by its adherents' strong support for FDR's New Deal. Irving Kristol, the "godfather of neoconservatism," states in his book *Neoconservatism: The Autobiography of an Idea* that neocons: ". . .accepted the New Deal in principle. . ." (x). Later in his book, Kristol writes:

In a way, the symbol of the influence of neoconservative thinking on the Republican party was the fact that Ronald Reagan could praise Franklin D. Roosevelt as a great American president-praise echoed by Newt Gingrich a dozen years later, when it is no longer so surprising. (379)

Why were neoconservatives so amicable towards the socialism of the New Deal? The answer is because Roosevelt's Marxist proclivities harmonized with the neoconservative variety of Technocracy. It is interesting to note that "godfather" Kristol was a Trotskyist in his youth. Kristol makes it clear that he is unrepentant: "I regard myself lucky to have been a young Trotskyist and I have not a single bitter memory" (13). The statist tradition found in Marxism is also carried on by the neocons. This is another point made clear by Kristol: "Neocons do not feel that kind of alarm or anxiety about the growth of the state in the past century, seeing it as natural, indeed inevitable" ("The Neoconservative Persuasion").

Marxist economic theory remains firmly embedded within neoconservative ideology. Several neoconservative ideologues have espoused socialist ideas. Former neoconservative Michael Lind admits:

The fact that most of the younger neocons were never on the left is irrelevant; they are the intellectual (and, in the case of William Kristol and John Podhoretz, the literal) heirs of older ex-leftists. The idea that the United States and similar societies are dominated by a decadent, postbourgeois "new class" was developed by thinkers in the Trotskyist tradition like James Burnham and Max Schachtman, who influenced an older generation of neocons. The concept of the "global democratic revolution" has its origins in the Trotskyist Fourth International's vision of permanent revolution. The economic determinist idea that liberal democracy is an epiphenomenon of capitalism, promoted by neocons like Michael Novak, is simply Marxism with entrepreneurs substituted for proletarians as the heroic subjects of history. (No pagination)

Arch-neocon William F. Buckley provides a prime example of a neoconservative ideologue who has espoused socialist ideas. Buckley has written:

"Congress shall appropriate funds for social welfare only for the benefit of those states whose per capita income is below the national average" (qutd. in Epperson 49).

Commenting on Buckley's statement, researcher Ralph Epperson writes:

This writer [Buckley] advocated a newer brand of Marxism: "From each *state* according to its ability, to each state according to its needs" (emphasis added). This writer advocated that the national government divide the wealth, taking it from the wealthier states and giving it to the less productive. Pure Marxism, except the writer involved both the state and the federal governments rather than just the federal government as Marx envisioned. This is only expanding Marx one step: the result is the same. Property is distributed by the government just as before. The shock is that this new thought came from the pen of William F. Buckley, Jr., hardly a paragon of Marxism. But notice that Buckley's intent is the same as that of Marx:

to use government to redistribute Consumption and Capital Goods. (Epperson 49)

No doubt, these Marxist proclivities were a consequence of neoconservativism's technocratic heritage. Of course, there are those who would argue that the technocratic tradition has been at variance with Marxism. Indeed, technocratic and Marxist theoreticians have feuded on occasion. Yet, the common thread of state socialism binds both, as is evidenced by their closely aligned economic policies and virtually identical outcomes.

In fact, some sects of the early technocratic movement "mingled socialism, managerialism, and assorted beliefs" with relative ease (Akin 83). Many were also "anticapitalistic and revolutionary in their rhetoric" (82). William Akin further examines the continuity of Marxist thought endemic to the early technocratic movement:

In California, adherents visualized an efficient totalitarian society in which "the individual must subordinate himself to the community" and which would function automatically under the guidance of engineers. The American Technocratic League of Denver was openly socialist. From Chicago the All American Technological Society proposed the use of the Soviet Union as a model for the new order. (83)

The petty differences between theoreticians become inconsequential. Technocracy was a logical outgrowth of earlier variants of socialism. The ideational continuum appears to have been a drift from Bellamyism to Technocracy to New Deal socialism. Neoconservativism is the latest segment in this larger continuity of thought. As neo-Jacobins and technocrats, they remain committed to the erection of a scientific dictatorship.

Chapter Two:

The Coming Clash of Scientific Dictatorships

Perpetual War for Perpetual Evolution

In his book *Evolution and Ethics*, Darwinian Sir Arthur Keith wrote:

> If war be the progeny of evolution--and I am convinced that it is--then evolution has "gone mad," reaching such a height of ferocity as must frustrate its proper role in the world of life--which is the advancement of her competing "units", these being tribes, nations, or races of mankind. There is no way of getting rid of war save one, and that is to rid human nature of the sanctions imposed on it by the law of evolution. Can man render the law of evolution null and void? I have discovered no way that is at once possible and practicable. (105)

In characterizing war as the "progeny of evolution," Keith makes it clear that conflict is the natural corollary of man's purported Darwinian ascent. Of course, Darwinism rests heavily upon the concept of natural selection, which depicts life as an enormous struggle for survival. On the microcosmic level, this struggle is bodied forth by the competition between species. On a macrocosmic level, this struggle manifests itself as war between nations. Thus, Keith establishes the centrality of war to evolutionary development. Jacques Barzun eloquently reiterates this contention:

> Darwin did not invent the Machiavellian image that the world is the playground of the lion and the fox, but thousands discovered that he had transformed political science . . . War became the symbol, the image, the inducement, the reason, and the language of all human beings on the planet. No one who has not waded through some sizable part of the literature of the period 1870-1914 has any conception of the extent to which it is one long call for blood . . . (100)

Indeed, Darwinism totally altered political science. Now, questions of governance and the administration of human affairs assumed an evolutionary character. In turn, the political terrain of the modern world is dominated by a criminal elite, which perpetuates a Darwinian power structure. Their supranational machinations, such as the IMF and World Bank, are designed according to evolutionary principles.

Because the paradigmatic character of governance is now distinctly evolutionary, national governments follow an inherently Darwinian course. Commensurate with this course is the continuous provocation of wars. In short, the primacy of evolutionary theory in politic science has predisposed the nations of the world to perpetual war.

The ruling class has a stake in engineering global conflicts. Not only do wars cull "surplus populations," thus satisfying the Malthusian precepts of Darwinism, but also they tangibly enact the dialectical framework intrinsic to evolutionary theory. This framework is Hegelian. The organism (thesis) comes into conflict with nature (antithesis) resulting in a newly enhanced species (synthesis), the culmination of the evolutionary process (Marrs 127). It is the hope of the elite that, through the continuous promulgation of warfare, this harmonious synthesis shall be tangibly realized. In this context, war serves an alchemical function.

Over the years, the various scientific dictatorships of the world have assumed convergent trajectories. While making war with one another, these scientific dictatorships simultaneously synthesize and birth new authoritarian regimes. Why? In reality, these scientific dictatorships have been little more than variants of the same socialist totalitarian system. Thus, the ostensible conflicts among these competing socialist regimes actually represent incremental phases in a process of coalitional integration. The final Hegelian synthesis is intended to be a global scientific dictatorship.

War is integral to the elite's evolutionary script, facilitating the dialectical convergence of the many scientific dictatorships littering the globe. The Hegelian synthesis of the world's various scientific dictatorships into a global government stipulates continual war. In turn, such perpetual conflict requires the manufacturing of adversaries to engage in fraudulent skirmishes. In short, war is alchemy. By synthesizing metallurgy, physics, occultism, and several other fields of study, the alchemists hoped to ultimately achieve the transmutation of man himself. Today, the power elite strives towards the same end, but through the synthesizing of global powers instead. War is one of the chief means by which this synthesis is achieved.

The Report from Iron Mountain

The plans for perpetual warfare were most thoroughly delineated within *The Report from Iron Mountain on the Possibility and Desirability of Peace*. Released in 1966, this document purported to be the product of a Special Study Group of fifteen men whose identities were to remain secret. For several years, the authenticity of the report has been in question. This skepticism only intensified when the document's author publicly declared that the report was a "satire." However, other parties with substantial credibility have defended the authenticity of *The Report from Iron Mountain*. For instance, Colonel L. Fletcher Prouty, who acted as an advisor to President Kennedy, stated:

> During the Kennedy Years, people within the government discussed frequently and quite seriously many of the major questions phrased by Leonard Lewin in Report From Iron Mountain. I had been assigned to the Office of the Secretary of Defense before the Kennedy election and was there when the McNamara team of "Whiz Kids" arrived. Never before had so many brilliant young civilians with so many Ph.D.s worked in that office. It was out of the mouths of this group that I heard so frequently and precisely the ideas that Lewin recounts in his "novel." (287-88)

The Report from Iron Mountain became a proverbial hot potato and changed hands several times. Its authorship was even attributed to William F. Buckley, a veritable icon among neoconservatives. Irrespective of who authored the document, its precise delineation of ruling class tactics and its accuracy in prognosticating future events is difficult to deny. Whether the document is genuine or some elaborate hoax, its didactic value remains intact.

Questions of morality and individual freedom were not addressed in *The Report from Iron Mountain*. In fact, the report only briefly mentions the concepts of human liberty and ethics, regarding them as anachronistic concepts embraced by bygone generations. The study concerned itself solely with the perpetuation of an absolute State and an elitist power structure. The report stated:

> Previous studies have taken the desirability of peace, the importance of human life, the superiority of democratic institutions, the greatest "good" for the greatest number, the "dignity" of the individual, and other such wishful premises as axiomatic values necessary for the justification of a study of peace issues. We have not found them so. We have attempted to apply the standards of physical science to our thinking, the principal characteristic of which is not quantification, as is popularly believed, but that, in Whitehead's words, ". . .it

ignores all judgments of value; for instance, all esthetic and moral judgments." (Lewin 13-14; emphasis added)

Evident in this statement is the power elite's fanatically religious adherence to the doctrine of scientism. The doctrine of scientism rigorously promotes the ecumenical imposition of physical science upon all fields of inquiry. Researcher Michael Hoffman most succinctly revealed the inherent folly of scientism:

The reason that science is a bad master and dangerous servant and ought not to be worshipped, is that science is not objective. Science is fundamentally about the uses of measurement. What does not fit the yardstick of the scientist is discarded. Scientific determinism has repeatedly excluded some data from its measurement and fudged other data, such as Piltdown Man, in order to support the self-fulfilling nature of its own agenda, be it Darwinism or "cut, burn and poison" methods of cancer "treatment." (49)

The dominant epistemological paradigm of today is one governed by scientific determinism. Now, institutionally accredited science is the infallible dogma of a new theocracy. Human beings are reduced to quantifiable entities for the scrutiny of priests donning lab coats. Atop the highest plateau of this epistemological hierarchy sits the power elite, whose scientism selectively excludes any data that could be disproportionate with the ultimate agenda: complete social control.

According to *The Report from Iron Mountain*, "the desirability of peace, the importance of human life, the superiority of democratic institutions, the greatest 'good' for the greatest number, the 'dignity' of the individual, and other such wishful premises" are disproportionate with the "yardstick" of physical science. Since science "ignores all judgments of value; for instance, all esthetic and moral judgments," there is no place for "axiomatic values" in a global scientific dictatorship.

The document proceeds to examine the necessity of war, declaring:

The war system not only has been essential to the existence of nations as independent political entities, but has been equally indispensable to their stable political structure. Without it, no government has ever been able to obtain acquiescence in its "legitimacy," or right to rule its society. The possibility of war provides the sense of external necessity without which no government can long remain in power. The historical record reveals one instance after another where the failure of a regime to maintain the credibility of a war threat led to its dissolution, by the forces of private interest, of reactions to social injustice, or of other disintegrative elements. The organization of society for the possibility of war is its principal political stabilizer. It has enabled societies to maintain necessary class distinctions, and it has insured the subordination of the citizens to the state by virtue of the residual powers inherent in the concept of nationhood. (Lewin 39, 81)

With the ever-present threat of war, the absolute State could maintain a standing army and implement a policy of compulsory service for its citizenry. According to the study, this system of obligatory service would provide the socially maladjusted and economically disadvantaged elements of society with a function. Thus, these "potential enemies of society" could be placated and pacified. The report elaborates:

We will examine the time-honored use of military institutions to provide anti-social elements with an acceptable role in the social structure... The current euphemistic clichés--"juvenile delinquency" and "alienation"--have had their counterparts in every age. In earlier days these conditions were dealt with directly by the military without the complications of due process, usually through press gangs or outright enslavement. Most proposals that address themselves, explicitly or otherwise, to the postwar problem of controlling the socially alienated turn to some variant of the Peace Corps or the so-called Job Corps for a solution. The socially disaffected, the economically unprepared, the psychologically uncomfortable, the hard-core "delinquents," the incorrigible "subversives," and the rest of the unemployable are seen as somehow transformed by the disciplines of a service modeled on military precedent into more or less dedicated social service workers.
Another possible surrogate for the control of potential enemies of society is the reintroduction, in some form consistent with modern technology and political processes, of slavery. It is entirely possible that the development of a sophisticated form of slavery may be an absolute prerequisite for social control in a world at peace. As a practical matter, conversion of the code

of military discipline to a euphemized form of slavery would entail surprisingly little revision; the logical first step would be the adoption of some form of "universal" military service. (Lewin 41-42, 68, 70)

The concluding remarks of this excerpt are particularly interesting. According to the *Report*, the "code of military discipline" is easily adapted to the standard civilian cultural milieu. Central to the military model is what late philosopher Michel Foucault dubbed the "panoptic schema." It is this highly versatile constituent of the military system that makes the adaptation of a similar authoritarian model to non-combatant environs relatively simple.

Based on Jeremy Bentham's Panopticon, the panoptic schema is a mechanism that allows for complete surveillance of the subject (pan=all, optic=seeing). Originally a hallmark of the prison system, the panoptic schema can assume numerous forms. It can be the security camera, the barbed wire fence, and the armed guard. However, it does have much more subtle manifestations.

Panoptic schema can also be bodied forth through the rigid timetabling of activities, which allows for the effective chronemic regulation of a subject's day. It is tangibly enacted by the observational design of specific types of architecture, which readily lend themselves to the spatial regulation of the subject. Finally, it is within the mind of the subject himself, who has internalized all of these mechanisms of control and now monitors his own behavior.

Commenting on the metastasis of the panoptic schema, Foucault observes that it:

is polyvalent in its applications; it serves to reform prisoner, but also to treat patients, to instruct schoolchildren, to confine the insane, to supervise workers, to put beggars and idlers to work. It is a type of location of bodies in space, of distribution of individuals in relation to one another, of hierarchical organization, of disposition of centres and channels of power, of definition of the instruments and modes of intervention of power, which can be implemented in hospitals, workshops, schools, prisons. Whenever one is dealing with a multiplicity of individuals on whom a task or a particular form of behaviour must be imposed, the panoptic schema may be used. (205)

According to Foucault, the elasticity of the panoptic schema allowed for its mass diffusion throughout society: "The panoptic schema, without disappearing as such or losing any of its properties, was destined to spread throughout the social body" (207). With the metastasis of the panoptic schema, Foucault observes that "its vocation was to become a generalized function" (207). Given the polyvalence of the panoptic schema, the *Report From Iron Mountain* seems correct in asserting that society's conversion into a carceral culture involves "surprisingly little revision."

The final result, Foucault observes, is a virtually mechanized society:

We are neither in the amphitheatre, nor on the stage, but in the panoptic machine, invested by its effects of power, which we bring to ourselves since we are part of its mechanism (217).

The concept of the "panoptic machine" synchronizes very comfortably with Weishaupt's society of *Maschinenmenschen*. Ever-present is the machine motif, which was also a hallmark of the technocratic movement of the early thirties. It comes as little surprise that the Bush administration, which is governed largely by the technocratic neoconservatives, would develop panoptic machinations like the Patriot Act and the Total Information Awareness program. It is even less surprising that the ascendancy of these panoptic machinations was facilitated by the so-called "War on Terror." Indeed, the war system has proven to be most advantageous to the power elite.

The Total Information Awareness program was the most prevalent manifestation of this contemporary panoptic age. Daniel Schorr, a journalist for the *Christian Science Monitor*, examines this panoptic machination:

Deep in the recesses of the Pentagon is the Office of the Defense Advanced Research Projects Agency (DARPA). DARPA is where Vice Adm. John Poindexter (USN ret.) hangs out these days, working on TIA. TIA stands for Total Information Awareness. The project, which is budgeted at $10 million this year and expected to get more next year, has been getting bad press. That is in part because its Orwellian-sounding purpose is to create a centralized database of personal information about Americans.

Cutting-edge technology would be used to gather everything

that the computer age has to offer, from travel plans to pharmacy prescriptions. Pentagon officials say it's meant to be a tool in the war against terrorism, not an invasion of privacy of innocent citizens. Well, maybe. But that would sound more reassuring if it were not for the identity of the project manager. (No pagination)

Indeed, Poindexter is certainly not one of the most ethical people who have ever lived. His past is replete with scandal and fraud, more than enough to preclude him from such a sensitive position as project manager of a national security program. Schorr proceeds to unveil Poindexter's shady past:

Admiral Poindexter is probably better known for destroying information than for gathering it. Before a congressional investigating committee in 1986, he admitted that, as President Reagan's national security adviser, he destroyed evidence in connection with the Iran-contra affair. Specifically, he tore up the only signed copy of a document called a "presidential finding" that retroactively authorized shipment of arms to Iran in return for the release of American hostages in Lebanon.
He testified that he did this to avoid embarrassment to Mr. Reagan. Poindexter, like Oliver North, who reported to him, was convicted in federal district court of lying to Congress and of obstruction. The conviction was overturned on technical grounds by an appeals court majority of two Reagan-appointed judges, Douglas Ginsburg and David Sentelle, over the vigorous dissent of Carter-appointed judge Abner Mikva. (No pagination)

Yet, despite Poindexter's dubious past, the Bush Administration had no qualms about employing him in such a sensitive post. Schorr states:

The Bush administration has shown no inclination to alter Poindexter's sensitive assignment. Mr. Rumsfeld says: "I would recommend people take a deep breath. Nothing terrible is going to happen." (No pagination)

What was the nature of Poindexter's Total Information Awareness project? What was its true magnitude and scope? *Washington Times* journalist Audrey Hudson provides a glimpse:

> In what one critic has called "a supersnoop's dream," the Defense Department's Total Information Awareness program would be authorized to collect every type of available public and private data in what the Pentagon describes as one "centralized grand database." (No pagination)

This data would include: "e-mail, Internet use, travel, credit-card purchases, phone and bank records of foreigners" (no pagination). Further elaborating on the ominous scope of this centralized database, *New York Times* columnist William Safire wrote:

> "To this computerized dossier on your private life from commercial sources, add every piece of information that government has about you - passport application, driver's license and bridge toll records, judicial and divorce records, complaints from nosy neighbors to the FBI, your lifetime paper trail plus the latest hidden camera surveillance - and you have the supersnoop's dream: a 'Total Information Awareness' about every U.S. citizen." (Qutd. In Hudson, no pagination)

The official emblem of the Total Information Awareness program semiotically gesticulated towards the rise of the panoptic machine. Schorr concludes his examination of the program with the following statement:

> Outside Poindexter's Pentagon office is a logo showing an all-seeing eye on top of a pyramid and the slogan, "Scientia est potentia" ("Knowledge is power"). The question is: How much power over knowledge about us should be entrusted to an admitted destroyer of federal documents? (No pagination)

At this juncture, it is most appropriate to recall Hoffman's remarks:

> The doctrine of man playing god reaches its nadir in the philosophy of scientism which makes it possible the complete mental, spiritual and physical enslavement of mankind through technologies such as satellite and computer surveillance; a

state of affairs symbolized by the "All Seeing Eye" above the
unfinished *pyramid* on the U.S. one dollar bill. (50; emphasis
added)

The truncated pyramid capped by the "All Seeing Eye" is a
recurring motif throughout Masonic iconography. Having previously
established Masonry's role in the erection of scientific dictatorships,
it comes as little surprise that this symbol should appear again. It
semiotically gesticulates towards America's ongoing transformation
into a fully functional Technocracy. The neoconservative technocrats,
who represent the philosophical scions of the Illuminist-bred Jacobins,
currently dominate this emergent Technocracy.

Extensive press coverage of TIA led to the Pentagon shutting the
program down and Poindexter leaving the government (Bamford, no
pagination). However, the TIA's method of collecting vast heaps of
information, known in national security circles as "data mining," did
not disappear. Initiatives that elites cannot establish in plain view, they
usually sneak through the back door. The panoptic machine is certainly
no exception to this rule. It was recently revealed that, in the fall of
2001, President Bush secretly ordered the National Security Agency
(NSA) to sidestep a special court and carry out warrantless spying on
American citizens (Bamford, no pagination).

The legality of the President's actions is questionable. Under the
Foreign Intelligence Surveillance Act, before spying domestically on
Americans suspected of having terrorist ties, the NSA must first go
before a Foreign Intelligence Surveillance court and show probable
cause to obtain a warrant (Bamford, no pagination). Persuading a FISA
court to give the NSA the warrants it desired would have been relatively
simple, as James Bamford points out:

> The court rarely turns the government down. Since it was
> established in 1978, the court has granted about 19,000
> warrants; it has only rejected five. And even in those cases the
> government has the right to appeal to the Foreign Intelligence
> Surveillance Court of Review, which in 27 years has only heard
> one case. And should the appeals court also reject the warrant
> request, the government could then appeal immediately to a
> closed session of the Supreme Court. (No pagination)

Given the ease with which the NSA might have procured the
required warrants it is shocking that the President's secret program

bypassed the FISA court entirely. Perhaps the evidence suggesting probable cause was extremely flimsy. It would not be the first time that such was the case. J. Edgar Hoover was convinced that Martin Luther King Jr. was a subversive and tirelessly snooped into the famous Civil Rights leader's private life. Today it is known that, while King's private life left much to be desired, the case that he was somehow a conscious agent of Moscow was weak. Perhaps Bush did not want take a chance, no matter how low the odds, because his oligarchical upbringing has made it hard for him to take "no" for an answer.

However, there is another possibility. Perhaps the neoconservative faction of the elite are determined to establish the panoptic machine before this President's time in office is over. Such a task would call for circumventing every safeguard the law provides. The President may believe he has done just that by having the NSA spy on hundreds, perhaps even thousands, of Americans just for a few days or weeks at a time (Bamford, no pagination). Such fishing expeditions would allow the President to argue that the eavesdropping was short-term and that FISA does not apply because it is for long-term monitoring (Bamford, no pagination). Even still, Bamford contends that such a method is "precisely the type of abuse the FISA court was put in place to stop" (No pagination). The President may be establishing a dangerous precedent, the ends of which is a fully functional panoptic machine.

How was the NSA able to conduct a spy campaign on hundreds, if not thousands, of American citizens? Investigative journalist Jason Leopold answers that question:

> A clandestine National Security Agency spy program code-named Echelon was likely responsible for tapping into the emails, telephone calls and facsimiles of thousands of average American citizens over the past four years in its effort to identify people suspected of communicating with al-Qaeda terrorists, according to half-a-dozen current and former intelligence officials from the NSA and FBI. (No pagination)

Echelon is a network of listening posts scattered across the world (Bomford, no pagination). The capabilities of this network are nothing less than shocking. The *BBC's* Andrew Bomford elaborates:

> Every international telephone call, fax, e-mail, or radio transmission can be listened to by powerful computers capable

of voice recognition. They home in on a long list of key words,
or patterns of messages. (No pagination)

Technology such as that employed by the Echelon network is
making it possible for the rising technocratic system of government
to become what author Charles Stross calls a Panopticon Singularity.
Stross even refers directly to Echelon as one of the ten technologies
contributing to the creation of a Panopticon Singularity (no pagination).
The Panopticon Singularity is nothing less than a carceral state made
absolute in its power by technology. Stross elaborates:

> A Panopticon Singularity is the logical outcome if the
> burgeoning technologies of the singularity are funneled
> into automating law enforcement. Previous police states
> were limited by manpower, but the panopticon singularity
> substitutes technology, and ultimately replaces human
> conscience with a brilliant but merciless prosthesis.
> If a panopticon singularity emerges, you'd be well advised to
> stay away from Massachusetts if you and your partner aren't
> married. Don't think about smoking a joint unless you want
> to see the inside of one of the labour camps where over 50%
> of the population sooner or later go. Don't jaywalk, chew gum
> in public, smoke, exceed the speed limit, stand in front of
> fire exit routes, or wear clothing that violates the city dress
> code (passed on the nod in 1892, and never repealed because
> everybody knew nobody would enforce it and it would take up
> valuable legislative time). You won't be able to watch those old
> DVD's of 'Friends' you copied during the naughty oughties
> because if you stick them in your player it'll call the copyright
> police on you. You'd better not spend too much time at the
> bar, or your insurance premiums will rocket and your boss
> might ask you to undergo therapy. You might be able to read
> a library book or play a round of a computer game, but your
> computer will be counting the words you read and monitoring
> your pulse so that it can bill you for the excitement it has
> delivered. (No pagination)

Those who believe that snooping technology will not be used
to erect a Panopticon Singularity and will only be employed against
terrorists are being extremely naïve. Echelon has already been used for

purposes outside of fighting terrorism. These purposes were criminal in nature. Andrew Bomford reports:

> Journalist Duncan Campbell has spent much of his life investigating Echelon. In a report commissioned by the European Parliament he produced evidence that the NSA snooped on phone calls from a French firm bidding for a contract in Brazil. They passed the information on to an American competitor, which won the contract. (No pagination)

If the powers-that-be would use Echelon to give American companies an unfair advantage, they would most certainly use it to silence opposition to the rising global scientific dictatorship as well.

It is interesting to examine the new definition of peace presented in *The Report from Iron Mountain*. It reads: "The word *peace*, as we have used it in the following pages implies total and general disarmament" (Lewin 9). Under such conditions, resistance against tyranny is virtually non-existent. With the exception of combatants, whose behavior will be closely monitored by their superiors in the military and whose dominant concern shall be survival on the battlefields of the elite's perpetual war, no one else shall have the weapons with which they could resist tyranny. Indeed, "war is peace." The "peace" afforded for the New World Order at the expense of others shall mean perpetual "war" for the rest of humanity.

The Report from Iron Mountain also acknowledged the role of science "fiction" literature in the psychological conditioning of the masses:

> Up to now, this has been suggested only in fiction, notably in the works of Wells, Huxley, Orwell, and others engaged in the imaginative anticipation of the sociology of the future. But the fantasies projected in *Brave New World* and *1984* have seemed less and less implausible over the years since their publication. The traditional association of slavery with ancient preindustrial cultures should not blind us to its adaptability to advanced forms of social organization (Lewin 70)

The *Report's* mention of Wells is most appropriate, especially within the context of the war system. *The Shape of Things to Come* provided numerous semiotic intimations of the Second World War. Orwell's *1984* is also significant. In this literary classic, Orwell presented a world order where the chief element of societal stability was war. The "machine," which represented a nation's technical and industrial infrastructure, had been transmogrified into a strategic weapon against its own population. Shamefully wasteful governmental programs were enacted to keep the citizenry perpetually impoverished. This Hobbesian war of "all against all" was perpetuated by a small elite for the purposes of maintaining their power. That which the *Report* euphemistically calls "imaginative anticipation of the sociology of the future" qualifies as sci-fi predictive programming. Within such literary works are narrative paradigms that are politically and socially expedient to the power elite. Thus, when the future unfolds as planned, it assumes the paradigmatic character of the "fiction" that foretold it.

The Semiotics of Sci-fi Predictive Programming

The media circus surrounding 9-11 and the science fiction films that preceded it provide an excellent case study in semiotic warfare. The entertainment industry of the pre-September 11th world was instrumental in creating a cultural milieu that was already predisposed to war.

Few are not acquainted with the scene in *Independence Day* during which the White House is destroyed by a powerful energy beam from a hovering alien ship. In his semiotic analysis of this famous clip, Professor Elliot Gaines discerns "the narrative qualities that embody the paradigmatic character of the situation and images" surrounding 9-11(Gaines 123). This researcher would contend that such synchronicities were consciously engineered by the entertainment industrial complex. Intrinsic to the narrative characteristics of *Independence Day* was a paradigmatic template that the elite successfully imposed upon September 11th. Promulgated vigorously by Establishment media organs, *Independence Day* was instrumental in creating a cultural milieu that would be hospitable to future media manipulations. By the time of the WTC attacks, the collective subconscious of America was fertile with memes (contagious ideas) planted by *Independence Day*.

This memetic fertility is most effectively illustrated by the comments of MSNBC reporter Ron Insana. Insana witnessed the disintegration of the World Trade Center firsthand (Gaines 125). In an interview with Matt Lauer, Katie Couric, and Tom Brokaw, Insana vividly recounted his experience:

> "[A]s we were going across the street, we were not terribly far from the World Trade Center building, the south tower.

As we were cutting across a, a quarantine zone actually, the building began disintegrating. And we heard it and looked up and started to see elements of the building come down and we ran, and honestly it was like a scene out of Independence Day. Everything began to rain down. It was pitch black around us as the wind was ripping through the corridors of lower Manhattan." (Qutd. in Gaines 125)

Gaines identifies the Independence Day reference as semiotically significant (125). Given his distinction as a journalist before a global audience, Insana is thoroughly cognizant of the fact that his "intertextual reference to the film will be understood as a commonly known cultural text" (125). At this point, the previously dormant seeds of virulent thought implanted by *Independence Day* have been activated. Insana's invocation of this "commonly known text" has triggered the release of ideational spores within humanity's collective consciousness. Gaines reveals the semiotic effect of Insana's intertextual reference upon the percipient's mind:

The violence in Independence Day, coded as fiction, constructs a narrative binary opposition that clearly identifies good against evil. The available images representing the events of September 11th, using inferences drawn from Independence Day's sign/object relations, construct a narrative paradigm based upon the same themes, but coded as reality. (126)

Indeed, Insana's intertextual reference helped establish the paradigm of "good against evil" upon which the "War on Terrorism" would be premised. Suddenly, Arabs became analogous to the "alien invaders" of *Independence Day*. Simultaneously, the United States became analogous to the beleaguered "home world." Semiotically, Insana's intertextual reference prompted America's collective subconscious to re-conceptualize the relational dynamic between the West and the Arab world. "Good" humans against "evil" aliens, a narrative paradigm coded as fiction in *Independence Day*, suddenly recoded itself in the guise of reality. However, according to the elite's narrative paradigm for September 11th, being neither "good" nor "human" is part of the Arab's role.

It is not this researcher's contention that Insana consciously designed his intertextual reference to achieve such an end. However, it is this researcher's contention that Insana's intertextual reference is product of a larger semiotic deception. This larger semiotic deception is part of a program for cultural subversion known as "sci-fi predictive programming," a term coined by researcher Michael Hoffman. Elaborating on this concept, Hoffman states: "Predictive programming works by means of the propagation of the illusion of an infallibly accurate vision of how the world is going to look in the future" (205).

Innocuous though the genre may seem, science fiction literature has had a history of presenting narrative paradigms that are oddly consistent with the plans of the elite. *In Dope, Inc.*, associates of political dissident Lyndon LaRouche claim that the famous literary works of H.G. Wells and his apprentices, George Orwell and Aldous Huxley, were really "'mass appeal' organizing documents on behalf of one-world order" (538).

Such would seem to be the case with Gene Roddenberry's Star Trek, which presents a socialist totalitarian world government under the appellation of the Federation. Moreover, Roddenberry espoused a core precept of the ruling class religion: "As nearly as I can concentrate on the question today, I believe I am God; certainly you are, I think we intelligent beings on this planet are all a piece of God, are becoming God" (Alexander 568). This statement echoes the occult doctrine of "becoming," a belief promoted within the Masonic Lodge and disseminated on the popular level as Darwinism. According to this doctrine, man is gradually evolving towards apotheosis.

In *2001: A Space Odyssey*, Stanley Kubrick and Arthur C. Clarke presented a semiotic signpost for the next step in this chimerical evolutionary ascent. Michael Hoffman explains:

> *2001, A Space Odyssey*, directed by Stanley Kubrick and based on the writing of Arthur C. Clarke, is, with hindsight, a pompous, pretentious exercise. But when it debuted it sent shivers up the collective spine. It has a hallowed place in the Cryptosphere because it helped fashion what the Videodrome embodies today. At the heart of the film is the worship of the Darwinian hypothesis of evolution and the positioning of a mysterious monolith as the evolutionary battery or "sentinel" that transforms the ape into the space man (hence the "odyssey").
>
> Clarke and Kubrick's movie, 2001, opens with a scene of the "Dawn of Man," supposedly intended to take the viewer back to the origins of humanity on earth. This lengthy sequence is vintage Darwinism, portraying our genesis as bestial and

featuring man-like apes as our ancestors. In the film, the evolution of these hominids is raised to the next rung on the evolutionary ladder by the sudden appearance of a mysterious monolith. Commensurate with the new presence of this enigmatic "sentinel," our alleged simian progenitors learn to acquire a primitive form of technology; for the first time they use a bone as a weapon.

This bone is then tossed into the air by one of the ape-men. Kubrick photographs the bone in slow motion and by means of special effects, he shows it becoming an orbiting spacecraft, thus traversing "millions of years in evolutionary time."

The next evolutionary level occurs in "2(00)1" (21, i.e. the 21st century). In the year 2001, the cosmic sentinel that is the monolith reappears again, triggering an alert that man is on to the next stage of his "glorious evolution." (Hoffman 11-12)

The monolith or "sentinel" semiotically gesticulates towards the next epoch of man's "glorious evolution." Like the tabula rasa of human consciousness, the barren canvas of the monolith awaits the next brushstrokes of unseen painters. A new portrait of man is scheduled to be painted and the "glorious evolution" of humanity continues. "Coincidently," this semiotic signpost reappeared before the public eye in the actual year 2001. This reappearance is recounted in a BBC news article entitled "Mysterious monolith marks 2001":

A black, steel monolith nearly three metres high has mysteriously appeared in a park in the American city of Seattle.

The unmarked sculpture, planted on a grassy knoll in Magnuson Park, seems to have been put in place on New Year's Eve.

It is believed to be a reference to the black monoliths featured in the classic science fiction film 2001: A Space Odyssey, a collaboration between director Stanley Kubrick and writer Arthur C Clarke.

As in the film, the appearance of the black structure raises questions, the most obvious one: who is its creator? (No pagination)

Before this question could be answered, the monolith vanished. A follow-up BBC article elaborates:

Just as mysteriously as it appeared, the monolith which was found on a grassy knoll in Seattle's Magnuson Park on New Year's morning has vanished into thin air.
Baffled park officials discovered its absence on Wednesday when they went to check whether the structure was a danger to public safety.
All that remained of the nine-foot-tall (three-metre) smooth steel block was a rectangular hole where it had stood, a few candles and a broken-stemmed rose.
The black sculpture had confounded locals who found it. Devoid of markings or inscriptions it was presumed to be a reference to the monoliths featured in the classic science fiction film 2001: A Space Odyssey.
Although only a fleeting attraction, the monolith had already become a popular draw for the park, with visitors flocking to see and touch it. ("Seattle's mystery monolith disappears," no pagination)

That same year, the World Trade Center attacks took place and the Bush Administration began to erect a garrison state under the auspices of "national security." The chronically recapitulated theme of exchanging freedom for security is one of the most prevalent symptoms of this transformational period. However, semiotic intimations of this emergent garrison state may be discernible in the 1997 film *Starship Troopers*. Based on the sci-fi novel by Robert Heinlein, the film presents a socialist totalitarian world government that owes its very existence to a threat from "beyond." Synopsizing the theme of the film, literary critic Geoffrey Whitehall makes an interesting observation:

Against, yet within, its clichéd ontological galaxy, *Starship Troopers* mobilizes the beyond to critique this dominant us/them narrative. It seeks to reveal how identity/difference, a relation of fear, founds a political galaxy... fear is the order word of a security discourse. Historically, a discourse of fear bridged what it meant to be human in the world under Christendom (seeking salvation) and the emergence of modernity (seeking security) as the dominant trope of political life in the sovereign state. The church relied on a discourse of fear to 'establish its authority, discipline its followers and ward off its enemies,' in effect creating a Christian world politics. Under modern world politics, similarly, the sovereign state relies on *the creation of an*

external threat to author its foreign policy and establish the lofty
category of citizenship as the only form of modern human
qualification. (182; emphasis added)

It is interesting that, the very same year of *Starship Troopers'* release,
former national security advisor Zbigniew Brzezinski published *The Grand
Chessboard*. In this overtly imperialistic tract, Brzezinski delineated the
geostrategy by which America would attain global primacy. According
to Brzezinski, this period of American hegemony would represent little
more than a transitional period preceding her amalgamation into a one-
world government. In one of the most damning portions of the text,
Brzezinski reveals the catalyst for America's imperialist mobilization:

Moreover, as America becomes an increasingly multi-cultural
society, it may find it more difficult to fashion a consensus
on foreign policy issues, except in the circumstance of *a truly
massive and widely perceived direct external threat.* (211; emphasis
added)

A "truly massive and widely perceived direct external threat" did
appear. His name was Osama bin Laden. *Starship Troopers* was premised
upon the same thesis that would underpin American foreign policy four
years later . . .consensus facilitated by an external threat. Like Insana's
Independence Day analogy, the thematic similarities between Brzezinski's
Grand Chessboard and Heinlein's *Starship Troopers* reiterate the semiotic
concept of intertextuality. The various texts comprising human discourse
are not read in a cultural vacuum. On the level of consumption, "any
one text is necessarily read in relationship to others and . . .a range of
textual knowledges is brought to bear upon it" (Fiske 108). Likewise, "a
range of textual knowledges" was brought to bear upon September 11th.
Like Independence Day, Heinlein's Starship Troopers constituted part
of this body of "textual knowledges."

The centrality of an external threat to the formulation of foreign
policy, which thematically underpinned Brzezinski's geostrategy, was
semiotically communicated to the public through *Starship Troopers*. In
the elite's narrative paradigm for September 11th, the necessity of the
external threat was illustrated by the nationalistic fervor that followed
the WTC attacks. Suddenly, the appellation of "patriot," which was
previously a stigma assigned to tax protesters and members of militias,
regained its place in the cultural lexicon of reverential labels. The
removal of the pejorative connotations previously imposed upon the

"patriot" facilitated the semiotic deception that was to follow with the introduction of the Patriot Act. Connotatively, the very title of the Patriot Act suggested that those who opposed it constituted "unpatriotic" elements. Thus, acquiescence meant patriotism. This inference echoes the mantra presented in *Starship Troopers*: "Service guarantees citizenship." In the post-911 cultural milieu where the term "patriot" was as elastic as the term "terrorist," independent reasoning was subverted by a burgeoning epidemic of cognitive dissonance.

Starship Troopers also reiterated the narrative paradigm of "good" humans against "evil" aliens, a belief integral to the imperial mobilization of Brzezinski's geostrategy. The forces of "good," embodied by America, were mobilized against the forces of "evil," embodied by the Arab world. In keeping with the narrative paradigm of the elite, the media continued its standard practice of typecasting. Like the extraterrestrial "bugs" of *Starship Troopers*, Arabs were cast as hostile aliens. Meanwhile, Americans maintained their roles as humans.

Again, it is not this researcher's contention that Ron Insana was a conscious agent of this semiotic deception. Yet, as a part of the Establishment media, Insana acted as the perfect transmission belt for memes emanating from the ruling class itself. As the old adage goes, "No one knows who invented water, but you can bet it wasn't the fish." Immersed within the sea of Establishment-controlled media, Insana could not identify the larger semiotic manipulation in which he unwittingly played an integral role. Science fiction has been called "the literature of ideas." Insana's intertextual reference suggests that he had contracted an ideational contagion through exposure to sci-fi films like *Independence Day* and *Starship Troopers*.

Ferdinand de Saussure observed that "normally we do not express ourselves by using single linguistic signs, but groups of signs, organized in complexes which themselves are signs" (Saussure 1974, 128; Saussure 1983, 127). Indeed, isolated signs say very little, if anything at all. Communication and cogent thought are contingent upon the coalescence of signs. Such coalescence constitutes the complex social interchange called discourse. Likewise, the semiotic significance of a particular scene becomes evident only once the percipient has correlated all the constituent signs comprising it. This is syntagmatic analysis, the study of a text's structure and correlating signs.

Because they are narratives, films largely depend upon sequential configurations that produce the illusion of causal relationships. Likewise, the narrative paradigm that the power elite wished to impose upon September 11th was sequenced to create a false causal connection between the WTC attacks and the Arab world. During the interview with Insana, Couric abruptly announced an "upsetting wire that just came across the wire from the West Bank" (qutd. in Gaines 126). Couric proceeded to paint a disturbing portrait of militant Muslims celebrating the destruction of the Twin Towers:

> "Thousands of Palestinians celebrated Tuesday's terror attacks in the United States chanting 'God is great' and distributing candy to passers by even as their leader, Yasir Arafat said he was horrified. The U.S. government has become increasingly unpopular in the West Bank and Gaza Strip in the past year of Israeli-Palestinian fighting." (Qutd. in Gaines 126)

As the report continued, Couric read the same "upsetting wire" again, this time as a voice-over narrative to video footage of Palestinian demonstrators (Gaines 126). The footage was accompanied by a title card claiming that the event had occurred "EARLIER THIS MORNING" (Gaines 126). This researcher contends that the juxtaposition of this image with Insana's intertextual reference was intentional. It was designed to reinforce the paradigmatic template of "good" Americans against "evil" Arabs. Within the mind of the percipient, causal connections were already being made. "Behold, the face of the enemy," the subconscious declared. The syntagmatic structure of the NBC report was designed to achieve precisely this effect.

Upon closer examination, the semiotic deception grows even more sinister. Gaines elaborates on the unfolding sham:

> NBC later acknowledged that it had committed a breach of ethics by using archive footage with an unverified wire report. Only through convention do we assume the indexical nature of an image grounded by the text of news. The image was not actually acquired September 11th as an authentic Palestinian celebration of the attack against the US. The image was selected from an archive as a global sign to imply Islamic extremism as the enemy. (126)

Was this an accident or a consciously engineered psychocognitive assault? Given the distinct possibility of a conspiracy to orchestrate 9-11, one cannot help but wonder if the NBC report was designed to distract attention. Gaines states: "The stereotypical images of Arab, mid-eastern-looking people celebrating on a street could be falsely anchored to a specific people from a designated time and place" (127). With the eyes of the world firmly fixed upon Islamic extremism as the enemy, the true of criminals remained hidden behind a semiotic veil.

Citing Richard L. Lanigan, Gaines asserts: "Fiction and nonfiction are both mediated popular texts-the convergence of human experience expressed through technology" (127). That the chief means of deception is technological in nature is intentional. The word "technology" is derived from the Greek word techne, which means "craft." Moreover, the term "craft" is also associated with witchcraft or Wicca. From the term Wicca, one derives the word wicker (Hoffman 63). Examining this word a little closer, researcher Michael Hoffman explains: "The word wicker has many denotations and connotations, one of which is 'to bend,' as in the 'bending' of reality" (63). This is especially interesting when considering the words of Mark Pesce, co-inventor of Virtual Reality Modeling Language. Pesce writes: "The enduring archetype of techne within the pre-Modern era is magic, of an environment that conforms entirely to the will of being" (Pesce). Through the magic of electronic media, the post-September 11th environment seemed to conform entirely to the will of the elite.

The Druid magicians of antiquity used to carry wands, which were made out of "holly wood." Does this sound familiar? The famous Hollywood sign is but an enormous semiotic marker for an industry that specializes in illusion. Independence Day could be considered just one more of its spells. Given the public compliance to the illusion of the so-called "War on Terror," it would seem that the spell is working. Through the alchemical sorcery of electronic media, America's consciousness remains immersed within the semiotic mirage of the post-911 culture.

The War on Terrorism: Preemptive War in Disguise

According to the sociopolitical Darwinism of the global elite, preemptive warfare is central to the evolutionary development of humanity. Anisa Abd el Fattah explains:

The idea of preventive wars, which we now call preemptive strikes, became popular during the rise of Social Darwinism and Eugenics, and led to the mass killings of those deemed weak, handicapped, poor and of inferior races throughout Asia, Europe, and the European colonies in Africa. The idea of perpetual war, and disaster as a means by which to accelerate the evolution of the human species was also popular during that era, as it is now. (No pagination)

The neoconservative faction of the elite has made preemptive war the centerpiece in their plan to establish a scientific dictatorship before the other competing bluebloods beat them to the finish line. Moreover, they have created the perfect pretext for employing this method: the phony "War on Terrorism."

Unlike the many other manufactured foes throughout history, international terrorism is not centralized within the borders of nation-states or easily reducible to a single entity. In this sense, international terrorism exhibits a sinister synchronicity with the elite. Like the ruling class, terrorism is a supranational institution. Thus, it is an ideal machination of the technocratic conspiracy and is instrumental in the tangible enactment of the elite's occult Darwinian doctrine.

Unlike the many other manufactured foes throughout history, international terrorism is not centralized within the borders of nation-states or easily reducible to a single entity. In this sense, international terrorism exhibits a sinister synchronicity with the elite. Like the ruling class, terrorism is a supranational institution. Both qualify as "virtual states." Philip Bobbitt defines this term:

> The virtual state has many of the characteristics of other states (a trained standing army and intelligence cadre; a treasury and a source of revenue; a civil service and even a rudimentary welfare system for the families of its fighters) but is borderless; it declares wars, makes alliances with other states and is global in scope but lacks a definable location on the map. (No pagination)

Zbigniew Brzezinski, President Carter's national security advisor, played no small role in the engineering of the terrorist threat that resulted in September 11. His 1997 book, *The Grand Chessboard: American Primacy and Geostrategic Objectives* essentially constitutes an open admission of guilt. He begins his elitist tract with the following observation:

> The last decade of the twentieth century has witnessed a tectonic shift in world affairs. For the first time ever, a non-Eurasian power has emerged not only as a key arbiter of Eurasian power relations but also as the world's paramount. The defeat and collapse of the Soviet Union was the final step in the rapid ascendance of a Western Hemisphere power, the

United States, as the sole and, indeed, the first truly global power . . . (xii)

According to Brzezinski, this emergent American empire can only maintain its primacy as the sole global scientific dictatorship through the imperialistic extension of its power. This extension involves the seizure and consolidation of geostrategic resources. In particular, Brzezinski cites Eurasia as geostrategically axial in this campaign of imperialism:

But in the meantime, it is imperative that no Eurasian challenger emerges, capable of dominating Eurasia and thus of also challenging America. The formulation of a comprehensive and integrated Eurasian geostrategy is therefore the purpose of this book. (xiv)

However, Brzezinski identifies a distinct threat to this campaign: "The attitude of the American public toward the external projection of American power has been much more ambivalent" (24). Brzezinski reiterates this fear later and in much more elitist language:

It is also a fact that America is too democratic at home to be autocratic abroad. This limits the use of America's power, especially its capacity for military intimidation. Never before has a populist democracy attained international supremacy. But the pursuit of power is not a goal that commands popular passion, except in conditions of a sudden threat or challenge to the public's sense of domestic well-being. The economic self-denial (that is, defense spending) and the human sacrifice (casualties, even among professional soldiers) required in the effort are uncongenial to democratic instincts. Democracy is inimical to imperial mobilization. (35)

Evidently, America's constitutional republican system and its natal revulsion towards imperialism is a threat to the extension of the empire. Worse still, Brzezinski claims that a renewed adherence to the American anti-globalist principles could result in doomsday:

America's withdrawal from the world or because the sudden emergence of a successful rival – would produce massive international instability. It would promote global anarchy. (30)

Thus, global stability stipulates America's supremacy abroad:

Without sustained and directed American involvement, before long the forces of global disorder could come to dominate the world scene. (194)

In an America of such vast racial diversity, a foreign policy stipulating such a campaign against the "dysgenic races" abroad is not likely to prompt popular support. In fact, it would justifiably provoke moral outrage. At this pivotal juncture, Brzezinski presents a solution:

Moreover, as America becomes an increasingly multi-cultural society, it may find it more difficult to fashion a consensus on foreign policy issues, except in the circumstance of a truly massive and widely perceived direct external threat. (211)

Brzezinski was involved in the development of just such a "direct external threat" years before he penned these words. This much he candidly admitted in an interview with a French magazine called *Le Nouvel Observateur*:

Q: The former director of the CIA, Robert Gates, stated in his memoirs ["From the Shadows"], that American intelligence services began to aid the Mujahadeen in Afghanistan 6 months before the Soviet intervention. In this period you were the national security adviser to President Carter. You therefore played a role in this affair. Is that correct?
Brzezinski: Yes. According to the official version of history, CIA aid to the Mujahadeen began during 1980, that is to say, after the Soviet army invaded Afghanistan, 24 Dec 1979. But the reality, secretly guarded until now, is completely otherwise: Indeed, it was July 3, 1979 that President Carter signed the first directive for secret aid to the opponents of the pro-Soviet regime in Kabul. And that very day, I wrote a note to the president in which I explained to him that in my opinion this aid was going to induce a Soviet military intervention.
Q: Despite this risk, you were an advocate of this covert action. But perhaps you yourself desired this Soviet entry into war and looked to provoke it?
B: It isn't quite that. We didn't push the Russians to intervene,

but we knowingly increased the probability that they would.
(No pagination)

The Soviet invasion of Afghanistan provided a catalyst for
Brzezinski's terrorist manufacturing project. Under the pretext of
education, Afghan children were propagandized and transformed into
a generation of potential "direct external threats." This project was
exposed in an article in the *Washington Post*:

> In the twilight of the Cold War, the United States spent millions
> of dollars to supply Afghan schoolchildren with textbooks
> filled violent images and militant Islamic teachings, part of
> covert attempts to spur resistance to the Soviet occupation.
> THE PRIMERS, which were filled with talk of jihad and
> featured drawings of guns, bullets, soldiers and mines,
> have served since then as the Afghan school system's core
> curriculum. Even the Taliban used the American-produced
> books, though the radical movement scratched out human
> faces in keeping with its strict fundamentalist code. (Stephens
> & Ottaway, no pagination)

Various governmental and educational organizations were involved
in this project:

> Published in the dominant Afghan languages of Dari and
> Pashtu, the textbooks were developed in the early 1980s
> under an AID [Agency for International Development]
> grant to the University of Nebraska-Omaha and its Center
> for Afghanistan Studies. The agency spent $51 million on the
> university's education programs in Afghanistan from 1984 to
> 1994. (Stephens & Ottaway, no pagination)

The material circulated by this campaign was replete with violent
images and language:

> Children were taught to count with illustrations showing
> tanks, missiles and land mines, agency officials said. They
> acknowledged that at the time it also suited U.S. interests to
> stoke hatred of foreign invaders. (Stephens & Ottaway, no
> pagination)

According to the article's authors, the material shocked and disturbed some: "An aid worker in the region reviewed an unrevised 100-page book and counted 43 pages containing violent images or passages" (Stephens & Ottaway, no pagination). The article elaborates:

One page from the texts of that period shows a resistance fighter with a bandolier and a Kalashnikov slung from his shoulder. The soldier's head is missing.
Above the soldier is a verse from the Koran. Below is a Pashtu tribute to the mujaheddin [sic], who are described as obedient to Allah. Such men will sacrifice their wealth and life itself to impose Islamic law on the government, the text says. (Stephens & Ottaway, no pagination)

After Afghanistan's population was sufficiently radicalized, the county was used as a base of operations for the dissemination of this new violent form of Islam to the rest of the Arab world. Ahmed Rashid pointed this out in his article for *Foreign Affairs* magazine entitled "The Taliban: Exporting Extremism." In the article, Rashid writes:

With the active encouragement of the CIA and Pakistan's ISI, who wanted to turn the Afghan jihad into a global war waged by all Muslim states against the Soviet Union, some 35,000 Muslim radicals from 40 Islamic countries joined Afghanistan's fight between 1982 and 1992. Tens of thousands more came to study in Pakistani madrasahs. Eventually more than 100,000 foreign Muslim radicals were directly influenced by the Afghan jihad. (No pagination)

Many of the children who were radicalized by these textbooks were recruited by Al-Qaeda during adulthood. The head of this terrorist network is Osama bin Laden, the heir to a Saudi construction fortune. In 1979, Bin Laden went to Afghanistan to fight the Soviets (Moran, no pagination). Osama came to head the Maktab al-Khidamar, also known as the MAK (no pagination). This organization would act as a front through which money, arms, and fighters were supplied for the Afghan war (no pagination). According to MSNBC's Michael Moran, hidden puppeteers controlled the MAK:

What the CIA bio conveniently fails to specify (in its unclassified form, at least) is that the MAK was nurtured by Pakistan's state security services, the Inter-Services Intelligence agency, or ISI, the CIA's primary conduit for conducting the covert war against Moscow's occupation. (No pagination)

Even after the war in Afghanistan was over, Bin Laden was still regarded by the CIA as an admirable freedom fighter:

Though he has come to represent all that went wrong with the CIA's reckless strategy there, by the end of the Afghan war in 1989, bin Laden was still viewed by the agency as something of a dilettante-a rich Saudi boy gone to war and welcomed home by the Saudi monarchy he so hated as something of a hero. (Moran, no pagination)

In his article entitled "Bin Laden Comes Home to Roost," Moran suggests that Osama was propped up by the Agency for reasons other than doing battle with our Cold War nemesis:

The CIA, ever mindful of the need to justify its "mission," had conclusive evidence by the mid-1980s of the deepening crisis of infrastructure within the Soviet Union. The CIA, as its deputy director Robert Gates acknowledged under congressional questioning 1992, had decided to keep that evidence from President Reagan and his top advisors and instead continued to grossly exaggerate Soviet military and technological capabilities in its annual "Soviet Military Power" report right up to 1990. (No pagination)

The Agency wished to keep Osama in the game in spite of his irrelevance in the Cold War crusade against Communism. In fact, it was so important to the CIA that they were willing to present a fraudulent assessment of Soviet military capabilities to the President. With all pretenses removed from the picture, a disturbing reality emerges. The elite are following an evolutionary vision that thrives on conflict. When following such a script, one can never have too many enemies to thrust the narrative towards its climax. With the specters of fascism and communism supposedly vanquished, a suitable substitute had to be invented. Bin Laden supplied just such a substitute.

Bin Laden's terrorist network provided the neoconservatives with an opportunity to radically transform America's approach to national security. Through their figurehead, George W. Bush, the neoconservatives introduced a document that would make preemptive warfare an official policy option. Entitled *National Security Strategy of the United States*, this document states:

> The United States has long maintained the option of preemptive actions to counter a sufficient threat to our national security. The greater the threat, the greater is the risk of inaction—and the more compelling the case for taking anticipatory action to action to defend ourselves, even if uncertainty remains as to the time and place of the enemy's attack. To forestall or prevent such hostile acts by your adversaries, the United States will, if necessary, act preemptively. (No pagination)

The neoconservative doctrine of preemptive warfare was put into practice with the 2003 Iraq invasion. Looking at documents from the neoconservative Project for a New American Century (PNAC) reveals that Iraq had long been a target for military occupation. One example is a PNAC "open letter" to Bill Clinton in 1998. The letter states:

> We are writing you because we are convinced that current American policy toward Iraq is not succeeding, and that we may soon face a threat in the Middle East more serious than any we have known since the end of the Cold War. In your upcoming State of the Union Address, you have an opportunity to chart a clear and determined course for meeting this threat. We urge you to seize that opportunity, and to enunciate a new strategy that would secure the interests of the U.S. and our friends and allies around the world. That strategy should aim, above all, at the removal of Saddam Hussein's regime from power. We stand ready to offer our full support in this difficult but necessary endeavor. (No pagination)

The Iraq invasion is part of a larger neoconservative plan to achieve imperial dominance over the Gulf region. This was admitted in the September 2000 PNAC report entitled *Rebuilding America's Defenses*. The report states:

> Indeed, the United States has for decades sought to play a more permanent role in Gulf regional security. While the unresolved conflict with Iraq provides the immediate justification, the

need for a substantial American force presence in the Gulf transcends the issue of the regime of Saddam Hussein. (14)

Will the neoconservatives' imperial approach allow them to establish a global scientific dictatorship before the other factions of the power elite? There are some who are not convinced. One dissenting voice is former Secretary of State Colin Powell. Powell correctly observed that the neoconservatives' imperial approach would anger other excluded power elite factions. In an attempt to prevent such a situation, Powell opposed the Defense Secretary Donald Rumsfeld and tried to build a United Nations coalition for the invasion (Jeffery, no pagination). These efforts failed and the neoconservatives won the day.

Powell's failure to prevent American imperialism may have led him to fear a potential global war. As sensational as this may sound, some of Powell's behavior reinforces the contention. In a discussion with Jack Straw, Powell allegedly called Washington neoconservatives "fucking crazies" (Jeffery, no pagination). Simon Jeffery also recounts an encounter between Powell and foreign diplomat:

A foreign diplomat encountered the secretary of state on the eve of the Iraq war and recited a news report that the president was sleeping like a baby. Mr Powell reportedly replied: "I'm sleeping like a baby, too. Every two hours, I wake up, screaming." (no pagination)

Such a statement is certainly suggestive of apprehension on Powell's part. That apprehension should not be taken lightly. The imperialistic features of the neoconservative strategy may not result in an American empire. Instead, those features may instigate hostility from countries with the military clout necessary to face the United States in a war. Like an acrobat performing without a net, the neoconservatives are playing a dangerous game.

The outcome of the neoconservatives' power bid notwithstanding, Osama bin Laden will most likely never be captured. An elusive Bin Laden is more profitable to the power elite than a Bin Laden facing a war crimes tribunal. At least one member of the ruling elite, former CIA executive director A.B. "Buzzy" Krongard, has publicly stated that Bin Laden's capture "might prove counter-productive" (Allen-Mills, no pagination). Krongard tried to justify his position by claiming that Bin Laden's capture would result in a power struggle among Al-Qaeda subordinates that would unleash a wave of terrorist attacks (Allen-Mills, no pagination).

However, Krongard's argument strains credulity to the breaking point. In all likelihood, Osama's apprehension would result in a terrible loss of morale for Al-Qaeda and disorganization stemming from a serious leadership void. Truth be told, a scientific dictatorship needs enemies to provide a pretext for suppressing people at home and abroad. An invisible Bin Laden, who can be everywhere and nowhere at the same time, provides just such a pretext.

Red China

For many years now, the United States government has taken the engagement approach to Communist China, hoping that gestures of appeasement would result in China's volitional liberalization. Two authors who argue for this position are Daniel Burstein and Arne J. De Keijzer. In their book, *Big Dragon*, Burstein and Keijzer propose a policy of "Dynamic Engagement" (355). The authors contend that it will end the cold war with China (355). The two then enumerate supposed beneficiaries of such engagement.

First on the list of those who would supposedly benefit is American business, which will have better access to the China market. The atmosphere will change from one of hostility to one of cooperation and mutual respect. China will someday have the largest economy in the world. Therefore, to be a partner with and play a role in the development of China's economy will lead to great gains when China reaches this status (355).

America will not only gain economically, but politically as well. Washington's ability to persuade and influence China will increase considerably, especially when it comes to matters such as Asian and global security. In so doing, America will ease China's adjustment to "the very positive emerging world order of recent year" (356). "Dynamic engagement", contends Burstein and Keijzer will also benefit the Chinese people. A more market-oriented China will somehow lead to a more democratic China, because with Western money and business will come Western values concerning political liberties and human rights. Burstein and Keijzer hold that the Chinese are already very receptive to Western ideas and concepts. They write:

Despite political rhetoric of recent years, the Chinese people admire the freedom and creativity of American culture and lifestyles, our pioneering spirit, our open society, and many of our ideals. (356)

Because of this deeply embedded affinity for America, China holds the innate potential of someday becoming a truly democratic society (356). However, shifting China's paradigm can only be accomplished through bilateral engagement.

Finally, Keijzer and Burstein reiterate the all-purpose mantra of global "peace and prosperity," claiming that the dormant synergies and potentially rewarding prospects of U.S.-China relations are hindered by the "politics of confrontation" (356-57). Resuscitating the traditional alarmist sentiments of the 60s anti-war movement, the authors claim that U.S.-China relations have assumed an escalatory trajectory towards "a new cold war– bordering on a hot one" (357). In order to avoid this inexorable descent into warfare, Americans must "reconceptualize" their relationship with China and abandon the ugly notions promulgated by the "China Threat school" (357). In other words, America's national security concerns must be ignored and the Chinese must be appeased . . .or else.

Ironically, Red China's rise to power is the result of such policies of dynamic appeasement. In fact, the country's transformation into a scientific dictatorship is directly attributable to the Truman administration's attempts to appease Josef Stalin. Six days before Japan's surrender, the Soviet Union was permitted entry into the Pacific theatre. "Uncle Joe" Stalin was promised the Northern Chinese Province of Manchuria in return for the Soviet Union's entry. When great amounts of Japanese military hardware were captured, they were handed over to Mao and his communist guerrillas. Manchuria was to become a staging ground for the communist acquisition of China. Three months before Potsdam, Truman was advised by fifty top Army intelligence officers through General George C. Marshall against just such an action. They stated:

"The entry of Soviet Russia into the Asiatic war would be a political event of world-shaking importance, the ill effects of which would be felt for decades to come.... [It] would destroy America's position in Asia quite as effectively as our position is now destroyed east of the Elbe and beyond the Adriatic.

"If Russia enters the Asiatic war, China will certainly lose her independence, to become the Poland of Asia; Korea, the Asiatic Rumania; Manchukuo, the Soviet Bulgaria. Whether

more than a nominal China will exist after the impact of the Russian armies is felt is very doubtful. Chiang may have to depart and a Chinese Soviet government may be installed in Nanking which we would have to recognize.
"To take a line of action which would save few lives now, and only a little time-at an unpredictable cost in lives, treasure, and honor in the future-and simultaneously destroy our ally China, would be an act of treachery that would make the Atlantic Charter and our hopes for peace a tragic farce.
"Under no circumstances should we pay the Soviet Union to destroy China. This would certainly injure material and moral position of the United States in Asia." (Hoar 254)

Instead of listening to the intelligence team, Truman allowed himself to fall under the influence of Owen Lattimore, whose concepts made up U.S. policy concerning post-war China (254-55). Lattimore would later be identified by an investigating Senate Subcommittee as a communist subversive (76). Besides this, he was also a member of the Institute of Pacific Relations, a subversive outfit that received millions of dollars from the Rockefeller and Carnegie foundations (76). The Institute consistently depicted Chiang as a dictator. Mao was played up as an "agrarian reformer" and not a communist. One who obviously fell for this line was George Marshall, who stated: "Don't be ridiculous. These fellows are just old-fashion agrarian farmers" (Flynn 14). Nothing could be further from the truth. In fact, Mao considered himself a full-fledged Marxist in late 1919. In 1921, he organized a small communist group in Changsha. In addition, that year Mao participated in the First National Party Congress of the Chinese Communist party. It was at this meeting that the Party was formally brought into existence (Dietrich 19).
The Institute's leaders also published a magazine called *Amerasia*. The FBI conducted a raid on the magazine's offices and found no less than 1800 stolen government documents. An investigation by the Senate Committee on the Judiciary led to the following declaration:

The Institute of Pacific Relations was a vehicle used by the Communists to orient American Far Eastern policies toward Communist objectives. Members of the small core of officials and staff members who controlled IPR were either Communist or pro-Communist. . . (Courtney 51)

Standing firmly against the communists was Chiang Kai-shek, disciple of Sun Yat-sen and one of those who were instrumental in the overthrow of the corrupt Manchu dynasty. In 1923, Chiang was sent by Dr. Sun Yat-sen to the Soviet Union to study the Bolshevik system. Chiang's first-hand experience compelled him to write: "I became more convinced than ever that Soviet political institutions were instruments of tyranny and terror" (Perloff 36). Chiang became dedicated against the communist cause. Initially his crusade against the communists was successful. In 1946, the Nationalists were winning against the communists. If allowed to continue, the Nationalists would, in a very short time, wipe the communists clean from China. However, General Marshall was dispatched by Truman to China to make sure that this was not the case. Marshall forced Chiang to agree to a cease-fire and let Mao and his forces retain what they had acquired in Manchuria (Hoar 255). Chiang would find that Marshall was quite antagonistic to the Nationalist cause. In one of his diary entries, he observed that Marshall:

> "continues to try to accommodate the Communists in every possible way and force us to make concessions. He doesn't seem to care whether China survives or perishes. This indeed is a most painful situation." (Perloff 40)

Chiang found out just how little Marshall did care in July of 1946 when the General clamped an embargo on the sales of ammunition and arms to China (Hoar 255). In 1948, when the China situation had almost reached the peak of desperation, Congress voted $125 million in military aid to Chiang. However, it was all for not, as the Truman administration successfully delayed its execution a full nine months, during which time China collapsed (Utley 44-45). These treacherous actions did not go unnoticed. On January 25, 1949, then Congressman John F. Kennedy stated before the House of Representatives:

> Mr. Speaker, over this weekend we have learned the extent of the disaster that has befallen China and the United States. The responsibility for the failure of our foreign policy in the Far East rests squarely with the White House and the Department of State. The continued insistence that aid would not be forthcoming, unless a coalition government with the Communists was formed, was a crippling blow to the National Government. (Burns 80)

China's transformation into a scientific dictatorship had begun. Over the next few decades, the size and strength of her military infrastructure would steadily grow. This growth would be fostered, in large part, by the sociopolitical Darwinians of the American Establishment. Many of these globalists were members of the infamous occult secret society, the Order of Skull and Bones. Sutton provides the rationale for the Order's support of Red China:

> By about the year 2000 Communist China will be a "superpower" built by American technology and skill. It is presumably the intention of The Order to place this power in a conflict mode with the Soviet Union. (*America's Secret Establishment* 181)

During the Cold War, the Order hoped to use China as a geopolitical counterbalance to the growing Soviet scientific dictatorship. Within the dialectical climate of the time, there was the very real possibility that Russia would become too difficult to manage. The Order believed that, should the Soviets decide to assert themselves militarily against the West, China would cast her lot in with the United States.

The Korean War saw the Truman administration's same mismanagement and irresponsibility. When South Korea was invaded, Truman announced:

> . . .I have ordered the Seventh Fleet to prevent any attack on Formosa. As a corollary of this action, I am calling upon the Chinese Government on Formosa to cease all air and sea operations against the mainland. The Seventh Fleet will see that this is done. (*American Foreign Policy, 1950-55: Basic Documents* 2468)

General Douglas MacArthur explained what this brought about:

> The possibility of Red China's entry into the Korean War had existed ever since the order from Washington, issued to the Seventh Fleet in June, to neutralize Formosa, which in effect protected the Red China mainland from attack by Chiang Kai-shek's forces of a half a million men.
> This released the two great Red Chinese armies assigned the

coastal defense of central China and made them available for transfer elsewhere. (Hunt 380)

To prevent Chinese entry into the war, MacArthur ordered the bombings of the bridges across the Yalu. This would have effectively kept the Chinese from crossing over into Korea. However, General Marshall came to the rescue for the Chinese by reversing the order. This led to MacArthur stating:

> I realized for the first time that I had actually been denied the use of my full military power to safeguard the lives of my soldiers and the safety of my army. To me, it clearly foreshadowed a future tragic situation in Korea, and left me with a sense of inexpressible shock. (Willoughby and Chamberlain 402)

The commander of the Chinese force, General Lin Piao, would go on to state:

> I never would have made the attack and risked my men and my military reputation if I had not been assured that Washington would restrain General MacArthur from taking adequate retaliatory measures against my lines of supply and communication. (MacArthur 375)

Nixon, too, would continue the trend of collaboration with the communist Chinese. In 1971, Henry Kissinger began secretly negotiating with Beijing to arrange a trip for Nixon to the communist-dominated country. Just a week after the negotiations, Nixon announced that he would soon visit China. As Dietrich points out: "Nixon, the fierce anti-communist, and Mao, the archfoe of capitalism—had executed a dramatic about-face" (211). Nixon would be meeting with Chou En-lai. This same man admitted to China's twenty-year plan to spread drug addiction in the United States in a 1965 conversation with Egyptian President Nasser. Mohammed Heikal provides a direct quote in *The Cairo Documents*:

> Some of them [American soldiers in Vietnam] are trying opium. And we are helping them. . .Do you remember when the West imposed opium on us? They fought us with opium. And we are going to fight them with their own weapons. .

.The effect this demoralization is going to have on the United States will be far greater than anyone realizes. (306-07)

The paragons of political correctness and the self-anointed "experts" of orthodox academia largely hold that Chinese involvement in the drugging and demoralization of America is either Taiwanese propaganda or baseless "conspiracy theory". However, the evidence is imposing. In his book *Red Cocaine: The Drugging of America*, Joseph D. Douglass convincingly argues the case using information given to him by Czechoslovakian defector Jan Sejna. Commenting on *Red Cocaine*, former Deputy Director for Intelligence Dr. Ray S. Cline stated:

Dr. Joseph Douglass, the author of this book, is not selling a theory but instead calling attention to evidence. He has marshalled his facts carefully, presents them responsibly and cautiously, and offers a wealth of soberly documented data. That data describes in detail the efforts of China, the Soviet Union, and its many surrogates, to use drugs over many decades as weapons designed to damage and weaken-if not destroy-the stability of Free World countries. The top target is and always has been, of course, the United States. (xvii)

Why, then, would the United States government hold a friendly dialogue with the communist government of China? Dr. Cline explains:

If we are serious about winning this war on drugs, we must know, too, to what extent it is true--as this book argues-- that top officials in our government have had access to this evidence for many years, but preferred to hush it up out of concern for what public disclosure would do to U.S.-Sino/ Soviet relations. (xviii-ix)

In spite of the fact that the communist Chinese government had obviously taken an adversarial position towards the United States, Nixon and Kissinger decided to deal. Exposure of the PRC's drug trafficking operations would have jeopardized the Establishment's geopolitical objectives in China. The Red Dragon was supposed to act as a dialectical arm of the Western elite, countering the slightly unpredictable Soviet scientific dictatorship. The dialogue established by Nixon and Kissinger set the stage for establishing relations with the communist Chinese government in 1979. The sociopolitical Darwinians of the West were playing a dangerous game with their Eastern counterparts.

With the tributaries of open diplomacy cleared, America's self-immolating tradition of bilateral engagement continued its seamless procession into the 90's. This time, a former governor from Arkansas would maintain the custom of appeasement. President Bill Clinton provided the ideal catalyst for the latest and, arguably, the most damaging compromise of national security. A report issued by Senator Fred Thompson's Governmental Affairs Committee reveals the reasons why the President entertained a treasonous course of action:

> "On November 8, 1994, Americans shifted control of both houses of Congress to the Republican Party for the first time in 40 years. For a time, the election rendered President Clinton so weak in the polls that many experts questioned his 'relevance,' suggesting that he might face a primary challenge as he attempted to secure his re-election in 1996. The election results spurred great concern among the President's supporters that he might suffer a similarly disastrous defeat in 1996... The President and his advisors determined that the key to their success in the 1996 elections would be to wage immediately a massive television political advertising campaign of unprecedented cost." (Jasper, "Beijing Bailout," 9)

Fearing that the upset in both Congressional houses was an ill omen of things to come for his Presidency, Clinton realized that drastic measures had to be employed. According to the Senate report, Clinton and his strategists proceeded to develop "a legal theory to support their needs and proceeded to raise and spend $44 million in excess of the Presidential campaign spending limits" (9). Many of the monetary sources were illegal. Many of the illegal sources were foreign. Many of the foreign donors were channeling money into the Clinton-Gore fund from China.

One group of foreign donors was the Riady family. James Riady and his wife were the biggest contributors to the Clinton-Gore ticket in 1992, giving a whopping $450,000 dollars to the election effort (Timperlake and Triplett 7). As the campaign drew to a close, the Riady family, along with associates and executives in Riady companies, gave an additional $600,000 to the DNC and Democratic state parties (7). When it was time to celebrate, Riady and his employee John Huang each gave $100,000 for the cost of the Clinton-Gore 1993 inauguration (13).

Who exactly are the Riadys? They are ethnic Chinese whose center of operations lies in Indonesia (7-8). Their corporate flagship is the Lippo Group (9). The patriarch of the family empire is Mochtar Riady, the father of Clinton's biggest contributor, James Riady (7). Mochtar visited the United States frequently, and James was a permanent resident (7). Mochtar's other son, Stephen, was educated in the United States and worked in California in the early 1980s (7). However, all of the Riadys have mysteriously left the United States. Many Riady employees that had comprehensive information concerning the family's activities in the United States have also exited the scene. The only Riady operative with detailed data over the family left here in the United States is John Huang, and he is not talking. Huang has pleaded the Fifth Amendment, claiming that sharing what he knows would be tantamount to self-incrimination (7).

Why all the secrecy, one might ask. In their detailed and carefully documented book, *Year of the Rat*, member of the professional staff of the House Committee on Rules Edward Timperlake and former Republican counsel to the Senate Foreign Relations Committee William C. Triplett state: "The Riady's chief partners in China (including Hong Kong)-China Resources and the China Travel Service-are government-owned companies that accommodate or serve as an extension of Chinese military intelligence" (18). One of these arms of Chinese military intelligence, China Resources, came to the Riady's rescue when their bank, LippoLand, was about to go belly-up and bring the entire Riady empire crashing down like Humpty Dumpty (17). Timperlake and Triplett state:

> What truly saved the bank was a timely purchase of Lippo shares by the Riady's chief Chinese partner, China Resources. The share purchase was not large-5 percent of LippoLand-but it was enough to restore confidence and bring in other investors. (17)

The CIA also provided information concerning the Riady's relationship with Chinese intelligence. The agency revealed the following to an investigating Senate Committee:

> The Committee has learned from recently acquired information that James and Mochtar Riady have had a

long-term relationship with a Chinese intelligence agency. The relationship is based on mutual benefit, with the Riady receiving assistance in finding business opportunities in exchange for large sums of money and other help.

Although the relationship appears based on business interests, the Committee understands that the Chinese intelligence agency seeks to locate and develop relationships with information collectors, particularly with close association to the U.S. government. (18)

A statement made by one ex-Lippo executive seems to indicate that the Riadys intended to fill the role of "information collector" for a Chinese intelligence agency: "Riady's goal was to sell his relationship with Clinton to two governments, Indonesia and China" (19). Do the Riadys have strong enough ties to the CCP to suggest that they were participating agents of the CCP and its intelligence service? The evidence already presented here is very compelling. Besides what was previously discussed, in his Hong Kong office, Mochtar Riady supposedly has a gold-framed picture of Chinese Politburo member Li Peng right next to one of the Clintons (19). However, while a picture may say a thousand words, this can hardly be considered proof, let alone evidence. Is there something we can look to that is more substantial?

The answer, unfortunately, is a resounding yes. In 1997 a senator posed a question to the CIA concerning relationships between the Riadys and Beijing officials. The CIA revealed that almost all of the Riadys joint ventures in China were "with local, regional and central governments in China." The CIA went on: "Lippo has substantial interests in China-about US$2 billion in the Riady's ancestral province of Fujian alone. These include real estate, banking, electronics, currency exchange, retail, electricity, and tourism". The CIA also stated: "Lippo has provided concessionary-rate loans to finance many of these projects in key [Communist] Party members' home areas." (16-17)

Lippo's top U.S. agent was John Huang. Huang's membership in Lippo was largely the result of all the right elements converging at once. In September 1983, Huang joined the Union Planters Bank of Memphis to facilitate a "correspondent relationship with LippoBank and other business ties to the Riadys" (24-25). Union Planters had assigned Huang the task of opening a representative office in Hong Kong (25).

During his assignment to Hong Kong, Huang made extensive sojourns throughout Asia, "broadening his contacts with officials in China, Japan, and Korea" (25). Yet, with little agricultural trade business to support Union Planter's Hong Kong office, John Huang soon found himself floating amidst the flotsam and jetsam of a disintegrated banking operation (25). The Union Planter's Hong Kong office closed, leaving Huang to the mercy of a marketplace devoid of prospects (25). It was at this precarious juncture of John's life that he was recruited by the Riadys (25).

The Riady's recruitment of Huang would prove to be an invaluable investment. John exhibited exceptional social skills, thus making him instrumental in the facilitation of "business developments" (25). During his stay at a law firm, Huang was dubbed a "rainmaker," a veritable lodestone attracting new business (26). Yet, this Lippo asset would most convincingly prove his weight in gold in March 1985, when the "rainmaker" shifted Hong Kong's attention towards a Bohemian Arkansas governor and Riady family friend (26). While escorting Riady clients to the Democratic National Convention in Atlanta, John would meet the governor again (26). Huang had identified the locus of the next major Riady project: Bill Clinton.

By September 1996, a time noticeably close to the U.S. presidential elections, the Los Angeles Times ran a story revealing Huang's illegal fund-raising activities on behalf of the DNC (26). Overall, Huang's financial harvest for Clinton and the Democrats exceeded $2.7 million, the majority of which was generated by illegal, foreign sources (Jasper, "Beijing Bailout," 11). The various contributors held connections to "organized criminal syndicates (Triads), narcotics trafficking, gambling, prostitution, the Chinese military, and all of Communist China's intelligence services" (Timperlake and Triplett 30). However, there are far more disturbing revelations surrounding this scandal.

Two years earlier in January 1994, under the pretext that Commerce Secretary Ron Brown urgently required the Lippo agent's assistance, Huang "received an interim 'Top Secret' security clearance" (30). According to the testimony of a Commerce Department security officer before the Senate Governmental Affairs Committee, Huang's acquisition of this clearance represented an unprecedented breach of protocol (31). The officer testified: "no other consultant on the Department of Commerce payroll was ever granted top security clearance" (31). Still, standard operating procedure was circumvented and Huang's "Top Secret" access continued with the blessing of Bill Clinton.

Yet, Huang 's career at Commerce Department did not begin for another five-and-a half months, a time period during which the Lippo representative still had legal access to classified material (30). From July

18, 1994 to early December 1995, Huang occupied a special position in the Commerce Department and enjoyed further access to highly classified information (30). Huang maintained this "Top Secret" access during his 1996 fund-raising campaign for the DNC, a time during which the Lippo agent was soliciting aid from illegal, foreign sources (30).

The full volume of sensitive information Huang collected and disseminated is not clear, but House Rules Committee Chairman Gerald Solomon revealed the following on June 11, 1997:

> "I have received reports from government sources that say there are electronic intercepts which provide evidence confirming that John Huang committed economic espionage and breached our national security by passing classified information to his former employer, the Lippo Group."
> (Jasper, "Beijing Bailout," 11)

A CIA witness, the identity of whom was protected, was asked about Solomon's revelation during the Thompson Committee hearings (Timperlake and Triplett 43-44). However, the witness could not answer the question "in open session" without imperiling sensitive "sources and methods" (44). Considering the fact that the agency could have simply dismissed Solomon's statement as an unsubstantiated claim, this reply suggests that at least some modicum of factual weight rests in the HRC Chairman's assertion.

Huang's penetration of the Commerce Department was only made easier by the policies of the Clinton Administration. Shortly after Clinton entered the Oval Office, his administration began to effectively eviscerate the existing security system (31). Serious background checks were removed and security clearances were generously dispensed to virtually anyone, including Huang (31). In short, the corpse of America's national security infrastructure became the very bridge across which the agent of a hostile power traversed detection.

However, Huang's appointment to an important post within the DNC was not so smoothly executed. DNC chairman Don Fowler cringed at the prospect of Huang's involvement with the forthcoming television advertising campaign (65). He had good reason to respond so negatively. Buried in the DNC files was a March 15, 1994 letter from Lippo consultant and Democratic activist Maeley Tom to Fowler's predecessor, David Wilhelm (66). When the letter was unearthed, it revealed that Riady wanted to assemble "business leaders from East

Asia" and galvanize them as "a vehicle to raise dollars from a fresh source for the DNC" (66).

The potential criminality of such a plan is obvious. Contributions to federal elections can only be made by American citizens and permanent residents (66). The likelihood of these Asian business leaders being either of these is doubtful. Participants in such a plan would face charges of conspiracy and substantial prison sentences. Despite these legitimate fears, President Clinton personally interceded on Huang's behalf and induced the DNC's compliance on November 13 (65-66). Lippo's agent was now a major Democratic fund-raiser.

Concerning the collection of vital information, Huang was literally a sponge. The man received 37 classified personal briefings from CIA officers (49). The Thompson committee comprehensively delineated the ten types of significant intelligence items that the Lippo agent acquired:

1. Business opportunities in Vietnam.
2. Economic issues confronting Taiwan and China.
3. Investment opportunities in China.
4. North Korean food shortage.
5. Succession of power in China.
6. China technology transfers.
7. Nuclear power industry in Asia.
8. Investments in the China auto industry.
9. Investment climate in Hong Kong.
10. Chinese government influence on investment in China and Taiwan. (50)

According to Timperlake and Triplett, item #6 would be of particular interest to the PRC (51). China's military modernization program is maintained with American technology and the "Er Bu" (military intelligence) would eagerly welcome the opportunity to know whom and what was the object of CIA surveillance (51). Such knowledge would insure the security of programs within China's military industrial complex. Unhindered, the communist Dragon would continue its inexorable march towards parity with the United States.

Item #5 would be extremely useful to the Ministry of State Security and the CCP's United Front Works Department (51). Timperlake and Triplett elaborate:

In 1994 and 1995 Chinese paramount leader Deng Xiaoping

was in failing health, and the various Chinese leaders were contending for position in the post-Deng era. Understanding the CIA's analysis of the situation would have allowed them to manipulate it to their advantage. (51)

The PRC would have also noticed Item #9. With the PRC's impending conquest of Hong Kong on July 1, 1997, concerns about the flight of western capital from the region began to emerge (51). Understanding the CIA's analysis of Hong Kong would allow the PRC to regulate its conduct in certain areas of strategic significance (51). In other words, the West would only see what China wanted it to see.

Additionally, Huang attended 109 meetings where classified information may have been disseminated, including many at the White House (109). Secret Service records reveal that, between March 15, 1993 and July 18, 1994, Huang had set foot in the White House at least forty-seven times (27-28). How much Lippo's top agent had learned and the extent of the damage done to America's national security is still being assessed.

Why did Kenneth Starr focus on sexual innuendo as opposed to the Chinese connection to the Clinton Whitehouse? After all, shady dealings with a foreign nation move one beyond the realm of poor character to the much more serious domain of treason. Assertions made by investigator and court reformer Sherman Skolnick may provide an explanation:

> There was a stand-off between Clinton and supposed "independent" Counsel Kenneth W. Starr. They are both master blackmailers against each other. Result: Starr's work dribbled down, no treason, just sex and Monica. Starr had as a PRIVATE law client Wang Jun, the reputed head of the Red Chinese Secret Police. Starr is also reportedly the UNREGISTERED foreign lobbyist of the Red Chinese government. From time to time, Wang Jun visited Clinton in the White House Clinton reportedly gave him U.S. industrial, financial, and MILITARY secrets. (No pagination)

After presenting this information, Skolnick asks the following questions:

> So who was going to arrest who? Starr arrested by Clinton's Justice Department? Or Starr to have arrested and prosecuted

Clinton for treason with Starr's private law client, Wang Jun, being the common factor? (No pagination)

The answer to these questions is obvious. Both sides would maintain a "hands off" policy as far as the Red China connection was concerned. America would only be exposed to lurid details of Clinton's sexual depravity. Meanwhile, behind the scenes, Red China would continue down the path to becoming a much more powerful scientific dictatorship.

Other assessments of enemy penetration have been running concurrently with the investigations into Chinagate. In July 1998, the Cox Committee was created for the express purpose of determining whether or not Loral Space and Communications Ltd. and Hughes Electronics Corporation compromised national security by assisting China's military technicians in the development of missile systems that could target American cities. The release of its findings provoked an aggressive Chinese response. In a January 6, 1999 article ran in the *People's Daily*, director of the Information Office of the State Council Zhao Qizheng called the Cox Report, "a farce to instigate anti-China feelings and undermine Sino-U.S. relations" (no pagination).

Is this the case? Is the Committee composed of nothing but a bunch of mean, bigoted Americans who want to destabilize U.S. relations with China? An August 11, 1999 article from *Capitol Hill Blue* seems to suggest the reverse. According to the news item, an update to the Cox report found the following:

"Events since the release of the select committee report have confirmed some of its most disturbing conclusions about the PRC espionage threat facing the United States, the weakness of our efforts to counter it, and the threats to our national security that have resulted from it. With the stolen US technology, the PRC has leaped, in a handful years, from 1950s-era strategic nuclear capabilities to the more modern thermonuclear weapons designs." (Morahan, no pagination)

Moreover, in June 1999, the PRC announced that it would test a new submarine launched ballistic missile with a range of 7,500 miles (no pagination). This announcement verified a prediction made the Cox Report (no pagination). The PRC's announcement that it now had a neutron bomb also validated the Cox Report's contention that

American nuclear technology had been stolen (no pagination). So much for Qizheng's contention that the Cox Report was a "farce."

Some argue that the Clinton sell out to China was merely motivated by greed and monetary profit. This may be true for Clinton, but it is not true for his hidden backers. While it may give democrats heartburn to hear it, the fact is that Clinton was merely a front man for several elites. This cozy relationship with the power elite probably began as early as Clinton's bid for governor of Arkansas. Joel Bainerman elaborates:

> When Arkansas' governor Bill Clinton was running for governor in October 1990, his campaign was in deep financial trouble. He called on Jackson Stephens, a member of the Bush 100 Club in Little Rock, and the prime mover behind Harken Energy's bid to win a lucrative oil-drilling contract in Bahrain. Stephens helped Clinton raise nearly $100,000, and receive a $2 million line of credit from the Worthen National Bank. This was done, it is believed, through one of Stephens' associates, Curt Bradbury, a former employee of his who is now chief executive officer of Worthen National Bank. Jackson Stephens is chairman of the bank. (305)

Elite support of Clinton continued with his bid for the Presidency. Pamela Harriman, the widow Bush's Brown Brothers associate Averell Harriman, funded Clinton's 1992 presidential campaign to the tune of $12 million (Sloan, no pagination).

For these elites, the Clinton sell out to China may have meant much more than just making money. Clinton temporarily appeased the Chinese. This kept the Chinese cooperating with the integration process necessary for the erection of a world government. However, the Chinese will eventually grow tired with being mere subordinates or junior partners in the New World Order. At that point, they will be the ideal enemy for the western elites to use as pretext for creating a world government. In the name of defeating the China threat, a worldwide scientific dictatorship may rise.

Russia

Russia presents yet another potential enemy of the 21st century. However, she is a foe that the Western elite unintentionally promulgated. Initially, America's secret Establishment had hoped that, through the gradual liberalization of the Russian government, she could eventually merge with the West in a global "scientific dictatorship." Concurrently, a Fabian process of societal transformation would be enacted in the United States, preparing the former constitutional republic for her comfortable amalgamation into a one world socialist totalitarian government. In 1954, Reece Committee staff director Norman Dodd met with CFR member and Ford Foundation head Rowan Gaither. Dodd recounts the astonishing revelations made during this meeting:

> Mr. Gaither said, "Mr. Dodd, we have asked you to come up here today because we thought that possibly, off the record, you would tell us why the Congress is interested in the activities of Foundations such as ourselves." And before I could think of how I would reply to that statement, Mr. Gaither then went on voluntarily and stated, "Mr. Dodd all of us that have a hand in making policies here have had experience either with the OSS during the war, or the European Economic Administration. After the war we have had experience operating under directives, and these directives emanate, and did emanate from the White House. Now we still operate under just such directives. Would you like to know what the substance of these directives is?" I said, "Mr. Gaither, I'd like very much to know" Whereupon he made this statement to me, namely, "Mr. Dodd, we here operate in response to similar directives, the substance of which is that we shall use our grant making powers so to alter life in the United States that it can be comfortably merges with the Soviet Union." (Qutd. in Zahner 92)

Ostensibly, the "scientific dictatorship" of the Soviet Union fell in 1991. However, this "fall" was actually an exercise in cheap theatrics orchestrated to instill the Western "scientific dictatorship" with a false sense of victory. Such a move was announced as far back as the 1930s, when Dimitri Manuilski stated:

> "War to the hilt between communism and capitalism is inevitable. Today, of course, we are not strong enough to attack. Our time will come in 30 to 40 years. To win, we shall need the element of surprise. The bourgeoisie...will have to be put to sleep. So we shall begin by launching the most spectacular peace movement on record. There will be electrifying overtures and unheard of concessions. The capitalist countries, stupid and decadent, will rejoice to cooperate in their own destruction.

They will leap at another chance to be friends. As soon as their guard is down, we will smash them with our clenched fist." (McAlvaney 196-97)

In November 1987, Mikhail Gorbachev reiterated this idea. According to Sir William Stephenson, head of Combined Allied Intelligence Operations during the Second World War, Gorby said the following in a speech to the Politburo:

"Comrades, do not be concerned about all you hear about glasnost and perestroika and democracy in the coming years. These are primarily for outward consumption. There will be no significant internal change within Russia other for cosmetic purposes. Our purpose is to disarm the Americans and let them fall asleep. We want to accomplish three things: One, we want the Americans to withdraw conventional forces from Europe. Two, we want them to withdraw nuclear forces from Europe. Three, we want the Americans to stop proceeding with Strategic Defense Initiative." (201)

Ex-KGB officer Anatoliy Golitsyn foretold the false "liberalization" campaign in 1984. In *New Lies For Old*, Golitsyn wrote:

If in a reasonable time "liberalization can be successfully achieved in Poland and elsewhere, it will serve to revitalize the communist regimes concerned. The activities of the false opposition will further confuse and undermine the genuine opposition in the communist world. Externally, the role of dissidents will be to persuade the West that the "liberalization" is spontaneous and controlled. "Liberalization" will create conditions for establishing solidarity between trade unions and intellectuals in the communist and noncommunist worlds. In time such alliances will generate new forms of pressure against Western "militarism", "racism", and "military industrial complexes" and in favor of disarmament and the kind of structural changes in the West predicted in Sakharov's writings.

If "liberalization" is successful and accepted by the West as genuine, it may well be followed by the apparent withdrawal of one or more communist countries from the Warsaw Pact to serve as the model of a "neutral" socialist state for the whole of Europe to follow. Some "dissidents" are already speaking in these terms. (336)

Europe has already witnessed this condition of alleged "liberalization." As the former Soviet "scientific dictatorship" feigned immolation, its various socialist machinations remained intact under the guise of "social democracy." Encouraged by the prospect of a more comfortable and non-violent merger with its eastern counterpart, the "scientific dictatorship" of the West pledged its support to this counterfeit "liberalization" movement. Golitsyn continues:

Political "liberalization" and "democratization" would follow the general lines of the Czechoslovak rehearsal in 1968. This rehearsal might well have been the kind of political experiment Mironov had in mind as early as 1960. The "liberalization" would be spectacular and impressive. Formal pronouncements might be made about a reduction in the communist party's role; its monopoly would be apparently curtailed. An ostensible separation of powers between the legislative, the executive, and the judiciary might be introduced. The Supreme Soviet would be given greater apparent power and the president and deputies greater apparent independence. The posts of president of the Soviet Union and first secretary of the party might well be separated. The KGB would be "reformed". Dissidents at home would be amnestied; those in exile abroad would be allowed to return, and some would take up positions of leadership in government. Sakharov might be included in some capacity in the government or allowed to teach abroad. The creative arts and cultural and scientific organizations, such as the writers' unions and Academy of Sciences, would become apparently more independent, as would the trade unions. Political clubs would be opened to nonmembers of the communist parties. Censorship would be relaxed; controversial books, plays, films and art would be published, performed, and exhibited. Many prominent Soviet performing artists now abroad would return to the Soviet Union and resume their professional careers. Constitutional amendments would be adopted to guarantee fulfillment of the provisions of the Helsinki agreements and a semblance of compliance would be maintained. There would be greater freedom for Soviet citizens to travel. Western and United

Nations observers would be invited to the Soviet Union to witness the reforms in action.

But, as in the Czechoslovak case, the "liberalization" would be calculated and deceptive in that it would be introduced from above. It would be carried out by the party through its cells and individual members in government, the Supreme Soviet, the courts, and the electoral machinery and by the KGB through its agents among the intellectuals and scientists. It would be the culmination of Shelepin's plans. It would contribute to the stabilization of the regime at home and to the achievement of its goals abroad.

The arrest of Sakharov in January 1980 raises the question of why the KGB, which was so successful in the past in protecting state secrets and suppressing opposition while concealing the misdemeanors of the regime, is so ineffective now. Why in particular did it allow Western access to Sakharov and why were his arrest and internal exile so gratuitously publicized? The most likely answer is that his arrest and the harassment of other dissidents is intended to make a future amnesty more credible and convincing. In that case the dissident movement is now being prepared for the most important aspect of its strategic role, which will be to persuade the West of the authenticity of Soviet "liberalization" when it comes. Further high-level defectors, or "official émigrés," may well make their appearance in the West before the switch in policy occurs.

If it [liberalization] should be extended to East Germany, demolition of the Berlin Wall might even be contemplated.

Western acceptance of the new "liberalization" as genuine would create favorable conditions for the fulfillment of communist strategy for the United States, Western Europe, and even, perhaps, Japan. The "Prague spring" was accepted by the West, and not only by the left, as the spontaneous and genuine evolution of a communist regime into a form of democratic, humanistic socialism despite the fact that basically the regime, the structure of the party, and its objectives remained the same. Its impact has already been described. A broader-scale "liberalization" in the Soviet Union and elsewhere would have an even more profound effect. Eurocommunism could be revived. The pressure for united fronts between communist and socialist parties and trade unions at national and international level would be intensified.

This time, the socialists might finally fall into the trap. United front governments under strong communist influence might well come to power in France, Italy, and possibly other countries. Elsewhere the fortunes and influence of communist parties would be much revived. The bulk of Europe might well turn to left-wing socialism, leaving only a few pockets of conservative resistance. (339-41)

Despite harsh criticism from Establishment-christened "experts" and "Sovietologists", many of Golitsyn's above predictions happened with frightening accuracy during the period of 1989-1991. The whole sham culminated with the August, 1991 Soviet coup. Several strange features of the overthrown suggest that whole event was staged. Donald McAlvany enumerates various oddities that are indicative of this thesis:

1. The U.S. and world press were warned about the coming coup for several days leading
up to August 19. Seldom is the world press given advance notice of such events.
Western intelligence sources knew of the coup several months in advance. Also curious
was the fact that in spite of the advance publicity of the coup, Gorbachev made no
moves to head it off or avert it.
2. All of the eight coup leaders were Gorbachev appointees and confidants.
3. Coup leader Gennady Yanayev referred to himself only as "acting president" and spoke of Gorbachev returning to power after recovering from "his illness."
4. The coup leaders did not cut the internal or international communication lines-something which is always done in a coup or revolutionary upheaval.
5. The coup leaders made no attempt to control the press- neither the Soviet nor the foreign press stationed in Russia- which had complete access to international phone lines throughout the coup.
6. Anti-coup leaders such as Yeltsin had access to international phone lines and operators throughout the coup.
7. Only minimal troops were used throughout the coup, and

troops loyal to Yeltsin were sent to surround Yeltsin in the parliament building.

8. The airports were all left open.

9. Utilities in the parliament building were never cut.

10. In a legitimate coup, the KGB would have killed Yeltsin, Gorbachev, and other reform leaders. No attempt was ever made to arrest Yeltsin, but the coup plotters did arrest Godiyan, a well-known enemy of Gorbachev's. (220-21)

There were so many phony characteristics to the coup that many expressed suspicions. McAlvany elaborates:

> The president of Soviet Georgia came out shortly after the coup and accused Gorbachev of having masterminded the coup, and 62 percent of the Soviet people (according to private polls) believe the coup was a fake. Even Eduard Shevardnadze (Gorbachev's former foreign minister) said that Gorby may have been behind the coup. (222)

All the suspicions aside, these theatrical overtures were successful. The western elites dropped their guard, believing that their version of a "scientific dictatorship" would be the one that would dominate the world. However, the Soviet Bear had not been vanquished. What appeared to be death was merely hibernation. The ascendancy of Vladimir Putin, formerly of the KGB, to the Russian Presidency may have marked the beginning of the slumber's end. The BBC's Bridget Kendall conducted an investigation into the background of this enigmatic political figure. She found that Putin's "burning ambition was always to be a Soviet secret agent" (no pagination). Kendall also reported that the Russian President "was devastated at the sudden and humiliating Soviet retreat from Eastern Europe and subsequent collapse of the Soviet Union" (no pagination). This hardly sounds like a reformer.

In 1999, Putin became Prime Minister of Russia. An incident during this period of his political career could just as easily been lifted out of a biography of Stalin. Kendall elaborates:

> In August of that year, he waged a brutal war against the Chechens after a series of explosions had ripped through tower blocks in Moscow and other cities. Thousands were killed, and Chechnya was all but obliterated. (No pagination)

It is very likely that the pretext for this war was generated by employing the Soviet tactic of state sponsored terrorism. On September 4, 1999, a bombing occurred in Buinaksk, Dagestan, claiming 62 people (Henry, no pagination). This attack was followed by another bombing in Moscow which cost the lives of 215 people (no pagination). Another bombing occurred on September 16 in Volgodonsk, killing 18 people (no pagination). While the government blamed Chechen rebels, "it has never produced evidence to back up this claim" (no pagination).

The Russian government's failure to produce any evidence implicating Chechen rebels leads one to consider the possibility of another culprit. Exiled media tycoon Boris Berezovsky leveled accusations at Putin and the FSB that suggested they were the real guilty party. Patrick Henry elaborates:

> Boris Berezovsky announced Tuesday that President Vladimir Putin "definitely knew" that the Federal Security Service was involved in four bombings that killed more than 300 people in Moscow and two other cities in the fall of 1999, as well as a foiled bombing attempt in Ryazan.
> "At a minimum Vladimir Putin knew that the FSB was involved in the bombings in Moscow, Volgodonsk and Ryazan," Berezovsky told reporters, adding that Putin's failure to order a full investigation of the attacks constituted a coverup. (No pagination)

Evidence supporting these allegations are true may lie in a failed bombing in Ryazan, which occurred on September 22, 1999:

> A bomb was discovered in the basement of a 12-story apartment building in Ryazan by local police. The device consisted of several bags of a white powder connected to a timer and a shotgun shell detonator. Investigators in Ryazan initially identified the powder as hexogen, a powerful explosive. But FSB chief Nikolai Patrushev quickly dismissed this finding, claiming that the whole incident was merely a training exercise with a dummy bomb, and that the bags contained sugar.
> According to Berezovsky, four explosives experts from Britain and France had examined the available evidence from the Ryazan incident-- including photographs of the explosive device made by investigators--and concluded that the bomb was authentic. All physical evidence from the Ryazan crime

scene has been classified and sealed for 75 years, he said. (No pagination)

One individual who reinforced Berezovsky's contentions was Nikita Chekulin, the former director of a research institute affiliated with the Education Ministry that deals with explosives (no pagination). Chekulin's claims were most telling:

> Chekulin claimed to have documentary evidence showing that the institute had purchased tons of the explosive hexogen from military installations in 2000. That hexogen was then falsely labeled and transferred to "various cover agencies in the regions," he said. An internal Education Ministry investigation led Minister Vladimir Filippov to ask for the FSB to get involved. Among those Chekulin said knew of this "possible terrorist activity" were Deputy Prime Minister Valentina Matviyenko, then-Deputy Prime Minister Ilya Klebanov, Patrushev, then-Interior Minister Vladimir Rushailo and then-Security Council Chairman Sergei Ivanov. "Mr. Patrushev forbade the investigation, and his deputy Yury Zaostrovtsev informed the Education Ministry of this decision," Chekulin said. (No pagination)

The FSB claimed that these allegations were "untenable and devoid of common sense" (Henry, 2002). However, it is interesting to note that the bombings provided Putin with the pretext to achieve objectives reminiscent of Stalin's agenda.

Putin also borrowed another page from the old Soviet playbook: suppression of media dissent. Kendall explains:

> But meanwhile, the independent television channel NTV was questioning the war in Chechnya. For Mr. Putin this amounted to betrayal.
>
> As part of his crackdown on corruption, he set about pursuing the channel's owner, Vladimir Gusinsky, one of the so-called Russian oligarchs who had allegedly exploited Russia's chaotic privatisation reforms to amass a personal fortune.
>
> Before long, his office had been raided by armed tax police, his journalists interrogated, and he had fled into exile where he was arrested on a Russian extradition warrant.
>
> Mr. Putin claimed this was just the Prosecutor's office doing its

job. But many worried it could be the first step in a crackdown on free speech and democratic freedoms.

"People are more afraid now," said one journalist we talked to. "Only influence from international leaders on Putin can protect Russia's democracy," said another. (No pagination)

Kendall also saw Putin's ascension as an enthronement of the infamous KGB. She states:

The KGB, or FSB as it is now called, is back at the heart of government. A plaque commemorating Russia's first KGB president, the Soviet leader Yuri Andropov, has been installed on Putin's orders to pride of place at the security service's headquarters in Moscow. (No pagination)

The Russian President's actions in 2003 certainly reinforce Kendall's contention. Consider the Russian President's recent restructuring of the government, reported by the BBC on March 12, 2003:

Russian President Vladimir Putin has restructured his government to extend the powers of the Federal Security Service (FSB).
The secret police will now absorb the border guards and the government agency for monitoring communications (Fapsi).
Liberal opposition politicians say the change amounts to the return of the KGB--the FSB's notorious predecessor. (No pagination)

The FSB is the successor to the infamous KGB. However, it never could boast the same tyrannical power possessed by its Soviet forerunner. According to the BBC, that could all be ending now:

The new powers given to the FSB by President Putin's decrees were enjoyed by its Soviet predecessor.
Post-Soviet reforms had gradually stripped the secret police of its control over the border guards—a force now numbering about 174,000 which still plays an important part in Tajikistan and other flash-points—and Fapsi.
The FSB's headquarters remain in the old KGB building on Lubyanka Square, a few streets away from the Kremlin. (No pagination)

It seems very suitable for the FSB to be located in the old KGB building, as it is becoming virtually indistinguishable from its Cold War precursor. It also suggests that history is about to repeat itself. The false liberalization campaigns of the past were always followed by the considerable strengthening of Russia's internal security organs. Russian communism, which now eschews the hammer and sickle, may be preparing to make a twenty-first century return. This means the war between communism and capitalism predicted by Manuilski could be just on the horizon. The Russian elite, previously known as the Soviet elite, has their own version of a "scientific dictatorship." They have never abandoned that model and will, if necessary, fight a war to see it implemented.

However, while this war was probably not intended by the western elites, it would still fit into their evolutionary script quite nicely. As researcher James Perloff has noted: "The Establishment has frequently exploited the native anti-Communism of the American people to inveigle them into destructive circumstances" (137). In this case, those destructive circumstances would be a socialist West. The Western elites have always offered up their own unique brand of socialism as a bulwark against the Russian threat.

The Sino-Russian Superstate

In *America's Secret Establishment*, deceased researcher Antony Sutton examined the role played by Yale's Order of Skull and Bones in the transformation of China into a formidable military power. This transformation, as portrayed by Sutton, was part of a larger project in dialectical manipulation:

> By about the year 2000 Communist China will be a "superpower" built by American technology and skill. It is presumably the intention of The Order to place this power in a conflict mode with the Soviet Union. (181)

Like all good sociopolitical Darwinians, the Order view war as an integral component in the world's political evolution. For years, several Bonesmen have endeavored to create a geopolitical climate that is predisposed to conflict. Within the petri dish of nation-states, Russia and China boast the greatest potential for a violent chemical reaction. Such a reaction could accelerate the world's alleged political evolution towards global governance. This is an objective long desired by sociopolitical Darwinian organizations like Skull and Bones. However, the experiment could eventually go awry:

> Yet, The Order has probably again miscalculated. What will be Moscow's reaction to this dialectic challenge? Even without traditional Russian paranoia they can be excused for feeling more than a little uneasy. And who is to say that the Chinese Communists will not make their peace with Moscow after 2000 and join forces to eliminate the super-super-power- -the United States. (181)

A Hegelian synthesis of the Russian and Chinese scientific dictatorships may already be underway. The alliance between the two powers predicted by Sutton took shape in the year 2000. *The People's Daily* reports:

> President Jiang Zemin and Russian President Vladimir Putin Tuesday signed a joint declaration pledging that the two countries will continue to develop their friendly relationship and promote all-round cooperation. The Beijing Declaration says the state leaders of China and Russia agree to deepen China-Russia relations in the 21st century. China and Russia, as strategic partners, will press ahead to strengthen their good-neighborly friendship and expand cooperation so that the two countries will grow and prosper. (No pagination)

On the surface, this alliance might appear benign enough. Certainly, there is nothing inherently wrong with two countries attempting to improve their diplomatic relations. However, many of the terms presented in the Beijing Declaration stipulate extensive military collaboration between China and Russia. This becomes evident with even the most preliminary analysis:

> Highlights of the declaration are as follows: -- All political documents signed and adopted by China and Russia serve as the solid basis for the healthy development of bilateral relations. The two sides will strictly abide by them and make continuous efforts to push the relationship to higher levels. China and Russia will maintain close and regular contact between the two state leaders, and departments of foreign

affairs, national defense, law enforcement, economy, science and technology will also maintain close contact.

China and Russia support in the international arena forces of peace, stability, development and cooperation, defy hegemonism, power politics and group politics, and oppose attempts to amend the basic principles of international law, to threaten others by force or to interfere in other countries' internal affairs.

As permanent members of the United Nations Security Council, China and Russia share the responsibility to safeguard the leading role of the UN and its Security Council in maintaining world peace and security, and to push forward multi-polarization of the world. The two state leaders are satisfied with the achievements of the Shanghai Five Summit held not long ago in Dushanbe, capital of Tajikistan. They agree that the cooperation among members of the Shanghai Five has reached a new level and should be further promoted. China and Russia have reached consensus on maintaining security and stability in their neighboring regions.

The aim of the joint statement on the anti-ballistic missile treaty signed during this summit is to consolidate global and regional strategic stability, to safeguard the existing system of arms control and disarmament treaties, to accelerate the non- proliferation process of weapons of mass destruction and their carrier vehicles, and to ensure the security of all countries, without exception. A look at the current world situation reveals the theory that the anti-ballistic missile treaty should be amended on the grounds that some countries are a missile threat is groundless.

China and Russia respect each other's independence, sovereignty and territorial integrity, and firmly oppose any attempts to split the country from within or outside the country. They understand and support each other's efforts to safeguard national unification, sovereignty and territorial integrity. National separatism, international terrorism, religious extremism and cross-border criminal activities have endangered the safety of sovereign countries and the peace and stability of the world. China and Russia are determined to take clear-cut measures to crack down on these problems both bilaterally and multi-laterally.

Russia reiterated its consistent principled stance on the Taiwan issue, saying that it recognizes the government of the People's Republic of China is the sole legitimate government representing China, and that Taiwan is an inalienable part of the Chinese territory. Russia will not support any form of Taiwan independence. It supports the People's Republic of China's stance on not accepting "two Chinas" or "one China, one Taiwan."

Russia opposes Taiwan's entry into the United Nations or into any international organization eligible only to sovereign states, and will not sell weapons to Taiwan.

China is truly grateful for Russia's faithful adherence to the "one China" principle on the Taiwan issue.

Both China and Russia believe that the Taiwan issue is China's internal affair. Both believe that no outside force should be allowed to interfere in resolving the Taiwan issue, and stress that such an attempt can only add to the tension in the Asia-Pacific region.

The further and comprehensive development of economic, trade, scientific and technological, and military-related technological cooperation between China and Russia is vital for the expansion of the Sino-Russian strategic partnership of cooperation based on equality and trust. The two heads of state said they were satisfied with the performance of prime ministers at regularly-held meetings, and think the regular-meeting system plays a major role in promoting bilateral cooperation in the areas of economy and trade, science and technology, national defense, energy (including oil and gas industry), transportation, nuclear industry, aviation and aerospace, and banking. China and Russia are committed to widening cooperation in specific areas to consolidate the strategic cooperation.

China and Russia will explore possibilities for Russia's participation in the development of China's western regions, including the joint development of oil and gas resources and gas pipeline laying. Both wish to strengthen cooperation in the sectors of science and technology, education, culture and sports.

The two countries believe that the Sino-Russian agreement signed on December 9, 1999, for the joint use of certain islets in border rivers and surrounding waters for economic purposes is unprecedented. The smooth implementation of the agreement marked a major step forward for the two countries to build their border into a bridge of friendship. In a constructive and pragmatic spirit, China and Russia will continue their talks to speed up resolution of disputes over areas still under negotiation. The status quo should be maintained for the areas until a solution is reached. China and Russia are satisfied with the initial implementation of a treaty signed by China, Russia, Kazakhstan, Kyrgyzstan and Tajikistan on deepening trust between their militaries, and reducing military forces in border regions.

Both believe the implementation of the pact will promote peace, tranquility, stability and prosperity in the border regions, and push forward the good-neighborly relations among all signatories.

China and Russia think that now is the time to study the possibilities of finding ways to promote trust between all these countries in military matters.

It is the common aspiration of the Chinese and Russian peoples to preserve their friendship for generations to come. To this end, tireless effort is required not only from the two governments, but also from the two peoples. The two countries agree to support the Sino-Russian Committee for Friendship, Peace and Development and to encourage other forms of non-governmental exchange. To build up a long-term and stable relationship between the two countries on the basis of good-neighborly friendship, mutual trust and mutual benefit, the two heads of state agreed to conduct negotiations on preparations for the reaching of the China-Russia Good-Neighborly Friendship and Cooperation Treaty. (No pagination)

Stripped of all its quixotic language, the Beijing Declaration established the type of cooperation that is necessary for the erection of a Sino-Russian Superstate. Certainly, the combined militaries of Russia and China would be a force to be reckoned with, even for the currently undefeated *Pax Americana*. Perhaps that is just what the Russians and Chinese have in mind. Neither superpower is enthusiastic about the neoconservatives' blueprint for world government. They have designs of their own and they do not intend to abandon them without a fight.

Soon, a newly formed scientific dictatorship in the East may halt the advance of the scientific dictatorship from the West.

In his article entitled, "The Chinese-Russian Alliance—Birth of a Superstate?" journalist Toby Westerman states:

> The union of Russia and Communist China is beginning to attract attention from the world's press, but the full—and dangerous—implication of the alliance is not acknowledged. The visit of Chinese Communist leader Hu Jintao to Russia prompted Fred Weir of the Canadian Press to write from Moscow and analyze the relationship between Russia and China. He recognized that Moscow and Beijing are forging the "world's next economic, military, and spacefaring superpower." (No pagination)

Westerman explains that this emergent alliance is militaristic in nature:

> Two basic facts are undeniable and demand the public's attention: "democratic" Russia supports Communist China's foreign policy, including Beijing's aggressive stand toward free Taiwan, and Moscow is the main supplier for Communist China's massive arms build-up, which has caused deep concern throughout the Asia-Pacific region.
> Sino-Russian ties have been close, long-term, and militarily oriented. (No pagination)

Evidently, Russia's provision of military infrastructure to China shows no sign of abating:

> In August 1994, Radio Moscow (known today as the Voice of Russia) declared that "Russia will remain a major source of Chinese weapons," and was supplying Communist China with "tanks, air-born radar, and the training of Chinese military officers in Russian military academies."
> Sixteen months later, in December 1995, Pavel Grachev, then-Russian Defense Minister, defined Russia as Communist China's "major partner" in the weapons trade, and that Moscow-Beijing cooperation was "an example of mutual trust and genuine friendship," according to a broadcast from the Voice of Russia World Service. (No pagination)

In fact, no questioning voices have been heard from within
Russia:

No politician in "democratic" Russia has called into question
Moscow's close ties with Communist China, just as Russia's
close association with every overt communist state, including
North Korea and Cuba, has been criticized. (No pagination)

In his final analysis, Westerman concludes that China and Russia
are coalescing into a superstate that could challenge the West:

By April 1997, then-Presidents of Russia and China, Boris
Yeltsin and Jiang Zemin, declared their intentions to establish
a "New World Order, " which would replace the purported
American domination of world affairs. (No pagination)

China and Russia took a giant leap forward in 2005 when the two
actually began conducting joint military exercises. The *China Daily*
reported on this event in an August 18, 2005 article:

Russia and China launched their first-ever joint military
exercises Thursday on a Chinese peninsula jutting into the
Yellow Sea — an eight-day event that symbolizes the two
countries' bolstered ties, the Associated Press reported.
"Peace Mission 2005" exercises include some 10,000 troops
from land, sea and air forces. (No pagination)

Needless to say, the title "Peace Mission 2005" is hardly fitting for
such an exercise. Establishing military cooperation is a vital step in the
erection of an empire. Both China and Russia are well aware of this fact.
The message to western factions is loud and clear: "We are powerful
enough to challenge you."
The sociopolitical Darwinians of the West have played an extremely
dangerous game, engaging their Eastern counterparts in a policy of
strategic engagement. The result has been the transmogrification of the
Chinese and Russian scientific dictatorships into a single technocratic
superpower. The governing principle of this Darwinian geopolitical
climate is "Survival of the fittest." The elite of America's Establish
are guaranteed no victory. Then again, a Sino-Russian super-state may
work to their advantage. After all, the threat of encroachments by

the communist world has always been used by the western elites as a pretext for world government. The American people, as well as the rest of the world population, may accept a global scientific dictatorship if it is offered up as a bulwark to the communist Sino-Russian empire.

The Elite-Sponsored Race War

Recall the dialectical framework intrinsic to evolutionary theory. This framework is Hegelian. The organism (thesis) comes into conflict with nature (antithesis) resulting in a newly enhanced species (synthesis), the culmination of the evolutionary process (Marrs 127). In *The Descent of Man*, Charles Darwin candidly ponders the "benefits" of a race war:

> The enquirer would next come to the important point, whether man tends to increase at so rapid a rate, as to lead to occasional severe struggles for existence; and consequently to beneficial variations, whether in body or mind, being preserved, and injurious ones eliminated. Do the races or species of men, whichever term may be applied, encroach on and replace one another, so that some finally become extinct? (No pagination)

A little later into *The Descent of Man*, Darwin answers this question:

> We shall see that all these questions, as indeed is obvious in respect to most of them, must be answered in the affirmative, in the same manner as with the lower animals. (No pagination)

As sociopolitical Darwinians, the power elite also seek to promulgate just such a dialectic among the various races of man. In fact, an elite-sponsored race war may already be unfolding.

On February 13, 2002, U.S. Representative Tom Tancredo (R-Colo.) and former San Diego congressman Brian Bilbray addressed a sizable audience concerned with illegal immigration issues ("The Puppeteer," no pagination). Hosted at the Cannon House Office Building in Washington, D.C., the event was the prelude to a two-day lobbying campaign (no pagination). The discerning researcher who investigates this event will automatically identify intimations of a racist Fifth Column, which threatens to subsume the Minutemen Project and subvert efforts to restore America's border integrity. A white supremacist newsletter entitled the Citizens Informer was made

available at the meeting (no pagination). This newsletter is published by
the Council of Conservative Citizens, a racist organization financed by
the Shea Foundation (no pagination).

More importantly, this assembly was "masterminded by
NumbersUSA," which is but one appendage of a larger organizational
octopus (no pagination). When NumbersUSA executive director Roy
Beck addressed the audience, he admonished activists to underplay
the organization's involvement in the lobbying effort on Capitol Hill
(no pagination). Beck candidly stated that the campaign "needs to
look like a grassroots effort" (no pagination). However, NumbersUSA
is anything but a grassroots organization. The Southern Poverty Law
Center *Intelligence Report* elaborates:

> Despite attempts to appear otherwise, it [NumbersUSA] is a
> wholly owned subsidiary of U.S. Inc., a sprawling, nonprofit
> funding conduit that has spawned three anti-immigration
> groups and underwrites several others, many of which were
> represented at the NumbersUSA conclave. (No pagination)

NumbersUSA is part of a "loose-knit" network of groups connected
to a man named John Tanton ("John Tanton's Network," no pagination).
Tanton has founded, co-founded, and financed numerous organizations
that are gradually redirecting the movement for immigration policy
reform towards an agenda of anti-immigration bigotry. The following
is a list of those groups:

American   Immigration   Control   Foundation
AICF, 1983, funded
American   Patrol/Voice   of   Citizens   Together
1992, funded
California   Coalition   for   Immigration   Reform
CCIR, 1994, funded
Californians   for   Population   Stabilization
1996, funded (founded separately in 1986)
Center   for   Immigration   Studies
CIS, 1985, founded and funded
Federation   for   American   Immigration   Reform
FAIR, 1979, founded and funded
N   u   m   b   e   r   s   U   S   A
1996, founded and funded

Population-Environment   Balance
1973, joined board in 1980
Pro   English
1994, founded and funded
P   r   o   j   e   c   t   U   S   A
1999, funded
The   Social   Contract   Press
1990, founded and funded
U.S.   English
1983, founded and funded
U.S.   Inc.
1982, founded and funded (no pagination)

Following the February 13 meeting in Washington D.C., other racist elements began to rear their ugly heads. The SPLC *Intelligence Report* reveals these dubious individuals and groups:

> Two weeks after the NumbersUSA lobbying trip to the offices of Tom Tancredo and a series of other congressmen, Glenn Spencer, head of the Tanton-funded anti-immigrant American Patrol, was one of the main speakers at a conference hosted by Jared Taylor of *American Renaissance* magazine.
> Joining Spencer, who warned his audience that a second Mexican-American war would erupt in 2003, was an array of key extremists:
> Mark Weber, a principal of the Holocaust-denying Institute for Historical Review;
> White power web maven, former Klansman and ex-con Don Black;
> Gordon Lee Baum, "chief executive officer" of the CCC; and several members of the neo-Nazi National Alliance. (No pagination)

Upon closer examination of this assembly, strange confluences become apparent. For instance, the Institute for Historical Review is actually connected with left-wing icon Noam Chomsky. Werner Cohn expands on this relationship:

> The IHR's publishing and bookselling arm is called Noontide Press. Holocaust-denying is only one part of the anti-Semitic menu of this supermarket of Nazism. The latest NP catalog is

dated 1995. Among its offerings we find Nazi-made movies that are banned in Germany because of their brazen propaganda (pp. 29, ff), as well as the notorious *Protocols of the Elders of Zion* (p. 10), books by Adolf Hitler and Joseph Goebbels (pp. 10 and 12), a book by the late Father Coughlin (p. 7). Chomsky is represented by five separate items: *The Fateful Triangle* (p. 16); *Necessary Illusions* (p. 11); and *Pirates and Emperors* (p. 12). Chomsky, according to the IHR, "enlightens as no other writer on Israel, Zionism, and American complicity" (p. 4). (No pagination)

What makes this association even stranger is the fact that Chomsky is Jewish. However, researchers like Cohn have found substantial evidence suggesting that Chomsky is a *hofjuden* (self-loathing Jew). Many icons of the left have belonged to this odd racialist tradition. Karl Marx is one case in point, as is evidenced by his book *A World Without Jews*. It is possible that Chomsky is yet another example.

Even more significant is Glenn Spencer, founder of the American Patrol Report and a supporter of the Minutemen Project. As was previously mentioned, Spencer forecasted a second Mexican-American war in 2003. While Spencer's prophecy was never fulfilled, his open advocacy of a race war certainly raises suspicions about his true agenda. These suspicions only intensify when one considers the fact that Spencer's operation is financed by Tanton.

It is possible that a conscious effort is being made to radicalize the Minutemen project, subvert efforts to restore America's border integrity, and manufacture a politically expedient race war. This effort could be financed and coordinated by factions of the power elite, which have a vested interest in the destruction of America's national sovereignty and the establishment of a one-world government. The immigration issue is of particular interest to these supranational oligarchs because it directly affects America's border integrity, which is integral to a country's national sovereignty. John Tanton is no small player in this conspiracy.

In 1990, John Tanton founded the Social Contract Press. Those who visit the organization's website will find links to Population and Sustainability, Carrying Capacity Network, and Negative Population Growth. Obviously, these organizations are guided by a common theme: population control. Social Contract Press shares this agenda, as is evidenced by the group's mission statement:

The Social Contract Press is an educational and publishing organization advocating open discussion of such related issues as *population size and rate of growth*, protection of the environment and precious resources, limits on immigration, as well as preservation and promotion of a shared American language and culture. (No pagination; emphasis added)

In fact, population control is one of the primary motives underpinning the Social Contract Press's immigration reduction agenda. According to the official website, the Press members "favor immigration," but desire "fewer admissions in order *to reduce the rate of America's population growth*, protect jobs, preserve the environment, and foster assimilation" (No pagination; emphasis added).

Evidently, population control is one of Tanton's preoccupations. A preliminary perusal of Tanton's resume further expands on this preoccupation. Social Contract Press's website states:

His [Tanton's] conviction that continued human population growth was a large part of the conservation problem led him to chair the National Sierra Club Population Committee (1971-74), and to the national board of Zero Population Growth (1973-78, including a term as president, 1975-77). In 1979, as immigration grew to be the significant part of the U.S. population problem, he organized the Federation for American Immigration Reform (FAIR) based in Washington, D.C. (No pagination)

This preoccupation represented a natural progression from Tanton's radical environmentalist passions. A Southern Poverty Law Center *Intelligence Report* elaborates:

Raising a family and practicing medicine in Petoskey, Mich., Tanton started out as a passionate environmentalist. In the 1960s and early 1970s, he was a leader in the National Audubon Society, the Sierra Club and other mainstream environmental groups.

But Tanton soon became fixated on population control, seeing environmental degradation as the inevitable result of overpopulation.

When the indigenous birth rate fell below replacement level in the United States, his preoccupation turned to immigration. And this soon led him to race. (No pagination)

A novel entitled *The Camp of Saints* reinforced Tanton's racialist contentions (no pagination). Authored by Jean Raspail, the book's narrative involved "an invasion of the white, Western world by a fleet of starving, dark-skinned refugees" (no pagination). Tanton was so enamored of the novel that he supported its publication in English (no pagination). The "prophetic argument" of *The Camp of Saints* succinctly encapsulates Tanton's messages (no pagination).

The cause of population control has been historically connected with the eugenics movement. Researchers Webster Tarpley and Anton Chaitkin synopsize the intimate relationship between the two:

> The population control or zero population growth movement, which grew rapidly in the late 1960s thanks to free media exposure and foundation grants for a stream of pseudoscientific propaganda about the alleged "population bomb" and the limits to growth," was a continuation of the old prewar, protofascist eugenics movement, which had been forced to go into temporary eclipse when the world recoiled in horror at the atrocities committed by the Nazis in the name of eugenics. By mid-1960s, the same old crackpot eugenicists had resurrected themselves as the population-control and environmentalist movement. Planned Parenthood was a perfect example of the transmogrification. Now, instead of demanding the sterilization of the inferior races, the newly packaged eugenicists talked about the population bomb, giving the poor "equal access" to birth control, and "freedom of choice." (203)

With this synopsis, the thematic continuity running throughout Tanton's work becomes clearer. It represents a continuation of the eugenics agenda. In fact, portions of Tanton's organizational network have been intimately involved in eugenics projects. FAIR, which Tanton co-founded, received $1.2 million from the Pioneer Fund between 1985 and 1994 ("The Puppeteer," no pagination). According to eugenics expert Barry Mehler, the Pioneer Fund qualified as a "neo-Nazi organization, tied to the Nazi eugenics program in the 1930s, that has never wavered in its commitment to eugenics and ideas of human and racial inferiority and superiority" (no pagination). Evidently, Tanton is keeping the eugenics tradition alive.

It comes as little surprise that Tanton also concerns himself with environmental issues. Radical environmentalism is yet one more vehicle for the population control agenda and, by extension, a eugenical agenda. It is even less surprising that Tanton was the former president of the northern Michigan branch of Planned Parenthood. This organization, which was founded by the racist Margaret Sanger, has had a long history of involvement in the eugenics movement. In her book, *The Pivot of Civilization*, Sanger revealed the eugenical motives underpinning the cause of birth control. She unabashedly declares:

> Birth Control, which has been criticized as negative and destructive, is really the greatest and most truly eugenic method, and its adoption as part of the program of Eugenics would immediately give a concrete and realistic power to that science. . .as the most constructive and necessary of the means to racial health. (Sanger, *The Pivot of Civilization*, 189)

Tanton belongs to this eugenical tradition. Recall his membership in Zero Population Growth. This organization's founder, Paul Ehrlich, wrote *The Population Bomb*. Published in 1968, Ehrlich's book presented the following prediction:

> The battle to feed all of humanity is over. In the 1970s the world will undergo famines—hundreds of millions of people are going to starve to death in spite of any crash programs embarked upon now. At this late date nothing can prevent a substantial increase in the world death rate. . . (xi)

To counter this plague of global starvation, Ehrlich prescribes overtly authoritarian measures: "We must have population control at home, hopefully through a system of incentives and penalties, but by compulsion if voluntary methods fail" (xi). Of course, the arrival of the 1970s thoroughly refuted all of Ehrlich's prognostications. However, his fraudulent eschatological claims acted as a pretext for the proposal of totalitarian policies. Of course, such policies, if implemented, would have proven politically and socially expedient for the power elite. This is the true agenda underpinning the carrying capacity myth. Whether or not Ehrlich actually believed the overpopulation fables that he peddled, they still provided the ruling class with a readily exploitable threat.

Ehrlich's wife is a member of the Club of Rome, an organizational machination of the power elite. The Club specializes in manufacturing hypothetical scenarios concerning overpopulation and environmental degradation. The organization typically compiles these apocalyptic forecasts and publishes them for public consumption. It should be fairly obvious why. The following excerpt from the Club's 1991 report, *The First Global Revolution*, reveals the motive:

> In searching for a new enemy to unite us, we came up with the idea that pollution, the threat of global warming, water shortages, famine and the like would fit the bill. . .But in designating them as the enemy, we fall into the trap of mistaking symptoms for causes. All these dangers are caused by human intervention and it is only through changed attitudes and behavior that they can be overcome. The real enemy, then, is humanity itself. (King and Schneider 115)

In making humanity the ultimate enemy, the oligarchs have the perfect pretext for world government. After all, only a massive supranational entity with unlimited powers could compel the nemesis of humanity to retard its "environmentally unsound" industrial and technological development. Thus, a global socialist totalitarian government could be erected in the name of "ecological preservation." This is the cause that men like Tanton and Ehrlich are perpetuating.

Although Ehrlich's false predictions should have qualified the man as a certifiable phony, his claims were still given credence by certain factions of the elite and government think tanks. During his Congressional career, George Bush Sr. founded and chaired the Republican Task Force on Earth Resources and Population (Tarpley and Chaitkin 199). Bush's task force subscribed to the same old eschatological claims of environmentalists. It members contended that "the world was already seriously overpopulated; that there was a fixed limit to natural resources and that this limit was rapidly being reached; and that the environment and natural species were being sacrificed to human progress. (199) The task force was thoroughly neo-Malthusian in character and provided Ehrlich with an audience. Tarpley and Chaitkin explain:

> Comprised of over 20 Republican Congressman, Bush's task

force was a kind of Malthusian vanguard organization, which heard testimony from assorted "race scientists," sponsored legislation, and otherwise propagandized the zero-growth outlook. In its 50-odd hearings during these years, the task force provided a public forum to nearly every well-known zero-growth fanatic, from Paul Ehrlich, founder of Zero Population Growth (ZPG), to race scientist William Shockley, to the key zero-growth advocates infesting the federal bureaucracy. (199-200)

Ehrlich suggested "a 'tough foreign policy' including termination of food aid to starving nations" (200). Naturally, people of a darker skin hue populated most of these nations. The scientific racism of such a "tough foreign policy" should be fairly axiomatic. Ehrlich's mandates for domestic population control included "the addition of. . .mass sterilization agents" to America's water and food supplies (200). Such ideas were given serious credence, as is evidenced by the pedigree of those comprising the task force.

Ehrlich's fellow traveler, William Shockley, was even less circumspect about his scientific racism. During the 60s, this race scientist had generated a substantial amount of controversy by promoting his already refuted thesis that black people were mentally and cognitively inferior to white people (200). In the same year that the GOP task force supplied him with a congressional platform, Shockley wrote:

"Our nobly intended welfare programs may be encouraging dysgenics--retrogressive evolution through disproportionate reproduction of the genetically disadvantaged. . .We fear that 'fatuous beliefs' in the power of welfare money, unaided by eugenic foresight, may contribute to a decline of human quality for all segments of society." (200)

To counter this tide of so-called "retrogressive evolution," Shockley proposed:

a program of mass sterilization of the unfit and mentally defective, which he called his "Bonus Sterilization Plan." Money bonuses for allowing oneself to be sterilized would be paid to any person not paying income tax who had a genetic deficiency or chronic disease, such as diabetes or epilepsy, or who could be shown to be a drug addict. "If [the government paid] a bonus rate of $1,000 for each point below 100 IQ, $30,000 put in trust for some 70 IQ moron of 20-child potential, it might return $250,000 to taxpayers in reduced cost of mental retardation care," Shockley said. (Tarpley and Chaitkin 200)

Naturally, such racist rhetoric prompted more than a few to raise suspicions of Shockley harboring Nazi sentiments. In 1967, the race scientist made a damning response to these charges: "The lesson to be drawn from Nazi history is the value of free speech, not that eugenics is intolerable" (Tarpley and Chaitkin 200). This is the sort of thinking that motivates Tanton. As a former member of ZPG and Planned Parenthood, he lays claim to a neo-Malthusian heritage. Neo-Malthusianism originated with the ideas of Thomas Malthus, who concluded that society should adopt certain social policies to prevent the human population from growing disproportionately larger than the food supply. Of course, these social policies were anything but humane. They stipulated the stultification of industrial and technological development in poor communities. With the inevitable depreciation of vital infrastructure, society's "dysgenics" would eventually be purged by the elements. According to Malthus, such sacrifice guaranteed a healthy society.

Tanton's anti-immigration racism synchronizes comfortably with neo-Malthusian thought. Neo-Malthusians harbor no small amount of disdain for pro-fertility belief systems. According to *Wikipedia* online encyclopedia, this disdain is directly related to the population question:

> In any group some individuals will be more pro-fertility in their beliefs and practices than others. According to neo-Malthusian theory, these pro-fertility individuals will not only have more children, but also pass their pro-fertility on to their children, meaning a constant selection for pro-fertility similar to the constant evolutionary selection for beneficial genes (except much faster because of greater diversity). According to neo-Malthusians, this increase in fertility will lead to hyperexponential population growth that will eventually outstrip growth in economic production. (No pagination)

Of course, Hispanic immigrants come from predominantly

Catholic countries. Catholicism is a pro-fertility belief system. Such a system encourages "careless breeding." Therefore, Hispanic immigrants represent a threat to neo-Malthusians. Moreover, neo-Malthusians contend that immigrants contribute very little to the nation's economy. *Wikipedia* online encyclopedia elaborates:

> Neo-Malthusians argue that although adult immigrants (who, at the very least, arrive with human capital) contribute to economic production, there is little or no increase in economic production from increased natural growth and fertility. Neo-Malthusians argue that hyperexponential population growth has begun or will begin soon in developed countries. (No pagination)

Herein is the true motive underpinning Tanton's promotion of limits to immigration. His ultimate objectives are the culling of surplus population and the eugenical regimentation of society. Tanton's network is gradually redirecting the Minutemen Project towards these objectives.

However, it would be wrong to consider Tanton the center of the onion. Behind him lurk the controlling hands of the Power Elite. While this fact can be seen through Tanton's connection to the Club of Rome by way of Ehrlich's ZPG, it can also be seen through the different tax-exempt foundations that supply a steady flow of capital to Tanton's network. Tax-exempt foundations allow the elite to shield their money from the income tax and conduct social engineering projects.

Two foundations funding the Tanton network are significant because of their connections to the U.S. intelligence community. The first is the Smith Richardson Foundation ("The Puppeteer", no pagination). The financing for Smith Richardson comes from the *Vicks Vaporub* fortune ("Smith Richardson Foundation", no pagination). In 1973, the Smith Richardson Foundation came under the presidency of R. Richardson Randolph (no pagination). With a mind-boggling net worth of 870 million dollars, the Richardsons are one of America's richest families. We are far off the trajectory of back hill bubbas and rednecks and have entered the realm of the power elite.

The origin of the Smith Richardson Foundation's involves a rogue gallery that includes the Bush family dynasty, the Yale secret society of Skull and Bones, and Nazi sympathizers. Tarpley and Chaitkin explain:

> The Bush family knew Richardson and his wife through

their mutual friendship with Sears Roebuck's chairman, Gen. Robert E. Wood. General Wood had been president of the America First organization, which had lobbied against war with Hitler Germany. H. Smith Richardson had contributed the start-up money for America First and had spoken out against the U.S. "joining the Communists" by fighting Hitler. Richardson's wife was a proud relative of Nancy Langehorne from Virginia, who married Lord Astor and backed the Nazis from their Cliveden Estate.

General Wood's daughter Mary had married the son of Standard Oil president William Stamps Farish. The Bushes had stuck with the Farishes through their disastrous exposure during World War II. . .Young George Bush and his bride Barbara were especially close to Mary Farish, and to her son W.S. Farish III, who would be the great confidante of George's presidency.

The H. Smith Richardson Foundation was organized by Eugene Stetson, Jr., Richardson's son-in-law. Stetson (Skull and Bones, 1934) had worked for Prescott Bush as assistant manager of the New York branch of Brown Brothers Harriman. (77)

Tarpley and Chaitkin reveal the Smith Richardson Foundation's connections to the U.S. intelligence community:

In the late 1950s, the H. Smith Richardson Foundation took part in the "psychological warfare" of the CIA. This was not a foreign, but a domestic, covert operation, carried out mainly against unwitting U.S. citizens. CIA Director Allen Dulles and his British allies organized "MK-Ultra," the testing of psychotropic drugs including LSD on a very large scale, allegedly to evaluate "chemical warfare" possibilities. In this period, the Richardson Foundation helped finance experiments at Bridgewater Hospital in Massachusetts, the center of some of the most brutal MK-Ultra tortures. These outrages have been graphically portrayed in the movies *Titticut Follies*.

During 1990, an investigator for this book toured H. Smith Richardson's Center for Creative Leadership just north of Greensboro, North Carolina. The tour guide said that in these rooms, agents of the Central Intelligence Agency and the Secret

Service are trained. He demonstrated the two-way mirrors through which the government employees are watched, while they are put through mind-bending psychodramas. The guide explained that "virtually everyone who becomes a general" in the U.S. armed forces also goes through this "training" at the Richardson Center.

Another office of the Center for Creative Leadership is in Langley, Virginia, at the Headquarters of the Central Intelligence Agency. Here also, Richardson's Center trains leaders of the CIA. (77-78)

The second contributor to the Tanton network is Richard Mellon Scaife. The SPLC *Intelligence Report* documents this relationship:

Tanton's most important funding source for the last two decades may well have been the Scaife family, heirs to the Mellon Bank fortune.

Richard Mellon Scaife, a reclusive figure, has been instrumental in establishing right-wing organizations like the Heritage Foundation and supporting causes like the "Arkansas Project," an effort to dig up dirt on President Clinton.

Scaife family foundations, including those controlled by Scaife's sister, Cordelia May Scaife, provided some $1.4 million to FAIR from 1986-2000. (No pagination)

Scaife's attacks on Clinton might lead many to believe he is a genuine crusader against elite criminality. However, Scaife's skirmishes with the Clintons amounted to little more than factionalism in the ranks of the oligarchs. Richard Mellon Scaife's elite pedigree is impeccable. Like most bluebloods, Scaife has a fixation with eugenics. Robert G. Kaiser and Ira Chinoy report:

Scaife has long favored abortion rights, to the chagrin of many of those he has supported. In the first years of his philanthropy he stuck to a pattern set by his mother and sister and gave millions to Planned Parenthood and other population control groups, though most such giving stopped in the 1970s. (No pagination)

In addition to supporting eugenics, Scaife also attended the 1964 Bohemian Grove retreat, a major blueblood gathering (Kaiser and

Chinoy, no pagination). Scaife is also connected to the U.S. intelligence community. Richard was, at one time, the head of the publishing organ Forum World Feature, which was publicly named as a CIA front organization ("Richard Mellon Scaife," no pagination). However, Scaife's intelligence connection is even deeper. Edward Spannaus goes into this connection:

> Dickie Scaife is what one might call a second-generation ``OSS brat." During World War II, Dickie's father, as well as a number of his father's close business and familial associates, occupied high positions in the Office of Strategic Services (OSS)--America's wartime intelligence service. Alan Scaife, his father, was a lieutenant colonel in the OSS. A number of cousins of Dickie's mother, Sarah Mellon Scaife, also had very high positions in the OSS.
>
> For example: Paul Mellon (a cousin of Dickie's mother and a rabid Anglophile) was recruited in London to the OSS by his brother-in-law, David Bruce. Paul trained with British troops, became a major in the OSS, worked under Allen Dulles in Berne, Switzerland, and commanded a unit responsible for conducting propaganda operations behind disintegrating German lines.
>
> David Bruce, husband of Paul Mellon's sister Ailsa Mellon Bruce, was designated by OSS head William Donovan to oversee all OSS operations in Europe from his base in London. (Although some say, with justification, that it was Bruce who was designated by the U.S. banking-establishment families to oversee Donovan.) Another OSS cousin was Larimer Mellon, who likewise worked on Allen Dulles's staff in Berne.
>
> David Bruce (a direct descendant of the Scottish Bruce dynasty) later divorced Ailsa and married his second wife, Evangeline, an OSS secretary whose father had been a special liaison to British intelligence from the U.S. State Department.
>
> It is reliably reported that these Anglophilic OSS circles around Scaife's father were the crucial influence on steering Dickie into intelligence-related ``philanthropy"--i.e., the private funding of joint British-U.S. intelligence projects which were commonly mis-identified as ``CIA" projects or fronts. (No pagination)

Funding on the part of the Smith Richardson Foundation and Scaife makes it quite possible that the Tanton network is an intelligence project meant to foment race war. Would intelligence groups such as the CIA be interested in manufacturing such chaos? During an interview with William Norman Grigg, former DEA agent Michael Levine recounted a discussion he had with a CIA spook. The discussion, which occurred in Argentina in 1979, suggested that this is the case:

> "There was a small group of us gathered for a drinking party at the CIA guy's apartment. There were several Argentine police officers there as well; at the time, Argentina was a police state in which people could be taken into custody without warning, tortured, and then 'disappeared.'"
> "At one point my associate in the CIA said that he preferred Argentina's approach to social order, and that America should be more like that country," Levine continues. "Somebody asked, 'Well, how does a change of that sort happen?' The spook replied that it was necessary to create a situation of public fear—a sense of impending anarchy and social upheaval. . ." (No pagination)

Such a "situation of public fear" could be incited by agent provocateurs within the ranks of white supremacists. If the idea of intelligence agents working within white supremacist groups seems foreign, consider the following report by John Hooper:

> Germany's most notorious postwar neo-Nazi party was led by an intelligence agent working for the British, according to both published and unpublished German sources.
> The alleged agent—the late Adolf von Thadden—came closer than anyone to giving the far-right real influence over postwar German politics.
> Under his leadership, the National Democratic party (NPD) made a string of impressive showings in regional elections in the late 60s, and there were widespread fears that it would gain representation in the federal parliament.
> Yet, according to a report earlier this year in the Cologne daily, the Kölner Stadt-Anzeiger, the man dubbed "the New Führer" was working for British intelligence throughout the four years he led the NPD, from 1967 to 1971.
> However, a former senior officer in German intelligence told the Guardian this week that he had been informed of a

much longer-standing link between Von Thadden and British intelligence. His recollection raises the question of whether the German far-right-winger was under the sway of MI6 when he and others founded the NPD in 1964.

Dr Hans Josef Horchem, who was the head of the Hamburg office of the Verfassungsschutz--the West German security service--from 1969 to 1981, said he received regular visits from British intelligence liaison officers.

"We held general discussions on security. At one of these--I think it was towards the end of the 70s- they said, 'Adolf von Thadden was in contact with us', and that that was in the 1950s". Mr Horchem did not know whether the links between the German and British intelligence had continued into the 60s and 70s.

According to the Kölner Stadt-Anzeiger, whose report passed virtually unnoticed when it was published, the neo-Nazi leader met his British contact at a hotel in Hamburg. (No pagination)

As the above example clearly illustrates, government control of radical groups is really nothing new. The evidence suggests that shadowy factions of the intelligence community are pulling Tanton's strings. As the Tanton network entangles itself with the Minutemen project, the very same sinister hands may soon hold what started as a legitimate attempt at immigration reform.

Richard Mellon Scaife's involvement in the equation suggested a neoconservative manipulation of the Minutemen movement. This suspicion was confirmed on the October 8th program of Michael Corbin's *A Closer Look* radio show. On this show it was revealed that Minutemen founder Jim Gilchrist's congressional campaign was being ran by political consultant Mary Parker Lewis. Early in her career, Lewis was a staffer with the neoconservative Hoover Institution ("Mary Parker Lewis," no pagination). Lewis was also a confidential assistant to neoconservative icon William Kristol, son of "the godfather of neoconservatism," Irving Kristol ("Mary Parker Lewis," no pagination). Mary Parker Lewis also served as Chief of Staff to former U.S. Secretary of Education and neoconservative William Bennett ("Mary Parker Lewis," no pagination). Clearly, neoconservative hands were guiding Gilchrist.

Some might consider the idea of neoconservatives manipulating patriots unthinkable. However, consider the case of Norio Hayakawa.

Hayakawa is one of the foremost experts on Area 51, a top-secret test site in the Nevada desert. Hayakawa has collected a large body of circumstantial, yet compelling, evidence that technology at Area 51 is being used for social engineering. Therefore, it came as a shock when Norio Hayakawa cast his lot in with the neoconservative crowd. In an article entitled "Why I am a Neo-Conservative," Hayakawa writes:

> I consider myself to be a Neo-Conservative because I am in total agreement with the basic tenets of the Neo-Conservative political philosophies as defined by the Project for the New American Century (PNAC). (No pagination)

Hayakawa even supports the neoconservative contention that America has a right to preemptively strike nations that have taken no military action against the United States:

> As a Neo-Conservative, I fully support George W. Bush's War against Terrorism.
> It is obvious that the greatest danger to this world today is the presence of Islamic extremists and terrorists whose main goal is to destroy Israel and to undermine, debilitate and ultimately destroy any nation that supports Israel.
> Therefore the Neo-Conservatives take the position that any nation that harbors terrorists should be a target for pre-emptive, surgical strikes if those become the only recourse left for the survival of America and the freedom-loving nations of the world. (No pagination)

Instead of recognizing the neoconservatives as one of many different elite factions competing for dominance, Hayakawa assumes a grossly oversimplified and dualistic perspective:

> Another misconception about the Neo-Conservatives is the allegation that the Neo-Conservatives' ultimate goal is to create a global government, or a New World Order.
> The Neo-Conservatives have no such goals.
> On the contrary, the Neo-Conservatives believe that a strong American leadership is exactly what is preventing any future takeover of this world by global elitists.
> It is my belief that European Union-based (or controlled) global elitists, together with a loosely-knit network of European

royalties (who would attempt to manipulate European Jewish financiers) will be the ones that would ultimately attempt to create a global government.

Thus there is a struggle at the present moment between the American Neo-Conservative movement and the European Union-based global elitists, each of which will do its utmost to rally the support of Jewish financiers for their causes. My support goes to the American Neo-Conservative forces. (No pagination)

As the case of Norio Hayakawa clearly illustrates, when the neoconservatives turn on the charm, some patriots become starry-eyed. The story is no different for Jim Gilchrist and the Minutemen.

While neo-Malthusian interests are co-opting the immigration reform movement, the power elite could be manufacturing its opposition. Throughout the American southwest, Chicano supremacist groups have been forming and growing. By creating this artificial adversary for the immigration reform movement, the oligarchs could engineer a politically and socially expedient race war.

The Chicano supremacist threat is the result of a conscious radicalization project. It is being accomplished, in part, through the promulgation of racist ideologies. Marginalized and impoverished as they are, Hispanic immigrants are already susceptible to being propagandized. Among one of the racist ideologies disseminated by the ruling class is the racial myth of Aztlan. According to this myth, the southwestern states—California, Arizona, New Mexico, and Texas—comprise the original homeland of the Aztec Indians (Grigg, "Revolution in America,"9). Many of the groups that adhere to this myth believe that these lands must be forcefully reclaimed by "*la Raza*," the Mexican race (9). Ominously enough, some of the Aztlan cult is "equipped with a paramilitary auxiliary" (9).

Automatically, one will recognize the Marxist concept of a people's revolution, which is farcical at best. Invariably, such revolutions result in the creation and maintenance of new class distinctions. The ostensible proletarian dictatorship is merely another oligarchy in disguise. Moreover, one might notice parallels between this myth and the Aryan myth of Nazi Germany. In sum total, the doctrine of Aztlan amounts to little more than a racialist variety of socialism. The Aztlan myth culminates with La *Raza's* reclamation of the southwestern states and the establishment of the Chicano Homeland (9). This "worker's paradise" would become a nation-state unto itself, separate from the

United States of America (9). Thus, the Chicano Aryan would have his *Lebensborn.*

This virulent racism was promoted through tax-exempt foundations, which insulate the wealth of the oligarchs from the income tax. While the common American is subject to the institutionalized theft of the IRS, the power elite receives a tax write-off. However, tax-exempt foundations are more than simple tax shelters. They allow the ruling class a channel for the dissemination of racist ideologies. Henry Santiestevan, who formerly headed the Southwest Council of La Raza, confessed as much when he stated: "It can be said that without the Ford Foundation's commitment to a strategy of national and local institutions-building, the Chicano movement would have withered away in many areas" (Jasper 35).

Santiestevan had good reason to express such gratitude. *New American* journalist William F. Jasper reveals that "a tabulation of Ford Foundation grants to the Hispanic radicals during the period of 1968 to 1992 came to over $31 million" (35). To make matters worse, Jasper adds: "Millions have been added since" (35). Jasper proceeds to enumerate various grants listed from the Ford Foundation's Summer/Fall 1995 report:

> National Council of La Raza, $160,000 and $75,000.
> Northern New Mexico Legal Services, $20,000.
> Mexican Academy of Human Rights, $20,000.
> Hispanic Leadership Opportunity Program, $2,325,000 (including $525,000 for MALDEF).
> Immigrant Legal Resource Center, $145,000.
> National Immigration Law Center, $335,000.
> National Immigration Law Forum, $130,000.
> Urban Institute Program for Research on Immigration Policy, $900,000.
> Hispanics in Philanthropy, $100,000. (35)

Ironically, the Aztlan movement may hold more in common with the white supremacist movement than one might expect. It certainly is no more popular with the Anti-Defamation League than the standard white hate group. In fact, the ADL contends that many of the Aztlan movement's doctrines are similar to neo-Nazi and white supremacy ideas. According to *Wikipedia,* these parallels include Holocaust-denial, homophobia, and anti-Israeli sentiments (no pagination). In fact, both the ADL and the Southern Poverty Law Center have accused the movement's official webzine, *La Voz de Aztlán*, of promulgating anti-Jewish conspiracy theories (no pagination). One such theory is that Jews plotted the infamous anthrax mailings of 2001 (no pagination). If the allegations leveled at *La Voz de Aztlán* are true, then the Aztlan cult is a certifiable counterpart of the white supremacy movement.

The founder of *La Voz de Aztlán* is Hector Carreon, who is a former member of the Brown Berets (no pagination). The Brown Berets were a community youth organization birthed from the tumult of the 60s counterculture ("Brown Berets," no pagination). The group was heavily influenced by several other radical revolutionary movements, particularly the Black Panthers (no pagination). Like the Aztlan cult of today, the Black Panthers were financed and controlled by the power elite. In fact, some Black Panther members may have been aware of the manipulation. One such Panther was Stokely Carmichael, who was also the leader of the SNCC. Former Communist Party member and FBI informant James Kirk made the following observations concerning Carmichael:

> Mr. Carmichael was obviously in the middle of something very important which made him more nervous and tense than in the past...He started speaking of things which he said he could not have said before because his research was not finished. . . He repeated the line from the song he liked so well, "Something is happening here, but you don't know what it is, do you, Mr. Jones?" He kept hitting on the theme that a very large monopoly capitalist money group, the bankers to be exact, was instrumental in fomenting (the) idea that the Jews are the ones actually behind the oppression of the blacks. . .In the agencies of this power, he cited banks, the chief among which were Morgan Guaranty Trust and Chase Manhattan. And the foundations connected with these monoliths. (Griffin 108)

Apparently, Carmichael's revelations presented a distinct threat to the hidden manipulators. According to researcher Des Griffin: "Within weeks Carmichael had been mysteriously removed from SNCC and the Black Panthers. He had learned too much!" (108) Evidently, this tradition of manipulation continues within the ranks of the Aztlan movement today.

Financed and propagandized by the elite, these radical Chicano organizations are helping to create a cultural milieu of neo-tribalism, which further contributes to the ongoing social fragmentation of

America. If such militant revolutionaries boast "a paramilitary auxiliary," then imagine the potential chaos should some branch of the Aztlan movement were to engage in armed conflict with a radicalized sect of Minutemen.

In fact, according to *Observer* journalist Paul Harris, just such a state of affairs is already developing:

> Michael Nicley, head of US Border Patrol in the sector where the Minutemen will operate, has called it a 'recipe for tragedy'. The Reverend Robin Hoover, of local relief group Humane Borders, said: 'It looks destined to deteriorate into some form of confrontation.'
>
> The Minutemen have become a great cause among white supremacists, including the notorious Aryan Nation. Though organisers screen all volunteers for links to extremists, there are fears some will descend on the area. The Hispanic criminal gang MS-13 has said it will try to attack the Minutemen. (No pagination)

White supremacist efforts to co-opt the Minutemen could represent a ruling class conspiracy in promulgating racial dialectics. If the ranks of the Minutemen project can be inculcated into some form of militant white supremacy, then they can be subsequently pitted against factions of the radical Marxist Aztlan movement. Such a manufactured race war would be extremely advantageous for the power elite.

Ground Zero: California

The most likely place for a race war to begin is California. It is in this state that two individuals possessing radical racist views have risen to political power. These individuals will be able to mobilize the different racist groups for a confrontation. They are Los Angeles mayor Antonio Villaraigosa and Governor Arnold Schwarzenegger.

Antonio Villaraigosa would lead the Mexican racists' charge. Roger D. McGrath describes Villaraigosa's racist background:

> Antonio Villaraigosa was known as Tony Villar when he was growing up on the Eastside. (He combined his surname with that of his wife, Corina Raigosa, to create the last name he uses now.) As Villar, he attended East Los Angeles Junior College and then transferred to UCLA in 1972 under an "affirmative action" program. By the time he left UCLA in 1975, he had

not graduated but had risen to a position of leadership in the campus chapter of Movimiento Estudiantil Chicano de Aztlan, better known by its acronym, MEChA. The group was organized in 1969 by Mexican-American college students, for the most part, who preferred to call themselves Chicanos and Chicanas. MEChA proudly proclaims that its mission is to reclaim California and the rest of the Southwest—an area MECha calls Aztlan—from "the foreigner 'gabacho' who exploits our riches and destroys our culture... [W]e declare the independence of our mestizo nation. We are a bronze people with a bronze culture. Before the world, before all of North America, before all our brothers in the bronze continent, we are a nation, we are a union of free pueblos, we are Aztlan." (12)

There is no evidence that Villaraigosa ever abandoned MEChA's racist ideology. Roger D. McGrath reports: "When asked if he still supported MEChA's mission and goals during an interview on a talk-radio station in Los Angeles, Villaraigosa refused to answer" (13). Villaraigosa's refusal to answer may as well be taken as an admission of guilt. The former MEChaista remains firmly dedicated to a thoroughly racist ideology. The problem is that Villaraigosa is not some low level street thug. On May 17, 2005, he was elected mayor of Los Angeles. This makes him a racist with authority and lots of political clout.

Mention the name of Schwarzenegger and most people think of ridiculous action films featuring Arnold jumping from one explosion to the next. The man's escapades in bodybuilding and the movie industry cause most people to view Arnold as just one more celebrity with a humorous back-story. However, closer examination reveals that many of Schwarzenegger's closet skeletons come dressed in Nazi uniforms. In *Arnold: The Unauthorized Biography*, Wendy Leigh reveals some of Schwarzenegger's darker aspects:

> According to Rick Wayne, who is black, when they discussed apartheid Arnold said he thought South Africa was right, saying things like "If you gave these blacks a country to run, they would run it down the tubes." However, Rick was accustomed to Arnold's reactionary views and quirky ways. He and Arnold had posed together in Munich. In his book, *Muscle Wars*, a study of bodybuilding politics, Rick recalled that after Arnold had "struck a pose reminiscent of the Nazi

salute," he received less applause from the German audience than he had expected. Arnold's response was to comment to Rick, "These people are nothing without an Austrian to lead them." (68-9)

Many close to Schwarzenegger viewed Arnold's Nazi mask as part of his body building façade. One such person was Dick Tyler. Tyler had given posing lessons to Schwarzenegger. According to Leigh: "Tyler commented that if Hitler had wanted to advertise the Aryan ideal, Arnold would have been its perfect representative"(87). Leigh continues:

Tyler's remarks may sound a trifle barbed; however, since 1977 rumors have circulated in the bodybuilding world that during the filming of *Pumping Iron*, the pseudo-documentary film that transformed him into a legend, Arnold said he admired Hitler. When contacted for a newspaper article in 1988, George Butler, the producer and director of the film and still a close friend of Arnold's today, admitted that during the filming of *Pumping Iron* Arnold definitely did say that he admired Hitler. Butler then conceded that the remark was cut from the final version of the film, adding that Arnold expressed his admiration of "Hitler and Kennedy in almost the same breath as people who were leaders." When asked why Arnold admired Hitler, Butler replied that the context in the film was that Arnold was saying he had "always wanted to be remembered like the most famous people in history, like Jesus and so on. . ." (87-88)

Schwarzenegger associate Manfred Thellig's account is also very telling:

Manfred Thellig, who worked with Arnold in Munich, offers a similar interpretation. According to Thellig, Arnold "definitely admires the Teutonic period of the Third Reich. He just loved those leftover relics of the Third Reich in Munich—those Teutonic statues." He added that Arnold would say, "If I had lived at that time, I would have been one of those Teutonic breeders" but explains, "Whenever he opened his mouth and it sounded like 'Oh, there is a neo-Nazi,' this was just playing Tarzan. It wasn't serious. . ." (88-9)

However, there is evidence that Arnold's Nazi antics were not so tongue and cheek:

> There are, nevertheless, witnesses over the years who have seen Arnold break into the "Sieg Heil" salute and play his records of Hitler's speeches. Arnold responded to this issue during a 1989 *Penthouse* interview with journalist Sharon Churcher. According to Churcher, a former associate of Arnold's during the seventies had heard from a mutual acquaintance that Arnold had Nazi paraphernalia in his apartment. According to the associate, Arnold's reaction at that time was to claim through *Pumping Iron* producer George Butler, that his interest "was only that of a student." Butler, professing to have forgotten the above exchange, says that he had never seen any Nazi paraphernalia at Arnold's house. (89)

Arnold's family background is also steeped in Nazism. Leigh elaborates:

> Arnold personified Aryan supremacy and Germanic strength of will. To top that, his father had been a member of the Nazi party. Both his heritage and his image were inescapable. Inescapable, but not ineradicable. Yet Arnold, far from underplaying his roots, embraced and advertised them. (89-90)

The Schwarzenegger family's Nazi background apparently had a profound influence of Arnold's life. Consider the testimony of black bodybuilder Dave DuPre:

> Black bodybuilder Dave DuPre, who would appear with Arnold years later in *Pumping Iron*, says Arnold declared one time while working out at Gold's. "Serge is your only black hope to beat me. Black people are inferior. You are not capable of achieving the success of white people. Black people are stupid." Black people weren't the only target of his venom, for as usual he was completely democratic in his heckling. According to DuPre, "He would make fun of Jews. If anybody looked Jewish, he would point it out and tell them that they were inferior." (101)

Schwarzenegger's past certainly makes him a candidate for mobilizing white supremacists for a race war. Even more disturbing is the fact that, as governor of California, Schwarzenegger has the political clout and influence to make this more than just a remote possibility. With two racist demagogues planted high up in Californian politics, the stage has been set for a race war. When the time is right, both Schwarzenegger and Villaraigosa will be in a position to mobilize the two racist camps for the fighting. All that is needed now is a pretext. A violent border incident involving radicalized Minutemen and illegals could provide just such a pretext.

The idea of an inevitable war with Aztlan beginning in California may already be spreading among white racist groups. One purveyor of the idea is Stephen McNallen. McNallen is the head of Asatru', a white racist group steeped in satanism and Norse paganism. Mattias Gardell describes McNallen's propaganda:

Alarmed by activities of radical Chicano separatists who want to establish an independent Aztlan Nation out of what is now California and the Southwest, McNallen believes that California soon will be a battleground." The spiritual descendants of the Aztec are looking northward," and Euro-Americans will either resign to a subordinate position or rise from their slumber to resist the conquest. McNallen suggests that accumulating ethnic tension is a reflection of deeper movements in the collective unconsciousness of Chicanos and northern Europeans. While admitting that the great majority of Mexican descendants are Christian, McNallen uses Jung to explain that the "old Aztec and Mayan deities never really went away, they simply went underground." What if Jung's analysis of the rise of national socialism as a manifestation of the Wotan archetype in the Germanic soul is applied to explain the emergent Hispanic nationalist scene? "Are Tonatzin and Tezcatlipoca moving among their folk, stirring them to conquest?" And who should better lead Euro-American resistance than their own archetypal deities? "Mighty psychic forces, and powerful religious impulses are on the move. The old Gods of Mexico, and the Gods of ancient Europe, are stirring their respective peoples." The spiritual "awakening" of the northern European folk is vital to accomplish lest they would follow Kennewick Man into obliteration. (283)

While not all racists may buy the Jungian and Norse pagan elements of McNallen's rhetoric, his presentation of California as battleground between Aztlan and white supremacist forces would appeal to most racists seeking an excuse to start shedding blood. This includes white supremacist infiltrants within the ranks of the Minutemen.

Upon closer examination, the dialectic of race merely camouflages a broader Hegelian manipulation. Marginalized and deprived of any substantial political capital, many ethnic and racial groups are predisposed to exploitation by socialist revolutionary movements. Race war invariably becomes class war and class war invariably results in some form of dictatorship. Whether the resulting dictatorship ostensibly belongs to the proletariat or bourgeoisie, absolute primacy is always held by a hidden oligarchy. An October 15, 2005 riot in Toledo, Ohio offers a prime case study in racial dialectics. The National Socialist Movement, a neo-Nazi organization, was scheduled to conduct a demonstration against so-called "black crime" ("Planned neo-Nazi march sparks violence," no pagination). The protest received a violent response. A CNN news report explains:

> Most of the violence happened when residents, who had pelted the Nazi marchers with bottles and rocks, took out their anger on police, said Brian Jagodzinski, chief news photographer for CNN affiliate WTVG.
> Video showed crowds at around 2:25 p.m. using bats to bring down a wooden fence as looters broke into a small grocery store.
> "The crowd was very. . .extremely agitated at the police. . .for doing this [making arrests in] the community when they should be doing this to the Nazis," Jagodzinski said.
> Around 3 p.m., crowds of young men pelted the outside of a two-story residence with rocks, smashed out the windows with wooden crates, ran inside and threw out the furniture and lamps from the upper-level windows to the sidewalk below. No police were on the scene. About 10 minutes later, the building's second story was in flames as a crowd of people watched. When police arrived, they used pepper spray on counter-demonstrators and shot tear gas containers into the crowd, Jagodzinski said. He added that his news van and a police car had windows smashed and doors bent back. (No pagination)

This civil unrest created a pretext for the implementation of authoritarian measures and considerable restrictions on personal liberties. The CNN news report elaborates:

> Toledo Mayor Jack Ford declared a state of emergency and asked for 50 highway patrol officers to reinforce Toledo police. A curfew came into effect at 8 p.m. for people "roaming around the streets," he said. (No pagination)

Several oddities surround this event, suggesting that it was orchestrated to create politically and socially expedient chaos. First, there is the question concerning the demonstration's venue:

> It's not clear why the National Socialist Movement chose north Toledo for its march, said Ford, himself African-American. "It is not a neighborhood where you have a lot of friction in the first place," he said. (No pagination)

In response to this question, the National Socialist Movement (NSM) issued an interesting allegation:

> A spokesman for the [NSM] group, Bill White, blamed the riot on Toledo police, saying the police intentionally changed the group's march route to make it collide with a counter-demonstration. (No pagination)

Given White's dubious affiliations with neo-Nazi and white supremacist interests, one could certainly question the voracity of his claim. However, his allegation gains more credence when one examines the "counter-demonstration":

> About 20 members from both the International Socialists Organization and One People's Project showed up, and some handed eggs to African-American residents to throw at the Nazi marchers, White said.
> Ford said that scenario was likely.
> "Based on the intelligence we received, that's exactly what they do—they come into town and get people riled up," Ford said. "I think that's a very common technique." (No pagination)

Herein is the second oddity suggesting Hegelian manipulation. It should not be lost upon the astute reader that the NSM demonstration and its counter-demonstration rival both represent some variety of socialism. The NSM represents national socialism or, more succinctly, fascism. The International Socialist Organization represents more of a traditional variety of Marxism, something more akin to communism. The distinctions, however, are superficial. This becomes clear when one contemplates the etymology of "communism" and "fascism." The appellation of "communism" comes from the Latin root *communis*, which means "group" living. Fascism is a derivation of the Italian word *fascio*, which is translated as "bundle" or "group." Both fascism and communism are forms of coercive group living, or more succinctly, collectivism. The only substantial difference between the two is fascism's limited observance of private property rights, which is ostensible at best given its susceptibility to rigid government regulation. In 1933, Hitler candidly admitted to Hermann Rauschning that: "the whole of National Socialism is based on Marx" (Martin 239). Nazism (a variant of fascism) is derivative of Marxism. The historical conflicts between communism and fascism were merely feuds between two socialist totalitarian camps, not two dichotomously related forces.

The dialectic of race merely disguises a larger Hegelian dialectic: communism against fascism. Ayn Rand probably provides the most eloquent summation of this dialectic:

> It is obvious what the fraudulent issue of fascism versus communism accomplishes: it sets up, as opposites, two variants of the same political system. . .it switches the choice of "Freedom or dictatorship?" into "Which kind of dictatorship?"--thus establishing dictatorship as an inevitable fact and offering only a choice of rulers. The choice--according to the proponents of the fraud--is: a dictatorship of the rich (fascism) or a dictatorship of the poor (communism). (180)

The Toledo riot represents a tangible enactment of this traditional Hegelian dialectic. The National Socialist Movement (thesis) superficially conflicts with the International Socialist Movement (antithesis) resulting in Toledo's transformation into a miniature garrison state (synthesis). Toledo is only a microcosm. On the macrocosmic level, the final Hegelian synthesis is a global socialist state.

If the Minutemen are successfully radicalized by some variety of National Socialism, then it can be set on a collision course with the radical Marxists of the Aztlan movement. Not only would such a dialectic effectively confuse the issues of border integrity and national sovereignty, but it would also facilitate America's Fabian migration towards the ultimate Hegelian synthesis: A scientific dictatorship.

The Altar of Warfare

As long as there is a sociopolitical Darwinian elite dominating the affairs of man, humanity can expect to experience a state of perpetual warfare. In *The Power Elite*, sociologist C. Wright Mills informs readers that the Power Elite "direct the military establishment" (4). This means the bluebloods are able to make sure that key positions within the military establishment are always occupied by Darwinian militarists. Darwinian militarists will almost always choose war over diplomacy. Richard Weikart elaborates:

Darwinian militarists claimed that universal biological laws decreed the inevitability of war. Humans could not, any more than any other animal, opt out of the struggle for existence, since—as Darwin had explained based on his reading of Malthus-population expands faster than the food supply. War was thus a natural and necessary element of human competition that selects the "most fit" and leads to biological adaptation or-as most preferred to think-to progress. Not only Germans, but many Anglo-American social Darwinists justified war as a natural and inevitable part of the universal struggle for existence. The famous American sociologist William Graham Sumner, one of the most influential social Darwinists in the late nineteenth century, conceded, "It is the [Darwininan] competition for life...which makes war, and that is why war has always existed and always will." (165-66)

Such an outlook holds tremendous ramifications for human society. Weikart explains:

By claiming that war is biologically determined, Darwinian militarists denied that moral considerations could be applied to war. In their view wars were not caused by free human choices, but by biological processes. Blaming persons or nations for waging war is thus senseless, since they are merely blindly following natural laws. Further, opposition to war and militarism is futile, according to Darwinian militarists, who regularly scoffed at peace activists for simply not understanding scientific principles. (166)

The famed General George MacArthur once stated that the soldier, above all other people, prays for peace. As a global scientific dictatorship is erected, such sane thinking is quickly disappearing. The sociopolitical Darwinian elite and their Darwinian militarists in the military establishment pray before the altar of perpetual war. It is painfully obvious that the common man is to be their sacrifice.

Chapter Three:

The Global Scientific Dictatorship and its Future

Pax Cosmica

According to Fischer, the Reagan administration's Strategic Defense Initiative qualified as a technocratic strategy (23). The "Star Wars" program, Fischer contends, represented an attempt to "sidestep the problematic social and political questions" inherent to traditional political and diplomatic processes (23). Fischer elaborates:

> In the case of Star Wars, the federal government has sought a technological space shield to deal with the problem of nuclear weapons. This is promoted by technocrats as preferable to the political alternative of sitting down with the Soviet Union and negotiating a mutual security treaty. (23)

SDI was certainly not the first technocratic strategy employed by factions of the elite to wrest control of space. Hitler's scientific dictatorship also had plans for cosmic expansionism. Such celestial ambitions were voiced in Hitler's *Mein Kampf*:

> The folkish philosophy finds the importance of mankind in its basic racial elements. In the state it sees on principle only a means to an end and it construes its end as the preservation of the racial existence of man... And so the folkish philosophy of life corresponds to the innermost will of nature, since it restores the free play of forces which must lead to a continuous mutual higher breed, until at last the best of humanity, having achieved possession of this Earth, will have a free path for activity *in domains which will lie partly above it and partly outside it.* (No pagination; emphasis added)

Of course, Nazi Germany would be defeated and such plans would never be realized. However, the same technocratic strategy of space militarization remains in vogue among the elites. Just as the Dutch "Low Countries" represented a pistol pointed at the heart of England during the Middle Ages, space represents a dagger hovering over the heads of major powers today. SDI was the Western elite's attempt to outwit its Eastern counterparts. Although the Reagan Administration failed to fully implement the plan, space militarization continues to resurface as a potential machination for circumventing the diplomatic processes of international relations. It comes as little surprise that the Bush Jr. Administration, which is a neoconservative scion of the Reagan Administration, has witnessed the reemergence of such plans.

In an article entitled "Revealed: US plan to 'own' space," Journalist Neil Mackay details the newest technocratic blueprint for cosmic expansionism:

> IT SOUNDS like the stuff of the darkest sci-fi fantasies, but it's not. The Air Force Space Command Strategic Master Plan is a clear statement of the US's intention to dominate the world by turning space into the crucial battlefield of the 21st century.
>
> The document details how the US Air Force Space Command is developing exotic new weapons, nuclear warheads and spacecraft to allow the US to hit any target on earth within seconds. It also unashamedly states that the US will not allow any other power to get a foothold in space. (No pagination)

Like its precursors, the new plan could potentially neutralize all international treaties and reconfigure the geopolitical landscape:

> The rush to militarise space will also see domestic laws and foreign agreements torn up. As the document warns: 'To fully develop and exploit [space] ... some US policies and international treaties may need to be reviewed and modified'. The Strategic Master Plan (SMP) changes the nature of war. No longer will battles be fought by ships, aircraft and ground forces. Instead the US will use its technology to dominate any theatre of war from space.
>
> The document also opens the door for the US to become the only global policeman. Control of space will give it uniquely instantaneous reach, capable of "worldwide military operations." (No pagination)

The language of the document is unabashedly expansionistic and imperialistic, articulating an agenda for cosmic supremacy:

> The first page of the document clearly spells out America's agenda. General Lance W Lord, of Air Force Space Command, writes in his foreword: 'As guardians of the High Frontier,

Air Force Space Command has the vision and the people to ensure the United States achieves space superiority today and in the future.'

The document also lays the groundwork for the development of '21st century space warriors' – a new military cadre tasked solely to fight 'from and in' space. The SMP says this Space Corps 'is just as crucial to the success of our vision as employing new technologies'.

Air Force Space Command operates from a base in Colorado and its mission is to "defend America through space and intercontinental ballistic missile operations." Its ultimate goal is to "project global reach and global power." Although little is known about Space Command in Europe, it is central to the US military machine and staffed by some 40,000 military and civilians.

General Lord says the strategy of the SMP "will enable us to transform space power to provide our nation with diverse options to globally apply force in, from, and through space with modern intercontinental ballistic missiles ... and new conventional global strike capabilities."

In gung-ho language, the foreword reads: "Precision weapons guided to their targets by space-based navigation -- instant global communications for commanders and their forces – enemy weapons of mass destruction held at risk by a ready force of intercontinental ballistic missiles -- adversary missiles detected within seconds of launch. This is not a vision of the future. This is space today!"

Lord adds: "Our space team is building capabilities that provide the President with a range of space power options to discourage aggression or any form of coercion against the United States."

The (SMP) says: "Effective use of space-based resources provides a continual and global presence over key areas of the world ... military forces have always viewed the 'high ground' position as one of dominance. With rare exception, whoever owned the high ground owned the fight. Space is the ultimate high ground of US military operations.

"Today, control of this high ground means superiority ... and significant force enhancement. Tomorrow, ownership may mean instant engagement anywhere in the world." (No pagination)

Mackay proceeds to enumerate the particulars of this technocratic strategy for space supremacy:

> The primary goal of the SMP is to give the US military "the capability to deliver attacks from space." The use of 'space power' would also let the US deploy military might instantaneously across the face of the earth and completely "bypass adversary defences."
>
> In order to "fully exploit and control space," the United States Air Force Space Command says it has to 'negate' the ability of foreign powers to develop their own space capabilities. The plan also demands that Space Command "focus on missions carried out by weapons systems operating from or through space for holding terrestrial targets at risk."
>
> The document proclaims US aspirations to "global vigilance, reach and power," and Space Command says its vision "looks 25 years into the future and is summed up as follows: space warfighting forces providing continuous deterrence and prompt global engagement for America ... through the control and exploitation of space."
>
> The aim, the SMP says, is to:
>
> "Extend the reach, precision and intensity of US military power and operations."
>
> "Ensure the ability to apply space forces when and where we need them and that our adversary understands the advantage we possess."
>
> "Use our space capabilities at our discretion while at the same time denying our adversaries access to space assets at their disposal."
>
> One of Space Command's key functions is the operation of America's arsenal of intercontinental ballistic missiles. The SMP details how the US wants to be able to fire either nuclear or conventional missiles from space, out of range of enemy weapons. "Such a capability will provide warfighting commanders the ability to rapidly deny, delay, deceive, disrupt, destroy, exploit and neutralise targets in hours/minutes rather than weeks/days," it adds.
>
> The SMP also shows how the US fears advances in space technology among other nations -- including its European allies. "Space capabilities are proliferating internationally,"

it says, "a trend that can reduce the advantages we currently enjoy." It points out that Space Command has no control over the European Galileo satellite system.

A list of strategies and objectives detail the goals of Space Command in the coming years. These include:

creating an instantaneous global strike force.

Total monitoring of the Earth by "real-time global situation awareness."

a nuclear arsenal in space.

the development of exotic new weapons.

the maintenance of US military dominance. The doctrine declares: "when challenged, pursue superiority in space through robust ... defensive and offensive capabilities."

a fully integrated "land, sea, air and space war-fighting system."

integrating civil and commercial space operations with military ones.

One of the exotic weapons in development is known as the Ground Moving Target Indicator (GMTI). This would be a tracking device, based in space, which could pinpoint and follow the smallest of targets on earth. GMTI, the document says, will improve the ability to "detect, locate, identify and track a wide range of strategic and tactical targets we currently have minimal ability to detect, such as nuclear, biological and chemical weapons and activities, hidden targets and moving air targets." (No pagination)

This veritable military juggernaut in space would be further compartmentalized, hosting numerous subsidiary agencies designed to micro-manage the many variables inherent to the burden of cosmic supremacy. Mackay explains:

The worldwide scope of Space Command's project is shown by the names of some of the units under its control: Global Strike, Air and Space Expeditionary, Global Response Task Forces and Global Mobility Task Force. Space Command is also setting up a wing of the intelligence services devoted to the militarisation of space. Space Command says it is "aggressively modernising our existing nuclear forces." (No pagination)

Summarizing his examination of this technocratic strategy, Mackay deduces that the plan is anything but hypothetical or innocuous:

> The conclusion of the SMP report leaves no doubt of how important these plans are to the US military and government: "Expanding the role of space in future conflicts ... produces a fully integrated air and space force that is persuasive in peace, decisive in war and pre-eminent in any form of conflict." (No pagination)

As *Pax Americana* asserts itself throughout the world, the sociopolitical Darwinians of the West are turning their eyes back toward the heavens. One can rest assured that other elite factions are doing the same. Although the strategies may differ, the technocratic character of such plans remains the same. In addition, they all echo the same ambitions. From the elite's egomaniacal viewpoint, man shall look up at the heavens and finally see "God."

Maschinenmenschen: From Autonomous to Automaton

With the popularization of Darwinism, physicalistic philosophies of the mind seem to dominate both the scientific and academic communities. This paradigm equates mental states with brain states, thus reducing the concept of the "soul" or "spirit" to a metaphysical fantasy. This view seems to pervade modern psychology as well. Ironically, the word "psychology" is derived from the word *psyche*, which meant "soul" in the original Greek. However, imposing the metaphysical doctrine of materialism upon psychology, Wilhelm Maximilian Wundt would expunge the soul from the halls of psychological research and enshrine the primacy of matter. Several years later, B.F. Skinner would continue the materialist-physicalist tradition of psychology. Dubbed behaviorism, Skinner's brand of psychology emphasized observable behavior as the primary indicator of mental states. Working from this premise, Skinner developed a "technology of behavior" by which human nature could be conditioned and manipulated. Skinner believed that, as desirable behaviors were promulgated within the human herd, the ideal society would eventually emerge.

Skinner presented his psychologically engineered Utopia as a *roman a' clef* entitled *Walden Two*. Characterizing *Walden Two* as an innocuous fiction, Skinner stated: "The 'behavioral engineering' I had so frequently mentioned in the book was, at the time, little more than science fiction" (vi). Yet, "behavioral conditioning" was much more than science fiction to dark forces with dark intentions. Thanks to a $5,000 grant from a group called the Human Ecology Fund, Skinner was able to pay for the secretary and supplies he needed during the writing of *Beyond Freedom and Dignity* (Marks 171). When approached about the grant and its origins, Skinner claimed to have no memory of the contribution (Marks 171). However, he did make the slightly suspicious comment: "I don't like secret involvement of any kind. I can't see why it couldn't have been open and aboveboard" (Marks 171).

When one examines the Human Ecology Fund closer, the reasons for the secrecy become clear. It was assembled in 1955 under the title of the Society for the Investigation of Human Ecology, which would later change to the Human Ecology Fund in 1961 (Marks 159). For the sake of convenience, researcher John Marks simply calls it the Society. The Society itself was funded and controlled by the CIA "for studies and experiments in the behavioral sciences" (158). In addition to behavioral research, the Society also entertained a preoccupation with the occult: "No phenomenon was too arcane to escape a careful look from the Society, whether extrasensory perception or African witch doctors" (173).

The Society's president was Harold Wolff, a neurologist involved in CIA research and operations (Marks 156). The vice president was Lawrence Hinkle, Wolff's colleague from Cornell Medical College in New York City (Marks 135, 167). According to one long-standing CIA associate, Wolff was:

"an autocratic man. I never knew him to chew anyone out. He didn't have to. We were damned respectful. He moved in high places. He was just a skinny man, but talk about mind control! He was one of the controllers." (Marks 161)

Evidently, the organization itself took on the character of its president. One of its board members, Adolf Berle, expressed concerns over the Society's mind control projects:

"I am frightened about this one," Berle wrote in his diary. "If scientists do what they have laid out for themselves, men will become manageable ants. But I don't think it will happen." (Marks 167)

Perhaps "manageable ants" was what the Society had in mind when it financed Skinner in his behavioral research. In *Beyond Freedom and Dignity*, Skinner candidly states:

> What is being abolished is autonomous man—the inner man, the homunculus, the possessing demon, the man defended by the literatures of freedom and dignity.
> His abolition has long been overdue. Autonomous man is a device used to explain what we cannot explain in any other way. He has been constructed from our ignorance, and as our understanding in creases, the very stuff of which he is composed vanishes. Science does not dehumanize man, it de-homunculizes him, and it must do so if it is to prevent the abolition of the human species. To man qua man we readily say good riddance. Only by dispossessing him can we turn to real causes of human behavior. Only then can we turn from the inferred to the observed, from the miraculous to the natural, from the inaccessible to the manipulable. (189-91)

This, the vision of a "de-homunculized" and "manipulable" man, was probably what prompted the Human Ecology Fund's $5,000 investment in Skinner's research. When *Walden Two* was released, many critics saw "shades of Aldous Huxley's *Brave New World*" in Skinner's fictional Utopia (Taylor 418). This analogy is very appropriate. Like its Huxlian kissing cousin, the *roman a' clef* of *Walden Two* is a reality in the making. Serious credence has been given to Skinner's behavioral theories. His methodology of behavioral tyranny has been employed in today's educational system. Researcher Ian Taylor elaborates:

> Nevertheless, the Skinner teaching techniques have been widely used for school children, although by use of a teaching machine rather than in a box with food pellets! In addition, by cooperation with drug companies, the effects of certain drugs to aid children with learning difficulties have been studied. Although new understanding has been gained, the whole idea of modifying human behavior in a purposeful way has not been an overwhelming success and the specter of crossing that fine line, from "aid" to "control" of tomorrow's society in today's classroom, has yet to become a total reality. (419)

Meanwhile, Skinner's method has also been applied under the guise of therapy. Taylor explains:

The vision of behavioral modification still has its enthusiasts. For example, in 1978 Sobell and Sobell reported a program to modify the behavior of a group of twenty gamma alcoholics. In this they used the electric shock "punishment" technique. These researchers believed that behavior therapy would enable hard-core alcoholics to become social drinkers, rather than having to become total abstainers. The experiment was widely reported to be successful, and the United States government began to invest considerable sums of money into this new approach. However, an independent study of the same twenty patients in a ten-year follow-up showed a totally different picture with only one success. This is another scandal, and the most charitable conclusion would be that... the theory in the minds of the Sobells assumed greater importance than the facts. (419)

Skinner's alter ego in *Walden Two* probably most succinctly voiced the rationale guiding such psychological engineering programs:

"I've had only one idea in my life--the idea of having my own way. 'Control' expresses it--the control of human behavior.
. . .it was a frenzied, selfish desire to dominate.
I remember the rage I used to feel when a certain prediction went awry. I could have shouted at the subjects of my experiments, 'Behave! Behave as you ought!'" (271)

For Skinner and those who carry on his tradition, humanity is little more than a lab animal to be conditioned and controlled. This authoritarian mentality becomes all the more evident when Skinner states: ". . .Russia after fifty years is not a model we wish to emulate. China may be closer to the solutions I have been talking about, but a Communist revolution in America is hard to imagine" (*Walden Two* xv). In other words, the communism of mass murdering Red China is preferable to the Russian variety of communism. Why? The Russian communists did not go far enough.

Skinner concludes *Walden Two* with the following contention: ". . .in the long run man is determined by the state" (257). Of course, the omnipotent State was also the god of Georg Wilhelm Friedrich Hegel.

According to deceased researcher Antony Sutton: "Both Marx and Hitler have their philosophical roots in Hegel" (118). This is also the intended result of the methods employed by Skinner and his adherents: the obliteration of the individual and the apotheosis of the State. Such a goal synchronizes comfortably with the vision of the power elite, which might be one of the main reasons that Skinner's methods have enjoyed widespread application today.

Of such a psychologically engineered society, C.S. Lewis writes:

> . . .many a mild-eyed scientist in a democratic laboratory means, in the last resort, just what the Fascist means. He believes that "good" means whatever men are conditioned to approve. He believes that it is the function of him and his kind to condition men; to create consciences by eugenics, psychological manipulation of infants, state education and mass propaganda. Because he is confused, he does not yet fully realize that those who create conscience cannot be subject to conscience themselves. But he must awake to the logic of his position sooner or later; and when he does, what barrier remains between us and the final division of the race into a few conditioners who stand themselves outside morality and the many conditioned in whom such morality as the experts choose is produced at the experts' pleasure? If "good" means only the local ideology, how can those who invent the local ideology be guided by any idea of good themselves? (81)

Indeed, when they speak of a psychologically engineered society, the "mild-eyed scientist" and the fascist mean exactly the same thing. They mean a socialist totalitarian society where the "many conditioned" are controlled by the "few conditioners." In short, they mean a scientific dictatorship.

Such a concept is nothing new. It finds its proximate origins with Auguste Comte, the "principal disciple" of Saint-Simon (Fischer 70). Comte would expand on his mentor's ideas and develop the conceptual models for the technocrats of the next century. Ian Dowbiggin states:

> Twentieth century liberals' statist and coporatist bent, as well as their confidence in reform, government interventionism, and technocratic elites, can be traced back to the Comtean tradition of the previous century. (11)

According to Fischer, Comte promoted the "development of a surrogate religion, an idea that Saint-Simon also contemplated" (71). In Comte's hypothetical theocracy, social scientists comprised the new priesthood. Fischer explains:

> . . .Comte advanced the concept of a "sociocracy," defined as a new "religion of humanity." Sociologists were to identify the principles of this new faith and to implement them through a "sociolatry." The sociolatry was to entail a system of festivals, devotional practices, and rites designed to fix the new social ethics in the minds of the people. In the process, men and women would devote themselves not to God (deemed an outmoded concept) but to "Humanity" as symbolized in the "Grand Being" and rendered incarnate in the great men of history. (71)

If sociocracy was to be the "religion of humanity," then Positivism was its theology. Positivism was vintage scientism, upholding the epistemological rigidity of radical empiricism and supplanting classical metaphysics with the scientific method. Ironically, radical empiricist claims, with their rejection of causality, required no less faith than mystical ones. Moreover, metaphysics was originally the province of religion. Positivism was but one more installment in an ongoing series of secular religions birthed by the Enlightenment. E. Michael Jones further comments on the religious nature of Positivism:

> Positivism might be called the Church of the Enlightenment, and through it, Comte attracted a following that would make a significant contribution toward turning sociology, in the broadest sense of the term, which is how Comte intended it, into a system of control which would become the world's dominant regime by the end of the twentieth century. Aldous Huxley would call Comte's Positivism "Catholicism minus Christianity," and in this it was similar to Weishaupt's appropriation of Jesuit spirituality in the service of Freemasonry. Both men took what they found appealing in the Catholic Church and ripped it out of its matrix and introduced it to a radically different context which changed its meaning completely. Both took what were essentially mechanisms of self-control based on Catholicism's understanding of the

moral order and turned them into essentially heteronomous instruments of social control whose goal was the betterment of "humanity" and whose validating principle was "science." (92-93)

Comte was one of the chief proponents of the "Positive State," a societal model premised upon the "hegemony of science and industry" (Fischer 71). Essentially, this new society envisioned by Comte amounted to a scientific dictatorship where "ideals of liberty and equality would eventually be supplanted by the technocratic values of order and progress" (71). This technocratic vision for society was probably a product of Comte's tutelage under Saint-Simon. Through Saint-Simon, Comte would be introduced to "what Marx and Engels would later call utopian socialism or critical utopian socialism" (Jones 93). Accompanying Saint-Simon's advocacy of socialism was the contention that science and industry held primacy, a hallmark of Comte's "Positive State." Jones explains:

> It was from Saint-Simon that Comte got the idea that Industrialism was to be the new form of social order that would replace the old order which had been swept irrevocably away by the revolution. The new order was to be based on science, not the now discredited religion, because no one could argue with science, which [Mary] Shelley has said, was based on fact, not hypothesis. "*Hypothesi non fingo*," Newton had written, and Shelley had quoted the passage in a footnote to *Queen Mab* as the marching orders for the New Man who would bring about heaven on earth. (94)

Thus, Comte's "Positive State" was a totally mechanized society and its citizenry was to be mechanized as well. After all, in a "totally scientistic society," all things are subject to quantification. That included man himself. It stood to reason that, if man were a quantifiable entity, then his mental and social behaviors could be guided through the predictive control of science. Following Saint-Simon's lead, Comte attempted to create a true "science of humanity." According to Fischer, Comte's ideas:

> proved to be very influential in the rise of modern sociology. Many call him the father of the discipline, a fact that also

underscores the technocratic origins of modern social science itself. (71)

Saint-Simon developed his precursory form of social science as a theoretical *"means for bringing an end to the revolution"* (Billington 212). The same rationale underpinned Comte's development and popularization of sociology. Understandably, both Saint-Simon and Comte were horrified by the excesses of the French Revolution. E. Michael Jones attributes the atrocities of the Revolution to the Enlightenment philosophy that underpinned it: "The Enlightenment appeal to liberty invariably led to the suppression of religion, which led to the suppression of morals, which led to social chaos" (15). Ending the very revolutions that brought them to power has been the chief aim of technocratic oligarchs throughout history. Invariably, the result has been a scientific dictatorship.

The scientific dictatorship is designed to create a society where, in the words of Aldous Huxley, "most men and women will grow up to love their servitude and will never dream of revolution." Huxley himself was a revolutionary and amorality was the catalyst for his cause:

> I had motives for not wanting the world to have a meaning. For myself, as no doubt for most of my contemporaries, the philosophy of meaninglessness was essentially an instrument of liberation. The liberation we desired was simultaneously liberation from a certain system of morality. We objected to the morality, because it interfered with our sexual freedom. We objected to the political and economic system, because it was unjust. The supporters of these systems claim that in some way they embodied the meaning—a Christian meaning, they insisted—of the world. There was one admirably simple method of confusing these people and at the same time justify ourselves in our political and erotic revolt. We could deny that the world had any meaning whatsoever. (*Ends and Means* 270)

Paradoxical though it may seem, the pattern of liberation dissolving into tyranny has been the hallmark of every socialist revolution throughout history. The excesses of radical libertarianism, typified by the violence of the French Revolution and the "free love" philosophy of the sixties counterculture, provided the pretext for rigid social regimentation. Jones explains:

Freedom followed by Draconian control became the dialectic of all revolutions, and, in this regard, the sexual revolution was no exception. Once the passions were liberated from obedience to the traditional moral law as explicated by the Christian religion, they had to be subjected to another more stringent, perhaps "scientific" form of control in order to keep society from falling apart. (15)

Moral law is central to the maintenance of a democratic government. C.S. Lewis most eloquently voiced this truth when he wrote:

The very idea of freedom presupposes some objective moral law which overarches rulers and ruled alike. Subjectivism about values is eternally incompatible with democracy. We and our rulers are of one kind only so long as we are subject to one law. But if there is no Law of Nature, the ethos of any society is the creation of its rulers, educators and conditioners; and every creator stands above and outside his own creation. (81)

In the absence of moral law, freedom cannot exist. Jones recapitulates: "Social control was a necessary consequence of liberation, something which the French Revolution would make obvious" (15). To achieve "social control," the state required a "science of control." Comte played no small role in the development of such a science:

It was the chaos stemming from the French Revolution, in fact, which would inspire August Comte to come up with the "science" of sociology, which was in its way an ersatz religion but most importantly a way of bringing order out of chaos in a world which no longer found the religious foundation of morals plausible. (15-16)

The technocratic social sciences filled the vacuum left by religion and the scientist became its ordained proselyte. Comte's vision for a scientific dictatorship synchronized with the vision of Adam Weishaupt, founder of the infamous Illuminati. Like Comte, Weishaupt developed "a system of control that proved effective in the absence of religious sanction" (16). This Illuminist incarnation of the "Positive State" became the "model of every secular control mechanism of both left and right for the next two hundred years" (16).

In examining the doctrine of "reason" promulgated by the Masonic lodges of the Strict Observance, Weishaupt correctly observed that: "[m]orals, cut off from their ontological source, became associated as a result with the will of the man who understood the mechanism of control" (16). Thus, Weishaupt's own moral cognizance became his "will to power." The irony of this fact is painfully illustrated by Weishaupt's own excesses, including sexual perversion and abortion.

The internal contention experienced by the lodges of the Strict Observance was a direct corollary of the Masonic conception of "reason," which "led more often than not to conflicting ideas of which program to take" (16). By contrast, Weishaupt's vision for the Illuminati was even more technocratic than Masonic philosophy. The Illuminist was supposed to be the supreme social engineer and ultimate psychological conditioner. Jones expands on Weishaupt's agenda of behavioral tyranny:

> . . .the Illuminist system had to take the law into its own hands and program behavior as its leaders saw fit. In this Illuminism followed the typical trajectory of every other form of Enlightenment social science which would come into being over the next two hundred years. As in the case of Comte's sociology, the old church was replaced with a new church. The older order, which was based on nature and tradition and revelation, was replaced by a new totalitarian order which was based on the will of those in power. (16)

Of course, the Illuminst program for a "Positive State" failed. However, the Bavarian authorities' publication of Weishaupt's work guaranteed the continuation of the Illuminist vision (16). Jones elaborates:

> Once released into the intellectual ether, the vision of machine people in a machine state controlled by Jesuit-like scientist controllers would capture the imagination of generations to come, either as utopia in the thinking of people like Auguste Comte or dystopia in the minds of people like Aldous Huxley and Fritz Lang, whose film *Metropolis* seemed to be Weishaupt's vision come to life. (16-17)

Jones enumerates the various forms this ideational contagion has assumed over the years:

In Illuminism we find in seminal form the system of police state spying on its citizens, the essence of psychoanalysis, the rationale for psychological testing, the therapy of journal keeping, the idea of Kinsey's sex histories, the spontaneous confessions at Communist show trials, Gramsci's march through the institutions, the manipulation of the sexual passion as a form of control that was the basis for advertising, and, via Comte, the rise of the "science" of behaviorism, which attempts, in the words of John B. Watson, to "predict and control behavior." (17)

Skinner's "technology of behavior" is really nothing new. It originates with the technocratic thought of Comte and its objective is the creation of Weishaupt's *Maschinenmenschen*. Throughout the 20th century and well into the 21st century, the social sciences have ascended to institutional dominance. Under the epistemological primacy of these institutions, society has undergone extensive technocratic restructuring. No longer does the judicial system arbitrate in matters of crime and punishment, but, under the watchful eye of the social scientist, it decides who is "normal" and "abnormal" as well. The political system, which is charged with the maintenance of a civil society, must exercise its prerogatives and issue its mandates within the technocratic parameters of the social sciences. In this state of affairs, the state is becoming the panoptic machine and its citizenry the *Maschinenmenschen*.

Darwin: Patron Saint of Sociocracy

Comte's philosophy of Positivism also had a significant impact on Charles Darwin. For Darwin, Comte's concept of a "theological state of science" was a "grand idea" (Desmond and Moore 260). Years later, two "Positive States" premised upon the theology of Darwin's "science" would arise. These were, of course, Nazi Germany and Communist Russia. Appropriating theoretical legitimacy to scientific dictatorships may have been the intended function of evolutionary theory from the beginning. The involvement of Freemasonry in Darwinism's popularization certainly reinforces this contention.

The social sciences, which were already devoted to the technocratic restructuring of society, were closely aligned with Darwinism. Harriet Martineau, a fanatical adherent of Comte's sociology, was even Darwin's dinner guest (Desmond and Moore 264). Her booklets, entitled *Poor Laws and Paupers Illustrated*, were recommended to Darwin by his sisters (153). Martineau's work had even drawn favorable attention from those in more esoteric quarters, as is evidenced by Freemason Erasmus Darwin's affinities for the sociologist. Writing to Charles during the *Beagle* voyage, his sisters commented on Erasmus' admiration of Martineau:

> "Erasmus knows her [Martineau] & is a great admirer & everybody reads her little books & if you have a dull hour you can, and then throw them overboard, that they may not take up your precious room." (Qutd. in Desmond and Moore 153)

Enamored of Comte's Positivism, Martineau would translate his work from French to English (261). In so doing, she would declare:

> "We find ourselves suddenly living and moving in the midst of the universe. . .not under capricious and arbitrary conditions. . .but under great, general, invariable laws, which operate on us as a part of a whole." (qutd. in Desmond and Moore 261)

This view of the cosmos being governed by an impersonal principle of predestination was consistent with the Masonic concept of the Great Architect, which was appointed scientific currency by Darwin's evolutionary determinism. Comte's Positivism only intensified this deterministic Weltanschauung (261). With his denial of free will, Darwin would attribute all human characteristics to heredity:

> He [Darwin] now routinely reduced thought and behaviour to cerebral structure, boiling it down to bits of the brain. If wishes are a consequence of neural organization-—evolving under the "circumstance & education"—then anti-social behaviour can be inherited. "Verily the faults of the fathers, corporeal & bodily are visited upon the children." (261)

Darwin's physicalist metaphysics and evolutionary determinism provided the rationale for rigid social engineering. The social sciences were designed to serve precisely this function. Since aberrations like "antisocial personalities" were genetically predetermined, man's evolutionary development had to be checked by the "Positive State." Of course, the socially and economically disadvantaged were bred within the dysgenic gene pool of the poor. Thus, the lower classes required the regulation of the technocratic social scientist.

Sociologists like Martineau viewed the eugenical regimentation of society as one method of social engineering. Not surprisingly, Martineau subscribed to Malthus' concept of carrying capacity, a myth that was central to Darwinian evolution. Astride Martineau's "edifying homilies," Malthus' theoretical eschatology enjoyed widespread exposure (153). Martineau's proselytizing was very effective. One pundit insisted that credence to Malthus' demographic prognostications promised to do "more for the country than all the Administrations since the Revolution" (qutd. in Desmond and Moore 154).

Characterizing the poor as the "gangrene of the state," Martineau endorsed the genocidal Poor Law Amendment Bill (153-54). In fact, Martineau received secret commission reports concerning the unpopular law from Lord Chancellor Henry Brougham (153). Her *Poor Laws and Paupers Illustrated* also did "more to pave the way for the new Poor Law than all of the government propaganda" (153). Arguably, Martineau's literature qualified as Malthusian propaganda itself. Martineau was a "darling of the Whigs," a political party favoring the Poor Law Amendment Bill (153). Martineau contended that the reforms would make the poor more self-sufficient (154). However, by immediately thrusting unskilled paupers into a competitive job market, the Whigs were actually "decreasing labour costs and increasing profits" (154). Evidently, Martineau's technocratic social agenda harmonized rather smoothly with corporate interests. A similar alliance exists between technocrats and Transnationalists today, as is evidenced by the techno-corporatism of the Trilateral Commission. At any rate, Martineau's contemporaries in the social sciences also seek to reconfigure society according to Malthusian designs.

The social Darwinian war against the impoverished continues to this very day, as is evidenced by the contemporary concept of "spatial deconcentration." This concept was devised by Anthony Downs, an Establishment-anointed "housing expert." Downs describes spatial deconcentration as an "adequate outmigration of the poor" from urban areas (176). This prescription for the "War on Poverty" bears eerie resemblance to the Khmer Rouge's tactics in Cambodia. As a part of campaign of eugenical regimentation, the Khmer Rouge force Cambodians out of the cities and into the rural areas. The end result was the infamous "Killing Fields." Anthony Downs went onto write the chapters over housing in the Kerner Report.

Herbert Spencer, a personal friend of Darwin's, would successfully integrate evolutionary theory with the already positivistic field of sociology. Among one of the evolutionary concepts most rigidly applied to sociology by Spencer was natural selection. John W. Burrow views sociology's assimilation of this evolutionary principle rather adversely:

Finally, there is the question of natural selection. In one sense, the influence of the theory of natural selection on sociology was enormous. It created for a while, in fact, a branch of sociology. It seems now to be felt that the influence on sociology of the doctrine of "survival of the fittest" was theoretically speaking, unfortunate, chiefly because it seemed to offer an explanatory short cut, and encouraged social theorists to aspire to be Darwins when probably they should have been trying to be Linnaeuses or Cuviers. As Professor MacRae points out, in sociology the principle explains too much. Any state of affairs known to exist or to have existed can be explained by the operation of natural selection. Like Hegel's dialectic and Dr. Chasuble's sermon on *The Meaning of Manna in the Wilderness*, it can be made to suit any situation. However, "Social Darwinism" was only a subspecies of the intellectual movement we are considering. Neither Maine, nor Tylor, nor McLennan made much use of the theory of natural selection and Spencer used it only as a garnish for a theory he had already developed. (115)

The coalescence of sociology and evolutionary theory, particularly the Darwinian principle of natural selection, has been anything but smooth. Donald G. McRae comments:

A peculiarity of Darwinism, both in biology and in other fields, is that it explains too much. It is very hard to imagine a condition of things which could not be explained in terms of natural selection. If the state of various elements at a given moment is such and such then these elements have displayed their survival value under the existing circumstances, and that is that. Natural selection explains why things are as they are: It does not enable us, in general, to say how they will change and vary. It is in a sense rather a historical than a predictive principle and, as is well known, it is rather a necessary than a sufficient principle for modern biology. In consequence its results when applied to social affairs were often rather odd. (304)

Indeed, it had been odd, if not absolutely frightening. The result has been the re-sculpting of society along the blood-stained contours of natural selection. Historical campaigns of eugenics, genocide, imperialism, state socialism, technological apartheid, and slavery all define the character of this conceptual integration. This prompts some interesting questions. What is so natural about natural selection if its social application results in such unnatural volumes of death? Moreover, if evolutionary theory were an immutable reality confirmed by "objective science," then why does it require human application at all? Adrian Desmond and James Moore may have already answered the question:

> "Social Darwinism" is often taken to be something extraneous, an ugly concretion added to the pure Darwinian corpus after the event, tarnishing Darwin's image. But his notebooks make plain that competition, free trade, imperialism, racial extermination, and sexual inequality were written into the equation from the start--"Darwinism" was always intended to explain human society. (xxi)

Darwinism itself was always a social theory, not a scientific one. It was designed according to Darwin's presuppositions, which were already oligarchical in character. Darwin was surrounded by aristocrats, technocrats, and other elitists. Freemason T.H. Huxley, who was involved in the establishment of the oligarchical Round Table groups, is just one case in point. The influence of such elements is evident in the Darwinian concept of natural selection itself. Ian Taylor observes that:

> the political doctrine implied by natural selection is elitist, and the principle derived according to Haeckel is "'aristocratic in the strictest sense of the word'" (411).

Darwinism facilitates the revolutionary dialectic of "[f]reedom followed by Draconian control." First, it appropriates currency to moral relativism, an economy of thought already bankrupted by self-refuting logical contradictions. H.G. Wells reiterates:

> If all animals and man evolved, then there were no first parents, no paradise, no fall. And if there had been no fall, then the entire historic fabric of Christianity, the story of the first sin, and the reason for the atonement collapses like a house of cards. (*The Outline of History* 616)

Subsequently, the architects of revolution establish their "sociocracy" over the thoroughly demolished "house of cards." Jane H. Ingraham explains:

"His [Darwin's] shattering 'explanation' of the evolution of man from the lower animals through means excluding the supernatural delivered the *coup de grace* to man's idea of himself as a created being in a world of fixed truth. Confronted with the 'scientific proof' of his own animal origin and nature, Western man, set free at last from God, began the long trek through scientific rationalism, environmental determinism, cultural conditioning, perfectibility of human nature, behaviorism, and secular humanism to today's inverted morality and totalitarian man." (Qutd. In Jasper, *Global Tyranny. . .Step by Step* 262-63)

William Jasper eloquently synopsizes this observation:

The rejection of Divine revelation and the sovereignty of God has resulted in the enthronement of man's "reason" as the ultimate source of truth and the apotheosis of the State as the supreme authority. (*Global Tyranny. . .Step by Step* 263)

In essence, Darwinism was an epistemological weapon for sociocratic revolution. As such, it was destined to merge with the rest of the technocratic social sciences. This was a prearranged marriage and one that was made in Hell. In the contemporary religious milieu of sociolatry, the golden calf of the Israelites has been exchanged for the golden ape-man of Darwinism.

The Social Scientific Dictatorship
In the *Science of Coercion*, Christopher Simpson writes:

Communication research is a small but intriguing field in the social sciences. This relatively new specialty crystallized into a distinct discipline within sociology-—complete with colleges, curricula, the authority to grant doctorates, and so forth—- between about 1950 and 1955. (5)

Indeed, communication research comprises a sizable portion of the standard "college- and graduate-level" curriculums (5). These academic programs produce "print and broadcast journalists, public relations and advertising personnel" and other closely aligned media experts that constitute the "ideological workers" of modernity (5). While the various professions in mass communication do not automatically qualify as technocratic vocations *per se*, the field's subsumption under sociology does predispose its occupations to technocratic applications. As a subsidiary of the technocratic social sciences, mass communication research has the capacity for being transmogrified into a weaponized form of semiotics. The military establishment is already acutely aware of this application:

> U.S. military, propaganda, and intelligence agencies favored an approach to the study of mass communication that offered both an explanation of what communication "is" (at least insofar as those agencies were concerned) and a box of tools for examining it. Put most simply, they saw mass communication as an instrument for persuading or dominating targeted groups. They understood "communication" as little more than a form of transmission into which virtually any type of message could be plugged (once one had mastered the appropriate techniques) to achieve ideological, political, or military goals. (5-6)

While the hawks sought to militarize communication research, the social scientists of academia hoped to use the field for their own technocratic purposes:

> Academic contractors convinced their clients that scientific dissection and measurement of the constituent elements of mass communication would lead to the development of powerful new tools for social management, in somewhat the same way earlier science had paved the way for penicillin, electric lights, and the atom bomb. (5-6)

Meanwhile, there were federal clients who viewed the "analysis of audiences and communication effects" as an instrument for the enhancement of "ongoing propaganda and intelligence programs" (6). Given this multiplicity of dubious parties that expressed an interest in communication research, it is reasonable to assume that the field had polyvalent applications in terms of the technocratic restructuring of society. The emergent scientific dictatorship in the West had discovered a new weapon. Predictably, the *modus operandi* was purely scientistic in character:

> Entrepreneurial academics modeled the scientific tools needed for development of practical applications of communication-as-domination on those that had seemed so successful in the physical sciences: a positivist reduction of complex phenomena to discrete components; an emphasis on quantitative description of change; and claimed perspective of "objectivity" toward scientific "truth." With a few exceptions, they assumed that mass communications was "appropriately viewed from [the perspective] the top or power center," Steven Chaffee and John Hochheimer put it, "rather than from the bottom or periphery of the system." (6)

Converted into an effective semiotic weapon, mass communication research played an active part in the Second World War. The mode of conflict was appropriately dubbed "psychological warfare" (24). This appellation was derivative of the German word *Weltanschauungskrieg*, a term cribbed from the conceptual lexicon of the Nazis in 1941 (24). The word literally means "worldview warfare" and connotes the "scientific application of propaganda, terror, and state pressure as a means of securing an ideological victory over one's enemies" (24). William "Wild Bill" Donovan, who was the director of the Office of Strategic Services in 1941, believed that the Nazis' psychological warfare methods could act as models for "Americanized" stratagems (24). Psychological warfare swiftly became part of the U.S. intelligence community's operational lexicon (24). Donovan believed the concept to be so significant that it would inevitably become "a full arm of the U.S. military, equal in status to the army, navy, and air force" (24).

Six organizations constituted the nucleus of U.S. psychological warfare research (26). These were the:

> (1)Samuel Stouffer's Research Branch of the U.S. Army's Division of Morale; (2)the Office of War Information (OWI) led by Elmer Davis and its surveys division under Elmo Wilson; (3) the Psychological Warfare Division (PWD) of the U.S. Army, commanded by Brigadier General Robert McClure; (4) the Office of Strategic Services (OSS) led by William Donovan; (5) Rensis Likert's Division of Program Surveys at the Department of Agriculture, which provided field research

personnel in the United States for the army, OWI, Treasury Department, and other government agencies; and (6) Harold Lasswell's War Communication Division at the Library of Congress. (26)

Of course, this wartime network was peopled heavily by "prominent social scientists" (26). In some instances, the same social engineers participated in two or more organizations (26). Simpson enumerates the various social scientists involved:

> The OWI, for example, employed Elmo Roper (of the Roper survey organization), Leonard Doob (Yale), Wilbur Schramm (University of Illinois and Stanford), Alexander Leighton (Cornell), Leo Lowenthal (Institut fur Sozialforschung and University of California), Hans Speier (RAND Corp.), Nathan Leites (RAND), Edward Barrett (Columbia), and Clyde Kluckhohn (Harvard), among others. (26)

The Army's Psychological Warfare Division was also largely staffed by social scientists, some of which being OSS officers as well (27). The OSS assigned Morris Janowitz (University of Michigan and Institut fur Sozialforschung), Murray Gurfein, Saul Padover (New School for Social Research), and W. Phillips Davison (Columbia and Rand) to the Psychological Warfare Division to employ their proficiency in "communication and German social psychology" (27). According to Art Kleiner, this wartime network:

> was generally an immense catalyst for social science in America (and England), because it pulled university researchers from their isolated posts. They worked together on real-world problems such as keeping up military morale, developing psychological warfare techniques, and studying foreign cultures. (33)

Indeed, the ascendance of the social sciences had begun. The OSS contributed substantially to this rise. Howard Becker (University of Wisconsin), Douglas Cater (Aspen Institute), Walter Langer (University of Wisconsin), Alex Inkeles (Harvard), and Herbert Marcuse (Institut fur Sozialforschung and New School for Social Research) were all "prominent OSS officers who later contributed to the social sciences"

(Simpson 27). However, OSS support extended beyond governmental channels. Simpson explains:

> OSS wartime contracting outside the government included arrangements for paid social science research by Stanford, the University of California at Berkley, Columbia, Princeton, Yale's Institute of Human Relations, and the National Opinion Research Center, which was then at the University of Denver. Roughly similar lists of social scientists and scholarly contractors can be discovered at each of the government's centers of wartime communications and public opinion research. (27)

During Senate hearings in early November 1945, OSS officer Brigadier General John Magruder adamantly maintained that:

> the government of the United States would be well advised to do all in its power to promote the development of knowledge in the field of social sciences. . .Were we to develop a dearth of social scientists, all national intelligence agencies servicing policy makers in peace or war would be directly handicapped. . .[R]esearch of social scientists [is] indispensable to the sound development of national intelligence in peace and war. (Qutd. in Simpson 32).

Given Magruder's prominence in the OSS, it is reasonable to assume that this contention represented the status quo within the fledgling intelligence organization. The consensus among those involved in psychological warfare was that the social sciences, which had been successfully tested during an exceptionally violent conflict, possessed equally promising potentials in times of peace. The weapon had become the surgical knife. Now, the incisions were to be made to the postwar psyche of the public mind.

The constellation of World War II psychological warfare programs provided its alumni with a "network of professional contacts" that proved to be "very valuable in their subsequent careers" (Simpson 28). In fact, many received influential positions within the tax exempt foundations of the power elite:

> Charles Dollard became president of Carnegie. Donald Young shifted from the presidency of SRCC [Social Science Research

Council] to that of Russell Sage, where he ultimately recruited Leonard Cottrell. Leland DeVinney went from Harvard to the Rockefeller Foundation. William McPeak. . .helped set up the Ford Foundation and became its vice president. W. Parker Maudlin became vice president of the Population Council. The later Lyle Spencer [of Science Research Associates]. . .endowed a foundation that currently supports a substantial body of social science research. (Qutd. in Simpson 28)

Of course, these tax exempt foundations play an integral role in the ruling class conspiracy. First, they provide tax shelters for the elite's wealth. In addition, they heavily finance socialist revolutionary movements, which provide a politically and socially expedient terrorist threat to the populace. Finally, they support further social science research, which provides the oligarchs with the necessary psychocognitive arsenal to wage their *Weltanschauungskrieg*.

Meanwhile, social engineers were also silently co-opting the mass media. Former members of the OWI became:

the publishers of *Time*, *Look*, *Fortune*, and several dailies; editors of such magazines as *Holiday*, *Coronet*, *Parade*, and the *Saturday Review*, editors of the *Denver Post*. New Orleans *Times-Picayune*, and others; the heads of the Viking Press, Harper & Brothers, and Farrar, Straus and Young; two Hollywood Oscar winners; a two-time Pulitzer prizewinner; the board chairman of CBS, and a dozen key network executives; President Eisenhower's chief speech writer; the editor of *Reader's Digest* international editions; at least six partners of large advertising agencies; and a dozen noted social scientists. (Qutd. in Simpson 29)

With the proselytes of sociocracy occupying strategically sensitive positions in the media, the power elite could re-sculpt public opinion. The semiotic deception following the events of September 11, 2001 is just one case in point (see Chapter Two: The Semiotics of Sci-fi Predictive Programming for further explication). The cultural milieu shaped by the mass media shortly after September 11th was one that was more amenable to technocratic restructuring, a fact painfully evidenced by the popular support for the draconian Patriot Act. Other examples of media manipulation are far too voluminous to enumerate. Ultimately, it is important to note the role of mass media in service to the purveyors of sociocracy.

Subsumed under the technocratic social sciences and refurbished for military applications, mass communication research has contributed to the epistemological primacy of the power elite's anointed priesthood. The field's development and application is now guided by a coterie of technocratic "experts." Legitimacy is bestowed only upon those theoreticians and ideologues that maintain the status quo. Simpson elaborates:

Government psychological warfare programs helped shape mass communication research into a distinct scholarly field, strongly influencing the choice of leaders and determining which of the competing scientific paradigms of communication would be funded, elaborated, and encouraged to prosper. The state usually did not directly determine what scientists could or could not say, but it did significantly influence the selection of who would do the "authoritative" talking in the field. (5)

Under the influence of the military establishment, the field of mass communication has become the orthodoxy of sociocracy. It has become an epistemological cartel, selecting the most socially and politically expedient "scientific paradigms of communication" that would be "funded, elaborated, and encouraged to prosper." It has become technocratic, maintaining a pseudo-meritocracy where the "authoritative" requires the sanction of the influential. In short, it has become a social scientific dictatorship.

The DSM Reconsidered

Understandably, Michel Foucault harbored great disdain for the social sciences and posited them as one of the chief facilitators for the rise of the panoptic machine. In the emergent carceral culture, Foucault argued, everyone was subject to the normalizing gaze of the social scientist. In order to know exactly just what "abnormalities" come under the scrutiny of this pervasive gaze, one need only read the American Psychiatric Association's Diagnostic Statistical Manual (DSM).

The DSM qualifies as a modern *Malleus Maleficarum* of sorts. Published in 1486, *The Malleus Maleficarum* (translated: *The Witch Hammer*) was an infamous text purporting to be an authoritative guidebook that could be used to identify practitioners of witchcraft. However, the book had more to do with snuffing out the Church's competition than it did with recognizing witches. At the time, herbal healers had more success curing people with alternative methods than did the priests with highly stylized rituals. Under the pretext of delivering the world from evil, innovation and eccentricity were criminalized. *The Malleus Maleficarum* played no small role in the process.

Likewise, the DSM has served a similar function in the marginalizing and, on occasion, incarceration of potential innovators. Now printed in four editions, the DSM is "the billing bible for mental disorders which commingles neurological diseases with psychiatric diagnoses" (O Meara, no pagination). While *The Malleus Maleficarum* stigmatized certain modes of thought and behavior as "witchcraft," the DSM stigmatizes them as "disorders." In an interview with *OMNI* magazine, R.D. Laing expands on the role of the DSM in marginalizing divergent paradigms:

In the later sixties it became apparent to the elite with the responsibilities for "control of the population" that the old idea of putting people in the proverbial bin and keeping them there for life--warehousing people--wasn't cost-effective. The Reagan administration in California was one of the first to realize this. So they had to rethink just what is said to the general public and what is practiced by the executive in control of mental health. The same problem prevails across Europe and the Third World.

To see what is happening, look at the textbook or manual called DSM-III: The Diagnostic Statistical Manual on Mental Disorders. Translated into economic and political terms, mental disorder means undesired mental states and behavior. The criteria for mental disorder in DSM-III include any unusual perceptual experience, magical thinking, clairvoyance, telepathy, sixth sense, sense of a person not actually present. You're allowed to sense the presence of a dead relative for three weeks after their death. After that it becomes a criterion of mental disorder to have those feelings.

. . .these are not exceptional examples out of DSM-III. The overall drift is what contemporary modern psychiatry, epitomized by this DSM manual translated into eighteen languages, is imposing all over the world - a mandate to strip anyone of their civil liberties, of habeas corpus; and to apply involuntary incarceration, chemicalisation of a person, electric shocks, and non-injurious torture; to homogenize people who are out of line. Presented as a medical operation, it is an undercover operation. (Liversidge 60-61)

Under the pretext of promoting mental health, the DSM has been instrumental in the stifling of cognitive dissent. Who, specifically, are the cognitive dissenters? In *The Architecture of Modern Political Power*, researcher Daniel Pouzzner presents an interesting assertion. Pouzzner contends that one of the power elite's greatest fears is chaos, more specifically the sort of chaos generated by innovation:

Fear of chaos is not unique to the power brokers. It is much more common than that. It is, in short, an important example of fear of the unknown—in practical terms, it is fear of the unknowable. This fear is a classic characteristic of small minds and of those of meager confidence. It is often observed that investors tend to hate uncertainty: today, roughly half of the value of US stock markets is held by individual investors, and 45 percent of American households own stock directly or indirectly. Chaos of the type introduced by innovators produces very serious uncertainty for these investors, and they hate it. Thus, because of fear and short-term interest, the bulk of mainstream first-worlders, being small-minded, tacitly supports the neutralization, or even extermination, of uncooperative innovators. In fact, the ordinary feel offended and disgraced by these innovators, and for that the innovators are resented like no other group. The small-minded must become larger-minded if they are to realize that they, too, are slated for enslavement and capricious extermination— except that they have, as a rule, already resigned themselves to obedient slavery in exchange for survival. The power brokers are the total enemies of the innovators and the masses alike, but the masses cower and bow, signalling their surrender. (Pouzzner, no pagination)

Because innovation abruptly reconfigures the socioeconomic playing field, the inventive personality is one of the greatest threats to the power of the ruling class. Innovators can potentially destabilize the elite's inequitable system of control and re-establish meritocracy. Innovators can introduce genuine competition to the marketplace, thus exposing the oligarchs' illusion of counterfeit capitalism and facilitating the emergence of a truly free enterprise system. As practitioners of usury, the parasitic ruling class cannot allow this to happen. The abatement of just such a shift in the power balance is precisely the function for which the DSM was designed. Pouzzner explains:

> The cultural prejudice against chaos is evident in contemporary language itself. Diseases of the mind are routinely referred to as ``disorders,'' whether or not they present themselves as, or are caused by, an imbalanced abundance of randomness. Dissociative Identity Disorder (DID), historically known as Multiple Personality Disorder (MPD), is not a disorder at all, but is in fact an additional level of ordered mental arrangement. In fact, most DSM-IV (American Psychiatric Association standard) mental illness involves minds and brains that are more ordered than healthy minds and brains. Chaos is healthy, and empowers consciousness. Order is morbid. An unusually regular and orderly electrocardiogram (EKG) is an indication of nascent illness; certain elements of chaos in heart rhythms are indications of good health. Another term that propels the prejudice is ``unstable,'' often used as a synonym for ``insane.'' This use of that term must be condemned with equal haste. As Ilya Prigogine (Nobel laureate and Clubber of Rome) observes, "over time, non-equilibrium processes generate complex structures that cannot be achieved in an equilibrium situation." (Pouzzner, no pagination)

The DSM is integral to civil commitment, one of the elite's legal instruments for the criminalizing of potential dissenters. Pouzzner elaborates:

> A more established institution in the same vein is civil commitment, which operates like civil forfeiture, with a reduced burden of proof, only the object seized by the state is an actual living human individual. Civil commitment is an extraconstitutional mechanism by which private citizens licensed by a committee of executive appointees cause the forcible imprisonment of individuals charged with no crime, with subsequent judicial review based principally on standards promulgated by the private American Psychiatric Association in its Diagnostic and Statistical Manual of Mental Disorders (the "DSM"). (Pouzzner, no pagination)

In short, all those who deviate from the Establishment's arbitrary criteria for mental health are incarcerated and assigned one of the APA's various stigmas... "unstable," "disturbed," or just plain "criminally insane." Typically, the recipients of such stigmas are the innovators who threaten the oligarchs' dominance. Worse still, the elasticity of such stigmas is increasing. According to Larry Akey, spokesman for the Health Insurance Association of America, "New mental illnesses are being included in DSM 4 all the time" (Porteus, no pagination). For every potential innovator, there is now a potential mental illness.

Observing the growing elasticity of qualifiers for mental illness, Kelly Patricia O Meara states:

A child who doesn't like doing math homework may be diagnosed with the mental illness developmental-arithmetic disorder (No.315.4). A child who argues with her parents may be diagnosed as having a mental illness called oppositional-defiant disorder (No.313.8). And people critical of the legislation now snaking through Congress that purports to "end discrimination against patients seeking treatment for mental illness" may find themselves labeled as being in denial and diagnosed with the mental illness called noncompliance-with-treatment disorder (No.15.81) (No pagination)

The list of so-called "mental disorders" continues to grow. Of course, if a divergent mode of thought or behavior does not find a corresponding "mental disorder" in the current DSM, the social engineers are always willing to invent a new one. Such is the case with the purported "mental disorder" of Attention Deficit Disorder. Already, this chimerical illness is drawing some healthy skeptical criticism. Kelly Patricia O Meara elaborates:

Fred Baughman, a San Diego neurologist and leading critic of the alleged mental illness called attention-deficit/hyperactivity disorder (ADHD), tells Insight the question that must be answered before a mental illness can qualify as a disease is this: "Where is the macroscopic, microscopic or chemical abnormality in any living patient or at death/autopsy?" (O Meara, no pagination)

Does ADHD even exist? The absence of any "macroscopic, microscopic or chemical abnormality in any patient" certainly drives one more nail into the coffin of this alleged "mental disorder." Yet, the imaginary illness remains part of the litany of stigmas and more await invention. Worse still, the standards for establishing mental illness have plunged into ambiguity. Baughman reiterates:

> "No one is justified in saying anyone is medically abnormal/ diseased until such time as they can adduce some such abnormality. This, by the way, would apply to a person suspected of having diabetes or cancer." The fact is, Baughman adds, "There is no psychiatric diagnosis for which any part of this question can be answered in the affirmative. In other words: no abnormality; no disease. There is no confirmation of abnormality in the brain in life or at autopsy for any of the psychiatric diagnoses. And they [in the psychiatric community] don't say this because it's part of the propaganda campaign to make patients out of normal people. The findings at autopsy would be very specific and would reveal whether it is a diseased brain and, if so, which disease it is. There is no proof in life or at autopsy of any of the alleged psychiatric mental illnesses, including schizophrenia, psychosis, depression, OCD or ADHD." (O Meara, no pagination)

For every independent thinker, there is a corresponding "mental disorder." If one does not currently exist, social engineers of the Establishment can always concoct one. Daniel Pouzzner provides an eloquent summation:

> The power brokers work to eradicate chaos both because of their own fear of it, and because they seek to eradicate the innovation it leads to (and the chaos which leads from innovation), insofar as that innovation and chaos directly threatens their hegemony. (Pouzzner, no pagination)

To prevent the innovators from toppling their epistemological cartel, the oligarchs of the Establishment work in tandem with the technocratic American Psychiatric Association. It is the proverbial church and state relationship. The Establishment dictates the prerogatives of the state and the church enforces them as the "will of God." However, the APA's "God" is hardly some benevolent spiritual entity. In a cultural milieu where materialism and physicalism rule, there can be nothing outside this ontological plane of existence. "God" is now the "exalted principle" of evolution and man is created in its image. The reflection of such a "God" amounts to little more than a soulless amalgam of behavioral repertoires.

Yet, humanity is not so easily reduced to a soulless amalgam of behavioral repertoires. Taylor explains:

> As in the case of biological determinism (nature), behavioral determinism (nurture) also denies the free will, since this says, in effect, that we are simply a product of our environment rather than a product of our genes. Clearly, both factors are important, but even then the human psyche involves far more than mere machine response to a combination of biological and environmental circumstances. It would be extremely difficult for humanistic psychology, however, based as it is on evolution, to acknowledge a spiritual dimension to man; this opens up a philosophical minefield involving the destiny of souls, for instance... the committed humanist cannot accept such a view. (419)

Humanistic psychology's rejection of the soul is a direct consequence of the field's characteristic physicalism. Physicalism rejects the soul because it does not conform to the developmental framework of evolutionary theory. How does the soul, an incorporeal entity, spontaneously generate itself within the purely corporeal processes of evolution? Yet, simultaneously, evolution posits a similar concept of spontaneous generation. The only difference is that the evolutionary concept of spontaneous generation, dubbed "abiogenesis," applies exclusively to physical organisms. This quasi-Gnostic doctrine of "self-creation" simply raises new questions. Moreover, like property dualism, it is no less mystical in character. In a universe where neo-Darwinism holds sway, what else is the organism but the veritable Kabalistic golem?

Physicalists criticize substance dualism because it cannot demonstrate mental causality. Yet, paradoxically, their own empirical epistemology forces them to relegate physical causality to the realm of metaphysical fantasy. What is perceived as A causing B could be merely a consequence of circumstantial juxtaposition. Although temporal succession and spatial proximity are axiomatic, causal connection is not. Affirmation of causal relationships is impossible. Given the absence of causality, all of a scientist's findings must be taken upon faith. Ironically, science relies on the affirmation of such cause and effect relationships.

This is all one can deduce while working under the paradigm of radical empiricism. Thus, causality, mental or physical, does not provide adequate grounds for rejecting the possibility of substance dualism.

Moreover, the physicalist ignores the semiotic character of behavior. Physicalism contends that behavior provides the only adequate indicator of "mental states." In fact, physicalism's reductionistic metaphysics virtually obliterate the "mental state," transforming it into a proverbial mirage imposed upon observable behavior by the percipient. However, observable behavior is not always an accurate indicator of "mental states," as is evidenced by the complex undercurrent of connotative meaning beneath the surface of behavior itself. Nonverbal gestures can semiotically communicate messages beyond what a behavior ostensibly denotes. For instance, the right hand over one's heart during the Pledge of Allegiance semiotically gesticulates towards more abstract and intangible concepts. . .patriotism, Americanism, etc.

Weishaupt, founder of the infamous Illuminati, was one individual who recognized the semiotic character of observable behavior. In understanding the correlation between visible gestures and abstract thought, Weishaupt was able to develop a system of control called "*Seelenspionage*," which means "spying on the soul" (Jones 14). This system allowed the Illuminist hierarchy to:

> . . .get access to the adept's soul by close analysis of the seemingly random gestures, expressions, or words that betrayed the adept's true feelings. Von Knigge, who was privy to the system, referred to it as a "*Smiotik der Seele*."
> "From the comparisons of all these characteristics," von Knigge wrote, "even those which seem the smallest and least significant, one draw conclusions which have enormous significance for knowledge of human beings, and gradually draw out of that reliable semiotics of the soul. (14)

While Weishaupt's system of semiotic manipulation was extremely powerful and pervasive, it still suffered from the same sort of gross oversimplification exhibited by physicalism. Although Weishaupt identified a semiotic dimension to observable behavior, he made the mistake of attempting to reduce its complex undercurrent of connotative meaning to simple schematicism. Such an approach is analogous to Delsartes' method of acting, which contended that truly great performers could mimic all the proper gestures to communicate the emotional state of their characters. Of course, the human soul is not so easily reduced to a "paint-by-numbers" schematic. Such reductionism underlies the operational protocol of every scientific dictatorship. The Illuminati's vision for society was certainly no exception.

Perhaps physicalism's greatest drawback is its inherent scientism. Scientism is, in essence, a form of epistemological imperialism. It stipulates the imposition of physical science upon all fields of inquiry. Those items and factors that fail to conform to the rigid parameters of scientific exactitude are automatically precluded. Thus, scientism concerns itself exclusively with quantifiable entities. In hopes of avoiding the ostensibly disjunctive relationship between natural and metaphysical phenomenon, scientism bestows absolute primacy upon naturalism. Anything that defies or falls outside of the realm of naturalistic explanation is selectively disregarded. Of course, the soul is one such unfortunate element. As a result, it is relegated to the realm of abstraction and metaphysical fantasy.

Such fetishization of science was a hallmark of Enlightenment humanism, which would result in the bloodbath of the French Revolution. It also underpinned the political and economic doctrines of both Nazi Germany and communist Russia. Scientism remains at the core of sociopolitical Utopianism. In turn, sociopolitical Utopianism remains at the core of all governmental aberrations. When extended beyond its legitimate fields of application, science becomes a rigid template to which even the most complex of entities, like man, must conform. Historian Richard J. Sutcliffe states:

> In the last hundred years or so, "scientific" views of history have become increasingly popular, for humanity as a statistical whole is thought of as being subject to analysis and prediction. In this thinking, once the motivations of the masses could be measured and tabulated, their response to economic or technological stimuli could be accurately predicted. Appropriate technology and education could then be adapted to engineer and control the desired society. Such theories are popular among both political rightists and leftists, neither of whom realize that they are advocating the same kind of society--a sort of "scientific totalitarianism" or "technocratic dictatorship." (No pagination)

Hoffman reiterates:

> The doctrine of man playing god reaches its nadir in the

philosophy of scientism which makes it possible the complete mental, spiritual and physical enslavement of mankind through technologies such as satellite and computer surveillance; a state of affairs symbolized by the "All Seeing Eye" above the unfinished pyramid on the U.S. one dollar bill. (50)

Contemporary totalitarianism is the outward expression of scientistic hubris. Every scientific dictatorship of the 20th century has attempted to reduce the complex creature of man into a quantifiable entity. Physicalism, which is scientistic at its core, is guilty of the same crime. Preoccupied with quantifiable entities, it precludes *qualia*. In overlooking *qualia*, it overlooks axiomatic values. Freedom and dignity, which are qualitative elements of the human condition, are eventually jettisoned because they are incompatible with the dominant criteria.

Cultivating Criminality

On the February 13, 2006 edition of MSNBC's *Live and Direct*, Rita Cosby examined the growing street gang known as La Mara Salvatrucha or MS-13 (no pagination). This criminal enterprise is transnational in scope, stretching from "El Salvador to Honduras to Guatemala to New Mexico, and now on U.S. soil" (no pagination). Infamous for their exceptionally violent methods, MS-13 has ascended to a prominent position in the criminal underworld. Rita Cosby elaborates:

> The majority of MS-13 members are foreign-born and are frequently involved in human and drug smuggling and immigration violations. Like most street gangs, MS-13 members are also committed to such crimes as robbery, extortion, rape and murder. They also run a well-financed prostitution ring.
> This notorious gang, best known for their violent methods, can now be found in 33 states, with an estimated 10,000 members and more than 40,000 in Central America. The FBI says MS-13 are the fastest growing and most violent of the nation's street gangs. So much so, even other gangs fear them. (No pagination)

The gang's membership also boasts a vicious array of skills:

> What makes MS-13 so deadly is their skill with the machete, and most have had extensive military training in El Salvador,

making them a double threat. The machete, typically used for cutting crops in El Salvador, is now the weapon of choice for this fearless gang. (No pagination)

Clearly, MS-13 is more than the average gang of thugs and miscreants. It is literally a terrorist network, peopled by skilled warriors and equipped with a paramilitary auxiliary. MS-13's growth and development is hardly some inexplicable social phenomenon. *Reader's Digest* writer Sam Dealy reveals the chief facilitator of MS-13's ascendance: "This is a problem that the federal government actually created" (no pagination). This is a very interesting claim. Just how did the United States create this burgeoning gang crisis? Dealy explains:

Our default policy throughout much of the past decade has been simply to, when you catch these guys, deport them. And they head back to Guatemala, or El Salvador, or Honduras, and weak states back there can't control them. (No pagination)

MS-13 is a threat fostered by America's own impotent immigration policies. No doubt, many categorize this situation as an instance of bureaucratic ineptitude. In a world ruled by the accidentalist perspective of history, this is a common explanation. However, it does not account for the factors that weakened America's border integrity in the first place. In this era of globalization, the public has been bombarded by talk of enacting open border policies. Many of those voicing this contention occupy lofty positions in the Establishment. It comes as little surprise that members of the ruling class would endorse such policies. Vanishing borders are a correlative of vanishing national sovereignty. This gradual subversion of the nation-state system is integral to the power elite's plan to establish a socialist totalitarian world government. Yet, the criminal culture resulting from this plan is hardly some unintended byproduct. MS-13 and other criminal enterprises produced by globalization are integral to the statist blueprint of a global scientific dictatorship.

Paradoxical though it may seem, deviance provides the power elite with an element of stability. This contention is premised upon the functionalist theories of sociologist Emile Durkheim. Durkheim believed that "deviance is not only normal but also beneficial to society because, ironically, it contributes to social order" (Thio 157). According to Durkheim, deviance serves four important functions. The first of these four functions is the enhancement of conformity (157). This function is premised upon the paradoxical notion that otherwise abstract concepts of criminal law can only be illustrated by their violation. Durkheim contends that, by committing crimes, the deviant tangibly enacts principles that are antithetical to the law. In so doing, the deviant supposedly makes the law "real." Once incarcerated and properly punished, the deviant is sacrificed on the altar of conformity for the education of the public. Because the letter of the law must be consistently reiterated for the common citizen, society requires an inexhaustible supply of deviants to act as examples.

The second function served by the deviant is the reinforcement of solidarity among "law-abiding" individuals (157). The deviant "promotes social cohesion" (157). The "collective outrage" generated by criminals unifies the citizenry and facilitates the stability of society (157). Thus, deviance, in the words of Durkheim, constitutes "a factor in public health, an integral part of all healthy societies" (qutd. in 157). In other words, society requires an enemy. The so-called "solidarity" induced by deviance is a solidarity of fear and paranoia, not of common dissent. There is no better example of this contention than the nationalistic fervor following the September 11th attacks. Bin Laden became the proverbial boogey man, a chimera invoked by the power elite to justify the erection of a garrison state. In this sense, deviance is analogous to Orwell's "Two Minutes Hate." It is an instrument for demonizing a nebulous adversary and apotheosizing the cult of personality.

The third function of deviance is the provision of a "safety valve" (157). Crime is a necessary cathartic exercise, allowing people to avenge themselves against the dominant social order (157). Fragmented deviance is a viable alternative to civil unrest. From the perspective of the power elite, individual criminal acts are far more desirable than movements unified by common dissent. Although deviance does induce a certain degree of so-called "solidarity," the resulting unity is one born of fear and paranoia. This unity should not be confused with genuine grass roots mobilization, which is born of legitimate social and/or political dissension. Criminality effectively atomizes society, stultifying grass roots opposition to the oligarchs. The rationale underpinning this third function inverts the classic mantra, "United we stand, divided we fall." Division becomes central to societal stability. Deviance, which fractures the social body by promulgating fear and paranoia among its members, becomes an agent of stability.

The fourth and most significant function of deviance is its role in the inducement of social change (157). Through shock and trauma, crime makes populations more tractable. Again, September 11th stands as a prime example. The WTC attacks provided the pretext for the introduction of the draconian Patriot Act. Traumatized as the population was, the post-911 cultural milieu began to entertain the legitimacy of omnipotent surveillance programs. America has come under the "normalizing gaze" of panoptic mechanisms like Echelon, Carnivore, and the Promis surveillance software. Public acquiescence to such authoritarian measures was made possible, in large part, by the perpetuation of deviance.

Durkheim's functionalist perspective on deviance seems to significantly influence the American legal system, a contention reinforced by the revolving doors of criminal justice. Murderers and rapists serve shorter sentences than non-violent drug offenders. They are subsequently unleashed back upon society. Members of radical organizations like MS-13 are temporarily incarcerated and deported. The same individuals eventually return to the United States with more initiates, drugs, and guns. Evidently, a cyclical pattern is taking shape.

Ultimately, Durkheim's functionalist criterion for social order facilitates the dialectic of mass criminality followed by authoritarian state control. This fact is illustrated perfectly by the 1992 L.A. riots. With racial tensions exasperated by the Rodney King beating, a pretext for the invocation of martial law and the further erosion of civil liberties was not difficult to concoct. In fact, several reports claimed that Police Chief Daryl Gates intentionally "held back his officers, some of whom literally cried as they watched the ensuing chaos" (Hoffman, no pagination). One such report surfaced in the *New York Times*:

> "Emerging evidence from the first crucial hours... provides the strong indication that top police officials did little to plan for the possibility of violence and did not follow standard procedures to contain the rioting once it began....
> The police... violated the basic police procedure for riot-control by failing to cordon off the area around one of the first trouble spots and not returning to that area for hours.
> Police 911 dispatchers attempted to send squad cars to the scene of the first violent outbreaks, but were repeatedly ignored or overruled." (Qutd. in Hoffman, no pagination)

Commenting on the inaction of hundreds of police officers and National Guardsmen, one Deputy Chief confessed to the *Los Angeles Times*: "This is alien to everything we're supposed to do in a situation like this" (qutd. in Hoffman, no pagination). With the chaos already in progress, all those in power needed to do was sit back and watch. The disaster became a pretext for the implementation of police state policies. As Los Angeles burned, so did the Constitution.

The L.A. riots graphically illustrate the centrality of deviance to the oligarchs' plan for a scientific dictatorship. All of the functions encapsulated within Durkheim's functionalist model were met. First, the aftermath of the riots witnessed the imposition of a curfew and the deployment of heavily armed federal authorities in Los Angeles. With these draconian elements in place, the populace quickly conformed to a totalitarian climate. Meanwhile, the rioters were swiftly rounded up and indicted during televised court sessions. For the sake of conformity, an example had to be made.

Second, there was an ostensible sense of solidarity among the "law-aiding" citizens, albeit a unity motivated by fear and paranoia. Citizens were joined in their support of the absolute State, which masqueraded as a savior from social upheaval. Third, the L.A. riots provided a "safety-valve" for the abatement of growing urban unrest. It allowed the inner city inhabitants, which was growing more and more dissatisfied with the corrupt LAPD, to cathartically expel their anger. Fourth, the L.A. riots induced social change. The event represents a major precedent in the invocation of police state mandates. Since then, law enforcement has become increasingly militaristic in nature. Within the Petri dish of society, a bacillus of totalitarianism was starting to grow.

Mass criminality is an instrument of cultural deconstruction. It facilitates the extension of the carceral system to the whole of the social body, creating a carceral culture. Just as prisoners require wardens and guards, a carceral culture requires a carceral state. As crime has steadily risen, society has witnessed the mass diffusion of panoptic schema. This is the ultimate function of deviance. It is an element of stability within the emergent scientific dictatorship.

The Global Skinner Box

The scientistic view of man as an easily "measured and tabulated" entity is best illustrated by Skinner's "box" experiment. Also dubbed the "Heir-conditioner," Skinner's box "enabled the environment to be controlled while the subject's behavior could be studied in terms of the conditioned reflex" (Taylor 418-19). Cloistered within this artificial environment, the subject's behavior was manipulated through a system of rewards and punishments (418). Skinner was so confident in the effectiveness of this system of behavioral modification that he made his infant daughter spend the first two years of her life in the conditioning box (419). His efforts to commercially market the "Heir-conditioner," however, were met with failure (419).

Yet, the power elite continue to carry out agendas of behavioral tyranny that closely parallel those of Skinner's. In fact, the society they are shaping is analogous to one enormous Skinner box. This is made evident by the social phenomenon of authoritarian hierarchicalization. Researcher Daniel Pouzzner expands on this phenomenon:

> When a superior determines to encourage, discourage, demand, or forbid among his subordinates a mode of action, thought, or awareness, those modes will tend to be encouraged or discouraged among everyone below him in the hierarchy. If that superior is a nuclear establishment leader, then these modes will tend to be encouraged or discouraged throughout most of society. In this case, only those not within the conventional hierarchy of civilized society escape the brunt of the behavioral tyranny. (17)

The power elite, who occupy the highest layers of societal strata, selectively deter or promulgate certain modes of thought and behavior. Ideational contagions emanating from those above are diffused and infect those below. Pouzzner continues:

> Authoritarian hierarchicalization is a memetic amplifier for people in higher echelons, and an attenuator for those in lower echelons. The memetic gain factor is not intrinsically correlated with the actual memetic aptitude of each individual; whatever characteristics favor ascension to higher echelons are the characteristics common to those positioned for high memetic gain factors. The characteristics are arbitrarily dictated by those who are already in the upper echelons of the hierarchy, and once those who exhibit them have ascended, the characteristics are themselves efficiently spread through society. (17-18)

The transformation of the United States under the Bush administration exemplifies authoritarian hierarchilization. In an article for *USA Today*, Alexandra Robbins examines President George W. Bush, "a loyal and particularly active member of Skull and Bones" (no pagination). Skull and Bones is a "mysterious, historically misogynist

Yale-based secret society" for the elite (no pagination). Out of loyalty to its ranks, Bush has appointed fellow society members to high-level positions (no pagination). Senior associate counsel on national security and General counsel of the Office on Homeland Security Edward McNally is one such Bonesman appointee (no pagination). Assistant Attorney General Robert McCallum is another (no pagination).

As a result of this discriminative staffing policy, Bonesmen now occupy hierarchical echelons that allow them to dictate the desirable modes of thought and behavior for the rest of society. Their lofty positions act as memetic amplifiers, intensifying the pervasive influence of their virulent ideas. Bush himself has literally transformed the American government:

He's [Bush] practically turning the government into a secret society—an old-boy, throwback establishment that even holds its secret spy-court proceedings in an elaborately locked, windowless room that sounds similar to the Bones' elaborately locked, practically windowless "tomb," or campus clubhouse. (No pagination)

Robbins notices a downward trickling of obscurantism from the highest levels of government to the lowest streets of the commoner:

Last month, Bush-appointed Assistant Attorney General Robert McCallum, a member of Bush's 1968 Skull and Bones class, filed pleadings in U.S. District Court seeking to extend executive privilege to any government official in pardon cases; the move makes information on presidential pardons more secret than it has ever been.

After 9/11, without initially telling Congress, Bush assembled a shadow government assigned to secret bunkers somewhere on the East Coast. He also tried to cut off some of the members of Congress from classified information about the anti-terrorist campaign.

The USA Patriot Act Bush eagerly signed lets the FBI—with the permission from a secret Washington "spy court"—view some customer records; store owners cannot reveal the review.

In October 2001, Attorney General John Ashcroft released

a memo encouraging federal agencies to withhold as much information as possible from the public.

A month later, just before documents from the Reagan-Bush administration were to be released, Bush signed an executive order severely hindering public access to former presidents' records.

Bush also signed legislation that jails or fines journalists who publish sensitive leaks, essentially reviving the Official Secrecy Act that President Clinton vetoed. (No pagination)

The neoconservative strategy of behavioral tyranny is working, as is evidenced by the Republic's transformation into an Empire. In this sense, Skinner's behaviorism and its theoretical progenies represent a form of alchemy. On a macrocosmic level, society begins to tangibly enact the occult dictum of "As above, so below." Meanwhile, the hidden alchemists continue to construct their global Skinner box.

Transhumanism: Techno-Eugenics and the End of Humanity

Today, Weishaupt's agenda of reducing humanity to a race of *Maschinenmenschen* has taken on new frightening dimensions with the World Transhumanist Association. However, equipped with nanotechnology and genetic engineering, this movement presents a technologically augmented form of eugenics. Richard Hayes, executive director of the *Center for Genetics and Society*, elaborates:

Last June at Yale University, the World Transhumanist Association held its first national conference. The Transhumanists have chapters in more than 20 countries and advocate the breeding of "genetically enriched" forms of "post-human" beings. Other advocates of the new techno-eugenics, such as Princeton University professor Lee Silver, predict that by the end of this century, "All aspects of the economy, the media, the entertainment industry, and the knowledge industry [will be] controlled by members of the GenRich class...Naturals [will] work as low-paid service providers or as laborers..." (No pagination)

Here is the vision of the Transhumanist movement: Huxley's *Brave New World* where the new class distinction is genetic. Yet, just how long shall the GenRich class tolerate the existence of its biological subordinates? Hayes continues:

What happens then? Here's Dr. Richard Lynn, emeritus professor at the University of Ulster, who, like Silver, supports human genetic modification: "What is called for here is not genocide, the killing off of the population of incompetent cultures. But we do need to think realistically in terms of the 'phasing out' of such peoples. . .Evolutionary progress means the extinction of the less competent" (No pagination)

This is a frightening proposition indeed. C. Christopher Hook delineates the philosophy underpinning Transhumanism:

That we are biological creatures is simply our current status, transhumanists believe, but it is not necessary for defining who we are or who we should be. Bart Kosko, a professor of electrical engineering at the University of Southern California, puts it more bluntly in his book *Heaven in a Chip* (2002): "Biology is not destiny. It was never more than tendency. It was just nature's first quick and dirty way to compute with meat. Chips are destiny."
British roboticist Kevin Warwick put it this way: "I was born human. But this was an accident of fate—a condition merely of time and place." This sounds startlingly reminiscent of what nihilist Frederick Nietzsche wrote in *Thus Spake Zarathustra*: "I teach you the overman. Man is something to be overcome." (No pagination)

Like Nietzsche's overman, the roboman of Warwick and Kosko represents yet another incarnation of Adam Weishaupt's "inner Areopagites: man made perfect as a god-without-God" (97). This is merely one more permutation of the occult doctrine of "becoming," which was disseminated on the popular level as the Gnostic myth of Darwinism. A central feature of Darwinism has been the belief in great extinctions. That belief remains firmly embedded within the crusade of the Transhumanist movement. Hook elaborates:

Katherine Hayles, a professor of English at the University of California, Los Angeles, says in *How We Became Posthuman* (1999) that "in the posthuman, there are no essential differences, or absolute demarcations, between bodily existence and computer simulation, cybernetic mechanism and biological organism, robot technology and human goals." She concludes her book with a warning: "Humans can either go gently into that good night, joining the dinosaurs as a species that once ruled the earth but is now obsolete, or hang on for a while longer by becoming machines themselves. In either case. . .the age of the human is drawing to a close." (No pagination)

According to the Darwinian doctrine of the Transhumanist movement, mankind is the next species slated for extinction. How does the GenRich class intend to regulate the rest of the "dysgenics" until their ultimate extinction? Transhumanist ideologue and Deputy Director of the National Science Foundation's Division of Information and Intelligent Systems William Sims Bainbridge provides the answer:

Techniques such as genetic engineering, psychoactive drugs and electronic control of the brain make possible a transformation of the species into docile, fully-obedient, "safe" organisms. (No pagination)

This aversion towards humanity echoes the precepts of an older religion. C. Christopher Hook elaborates:

Transhumanism is in some ways a new incarnation of gnosticism. It sees the body as simply the first prosthesis we all learn to manipulate. As Christians, we have long rejected the gnostic claims that the human body is evil. Embodiment is fundamental to our identity, designed by God, and sanctified by the Incarnation and bodily resurrection of our Lord. Unlike gnostics, transhumanists reject the notion of the soul and substitute for it the idea of an information pattern. (no pagination)

Both Transhumanism's inherent Gnosticism and eugenical character are semiotically expressed through the symbol of the Ourohazard. This symbol, which comprises some Transhumanist iconography, is a combination of the occult Ouroboros and the contemporary Biohazard insignia. The Ouroboros is commonly associated with Gnosticism and alchemy ("Transtopian Symbolism," no pagination). Transhumanism encapsulates both. In its derisive portrayal of the body, one finds Gnosticism. Meanwhile, the emergent techno-eugenics of genetic engineering provides the alchemy.

The Gnostic religion of Transhumanism is one more link in an ideational chain. It is a philosophical scion of the Ancient Mystery religion, which originated in Mesopotamia roughly 6,000 years ago. Of course, variants of the ancient Mysteries largely constitute the religious doctrines of the elite. Transhumanism is but a microcosm of the ruling class religious vision for man. A stratified society of rulers and slaves, eugenical regimentation, a technologically altered reality, the complete obliteration of all those things that define humanity...all these comprise the anatomy of the Eschaton they seek to immanentize. Chemically numbed and anesthetized, the "dysgenics" will resign themselves to extinction in the posthuman era. Meanwhile, the eugenical alchemists of the elite continue to write the final chapter of the evolutionary script and they have left no room for humanity in the last pages.

Domesticating the Anthropomorphic Ape

Recently, the London Zoo welcomed a new addition to its collection of animals: man. Sequestered within the zoo's bear enclosure, eight scantly dressed human beings "monkeyed around for the crowds" (Vinograd, no pagination). Affixed to the entrance to the exhibit was a sign reading: "Warning: Humans in their Natural Environment." *Associated Press* journalist Cassandra Vinograd describes the rationale underpinning the exhibit:

> Tom Mahoney, 26, decided to participate after his friend sent him an e-mail about the contest as a joke. Anything that draws attention to apes, he said, has his support.
> "A lot of people think humans are above other animals," he told *The Associated Press*. "When they see humans as animals, here, it kind of reminds us that we're not that special."
> Mark Ainsworth, 21, heard about the Human Zoo on the news.
> "I've lived in this country for nine years and have never come to a zoo," said Ainsworth. "This exhibit made us come to the zoo. Humans are animals too!" (no pagination)

This exhibit represented a semiotic assault, designed to inculcate the masses into the evolutionary Weltanschauung of the power elite. Darwinian transformism is simply the latest permutation of this Weltanschauung, which asserts that man is merely an animal. Freemason Isindag reiterates:

From the point of view of evolution, human beings are no different from animals. For the formation of man and his evolution there are no special forces other than those to which animals are subjected. (*Masonluktan Esinlenmeler*, 137)

The "human" exhibit at the London Zoo semiotically communicates this bestial message. Polly Willis, a spokesperson for the zoo, candidly admitted that this was the intended psychological effect: "Seeing people in a different environment, among other animals. . .teaches members of the public that the human is just another primate" (Vinograd, no pagination). This is a core precept of the ruling class supremacy doctrine. It was scientifically dignified by Darwin and, subsequently, disseminated on the popular level as evolutionary theory. Evolutionist and animal rights advocate Peter Singer summarizes the immediate consequences of this view's popularization:

Darwin's theory. . .undermined the foundations of the entire western way of thinking on the place of our species in the universe. He taught us that we too were animals, and had a natural origin as the other animals did. As Darwin emphasised in *The Descent of Man*, the differences between us and the nonhuman animals are differences of degree, not of kind. Nor did he rest his case on physical similarities alone. The third and fourth chapters of *The Descent of Man* show that we can find roots of our own capacities to love and to reason, and even of our moral sense, in the nonhuman animals. (171)

In this sense, Darwinism merely represents the latest incarnation of totemism. Totemism is the mystical belief that "kinship exists between humans and an animal" (Thio 326). In the case of Darwinism, the ape was designated man's kindred animal. In 1838, Darwin wrote:

"Man in his arrogance thinks himself a great work, worthy of the interposition of a deity. More humble and, I believe, true to consider him created from animals." (Qutd. in Singer 170)

Although Darwin was more circumspect about this contention in *Origin*, he openly declared it in *The Descent of Man*: "[M]an is the co-descendant with other mammals of a common progenitor" (no pagination). Examining this passage from *Descent*, Dowbiggin explains the ramifications of this belief upon Christianity:

With that statement, Darwin declared that the human mind—-and by implication the soul--rather than being prior to things, was emergent in nature. Mind was naturalized, and for philosophers such as Dewey, this was the most revolutionary message of the theory of evolution. After Darwin, it was perfectly logical to conclude that the moral instinct in human beings was a product of natural selection and there were no immutable laws governing ethical behavior. The roots of human moral consciousness lay in man's nonhuman ancestry. This, and much else, made the *Descent of Man* a greater threat to Judeo-Christian morality than the *Origin*. (10)

Downbiggin is correct. Observing the challenge Darwinism presented to Christianity, Singer rejoices over the supposed end to man's position as *imago viva Dei*:

No intelligent and unbiased student of the evidence could any longer believe in the literal truth of Genesis. With the disproof of the Hebrew myth of creation, the belief that human beings were specially created by God, in his own image, was also undermined. So too was the story of God's grant of dominion over the other animals. No wonder that Darwin's theory was greeted with a storm of resistance, especially from conservative Christians. Nor is it surprising that in these circles opposition to Darwin's theory continues to smoulder. (171)

Singer accuses Christianity of "speciesism," a term coined by Richard Ryder in 1970 (173). This term connotes the "discrimination against or exploitation of certain animal species by human beings, based on an assumption of mankind's superiority" (qutd. in 173). Singer agrees with Lynn White, Jr., a so-called "Christian" who feels that the faith is "the most anthropocentric religion the world has seen" (qutd. in 173). Such accusations grossly misrepresent Christianity.

First of all, Christianity hardly qualifies as an "anthropocentric" faith. In John 15:5, Jesus, who is God incarnate, states: "I am the vine; you are the branches. If you remain in me and I in you, you will bear much fruit, apart from me you can do nothing."If man can do nothing apart from God, then he can hardly claim to be the central element of the universe. In this sense, Christianity is Theocentric. Rama Coomaraswamy explains Theocentricism:

> . . .the traditional viewpoint is Theocentric. It is based on the principle of a "fall" from a state of grace in which man directly communicated with God, the need and gift of a Revelation by which means man can return to his primordial and sacred condition, and a metaphysic which explains the nature of God, Truth, Reality and the very essence of man. (No pagination)

In fact, Coomaraswamy characterizes "modern man" as "anthropocentric" (no pagination). This is the same "modern man" that is associated with terms like "scientistic," "rationalistic," "liberal," "democratic," "humanistic," and "relativistic" (no pagination). Such adjectives are inconsistent with Christ's depiction of man: a unique creature that is totally dependent upon God. According to *Wikipedia*, some Christians take issue with anthropocentricism:

> Some evangelical Christians have also been critical, viewing a human-centered worldview, rather than a Christ-centered or God-centered worldview, as a core societal problem. According to this viewpoint, a fallen humanity placing its own desires ahead of the teachings of Christ leads to rampant selfishness and behavior viewed as sinful. (No pagination)

Man's dominion over creation makes him no less accountable to the Lord. In fact, one could argue that it was man's abuse of his position that resulted in his fall in the first place. After all, it was humanity's anthropocentric ambition to become "like gods" that resulted in its expulsion from the Garden of Eden. Since his eviction from the Garden, man has had to suffer the consequences of his own subsequent excesses and abuses. The Lord has seldom withheld those consequences. Creation had come under the dominion of another. As a just and fair God, the Lord has observed the legal rights of man's usurper. As the antithesis of God, this landlord has been anything but fair or just. The new landlord even offered to return creation back to the Lord with one stipulation: "All these things will I give thee, if thou wilt fall down and worship me" (Matthew 4:9).

The Lord refused. Later, thanks to His propitiatory sacrifice at Calvary and His resurrection, the Lord regained the legal rights over creation. Subsequently, He returned these rights to man:

"And I [Jesus] will give unto thee the keys of the kingdom. . .and whatsoever thou shalt bind on earth shall be bound in heaven: and whatsoever thou shalt loose on earth shalt be loosed in heaven." (Matthew 16:19)

However, this dominion is reserved only for the believer. It cannot enjoyed by those who, like the original tenants of Eden, indulged their own hubris and aspired to become gods. Ironically, Darwinism was birthed by just such an anthropocentric tradition. Evolutionism was the outgrowth a "gospel of progress" that accompanied the age of science. William Jasper explains this secular evangel:

> The century now drawing to a close has witnessed man's greatest achievements in science, engineering, and technological progress. Our monumental advances in medicine, agriculture, communications, transportation, space exploration, and virtually every field of learning have far eclipsed the most ambitious hopes of those who lived a generation ago. So sweeping and breathtakingly rapid have these advances come that peoples everywhere have been seduced by the "gospel of progress," the beguiling doctrine of salvation through the all powerful cognitive powers of man. "Science" and "reason," this secular faith contends, will ultimately triumph over religious "superstition" and then usher in a new age of enlightenment, peace, prosperity, and continuous progress.
> The adherents of this "new" faith come in many stripes. Their "spiritual" lineage may be traced to Rousseau, Bacon, Hume, Descartes, Kant, Weishaupt, Marx, Lenin, Asimov, or a myriad of other masters. Darwin is certainly one of the leading points of light in this glittering firmament. (Japser, *Global Tyranny. . .Step by Step* 262)

The popularization of Darwinism created a scientifically legitimized caste system that elevated some men and, simultaneously, subordinated others. The waning dominance of the old oligarchs was revitalized and their progenies continued their elitist tradition under the various banners of the scientific dictatorship (e.g., communism, fascism, transnationalism, internationalism, etc.). The occult theocrats of antiquity were succeeded by the sociopolitical Darwinians of modernity. No longer were the lower classes referred to as "peasants."

Now, they were "dysgenics," "genetic inferiors," and "anthropomorphic apes." This was the new order

As Singer made clear, the differences between man and animals are "differences of degree" (171). This statement infers that the sanctity of human life requires a radical revaluation according to a hierarchical framework. If man did evolve, then some men are less human than others. This contention reiterates the elitist rationale satirized by Orwell in *Animal Farm*: "Some pigs are more equal than others." Thus, it would be reasonable to conclude that evolution is producing a race of "Over-men." Of course, in every instance where this belief was appropriated legitimacy, its adherents have argued for State intervention and eugenical regimentation to realize the production of such a race. It was toward just such an end that the scientific dictatorships of modernity directed their political policies and social programs.

It is interesting to note that Gnosticism played a significant role in the ascendancy of scientific dictatorships. Both Nazis and communist qualify as little more than secular Gnostics. Gnosticism portends the arrival of an adept. Jesus was but a mere "type" of this perfect man (*The Interruption of Eternity* 27). As a Gnostic myth, Darwinism portends the arrival of an evolutionary adept. As both secular Gnostics and sociopolitical Darwinians, the power elite impose selective breeding and technological apartheid upon humanity to facilitate the emergence of this adept. Again, the scientific dictatorship was designed according to the prerogatives and mandates of this myth.

Indeed, many contemporary scientific dictatorships have held aloft their own variety of *Ubermensch*. In Nazi Germany, he was called the Aryan. In communist Russia, he was called the Soviet Man. There can be little wonder why Darwinism received the sanction of occult secret societies and oligarchs. It scientifically legitimized their occult religion of control. The totalitarian theocracy of Babylonian antiquity had found a way to perpetuate itself.

This supremacy doctrine, which is evolutionary in character, exhibits a framework that is inherently hierarchical. The criterion for supremacy used to be racial and favored whatever race happened to occupy the societal capstone. Now, however, the elite's racism has expanded its borders. According to their evolutionary supremacy doctrine, the anthropomorphic apes are all those with the misfortune of being outside the oligarchs' insular clique. This was precisely the message that the "human" exhibit at the London Zoo was designed to semiotic communicate. Judging from the public response, this latest volley in the elite's ongoing *Weltanschauungskrieg* seems to have been successful.

This is, by no means, the first time human beings have been caged to promote an oligarchical agenda. George Herbert ("Bert") Walker, great grandfather of President George W. Bush, hosted a similar event. Webster Tarpley and Anton Chaitkin recount the details surrounding Bert's "Human Zoo":

> Back in 1904, Bert Walker, David Francis, Washington University President Robert Brookings and their banker/ broker circle had organized a world's fair in St. Louis, the Louisiana Purchase Exposition. In line with the old Southern Confederacy family backgrounds of many of these sponsors, the fair featured a "Human Zoo": live natives from backward jungle regions were exhibited in special cages under the supervision of anthropologist William J. McGee. (19)

Like Bert Walker's "Human Zoo" of 1904, the recent exhibit at the London Zoo is an outward expression of the power elite's supremacy doctrine. Accompanying the view of man as an animal is the advocacy of a hierarchical society. A central feature of Darwin's evolutionary theory is natural selection. Ian Taylor observes that "the political doctrine implied by natural selection is elitist, and the principle derived according to Haeckel is 'aristocratic in the strictest sense of the word'" (411). The caste system of such a political doctrine is racial and the criterion for superiority is arbitrarily determined by whatever race happens to hold socioeconomic primacy. In the past, it was the Anglo-Saxon who represented racial perfection.

This scientific racism stemming from Darwinism was especially evident in the case of Operation Fruehmenschen, an FBI project in political suppression of black Americans. John W. DeCamp elaborates:

> On January 27, 1988, then-Congressman Mervyn Dymally placed before the House of Representatives a shocking document. It was an affidavit sworn by an FBI agent, Hirsch Friedman, concerning an FBI policy named Operation Fruehmenschen (German for "primitive man"). According to Friedman's testimony, "The purpose of this policy was the routine investigation without probable cause of prominent elected and appointed officials in major metropolitan areas throughout the United States. It was explained to me that

the basis for this Fruehmenschen policy was the assumption by the FBI that black officials were intellectually and socially incapable of governing major governmental organizations and institutions."

Other evidence backed up Friedman's charges, including a 1987 book by Dr. Mary Sawyer, *Harassment of Black Elected Officials: Ten Years Later*, a follow-up to a 1977 report she had issued on the same subject.

The figures backed up Dymally and Sawyer's charges. Between 1983 and 1988, 14% of all political corruption cases targeted black officials, though they comprised only 3% of U.S. officeholders. From 1981-1983, roughly half of the 26 members of the Congressional Black Caucus were targets of federal investigation for indictments. In magnitude, this is as if 204 members of the (largely white) 435-member House of Representatives were under investigation at one time! (297-98)

The campaign of scientific racism has not only been conducted against black representatives. Several projects targeting regular black citizens have periodically surfaced. One such project was the Violence Initiative, which was to appear on the Alcoholism, Drug Abuse and Mental Health Administration's (ADAMHA) 1994 budget (Shipman 236). The man supposedly responsible for designing the Initiative was Dr. Louis Sullivan, the black director of the Department of Health and Human Services (236). However, it is possible that a black man was presented as the Initiative's architect to conceal a racist agenda. In his 1992 annual report, Sullivan correctly pointed out that there was a problem with violence among black youths:

"This increase [in homicide rates] is attributed, in large part, to a rising rate of homicide among young black men. Between 1985 and 1989," Sullivan noted gravely, "homicides were up 74 percent among young black males to reach the highest level ever." (236)

Minority children are exposed to extremely high levels of poverty. Criminals present criminal lifestyles to these children as the only escape from deprivation. Furthermore, children receive a daily dose of moral relativism in the public schools, thus rendering it impossible for their young, impressionable minds to differentiate between right and wrong. However, these factors were not taken into consideration when looking for the cause of violence among black males. Instead, it was held that these youths were genetically predisposed to violence. Working under this contention, those employed in the realm of public health would:

> . . .look for early predictors of future violence, by studying behavioral and biological markers, and try to establish a useful pattern of intervention once those predisposed to violence had been identified. (237)

Dr. Frederick Goodwin, the head of ADAMHA, "envisioned a target population of perhaps 100,000 inner-city youths" (237). What would be the "pattern of intervention" employed? Dr. Peter Breggin, an activist psychiatrist opposed to the Violence Initiative, charged that the program was a pretext for the same sort of pharmacological totalitarianism described in Huxley's *Brave New World*:

> There could never be any doubt that the proposed "intervention" was pharmacological, because that's what Fred [Goodwin] knows. This is what he does [for his own research]. He has systematically purged NIMH of all psychosocial research. There couldn't be anything else other than drugs, shock treatment, or incarceration; they don't promote anything else; it was a foregone conclusion. (243)

Goodwin denied any plans for biological intervention (244). To dodge such accusations, Goodwin prepared plans for counseling and special school programs (244). However, it is highly suspicious that, when he became director of the National Institute of Mental Health, Goodwin shifted research away from psychosocial forces and focused on biology (243). This seems to suggest that Goodwin saw a predominantly genetic or biological cause to violence. Such a contention holds that drugs, not counseling, are the remedy to violence. In front of the Congressional Black Caucus, Breggin also made an interesting observation:

> While not specifically discussing drugs, Goodwin focuses on the need to correct presumed imbalances in the serotonergic neurotransmitter system...Drugs are the only possible cheap, effective intervention into the lives of tens of thousands of children. . . (244)

Upon closer examination, one will find that the Initiative's agenda was premised upon the Darwinian contention that man is nothing more than a slightly higher form of primate. Following this bestial notion of man to its logical ends, one must conclude that apes and monkeys can provide a comparative model to explain human attributes. Goodwin held this belief, and compared those targeted by the Initiative with monkeys:

> I say this with the realization that it might be easily misunderstand, and that is, if you look at other primates in nature-male primates in nature-you find that even with our violent society we are doing very well.
>
> If you look, for example, at male monkeys, especially in the wild, roughly half of them survive to adulthood. The other half die by violence. That is the natural way of it for males, to knock each other off and, in fact, there are some interesting evolutionary implications of that because the same hyperaggressive monkeys who kill each other are also hypersexual, so they copulate more and therefore they reproduce more to offset the fact that half of them are dying. Now, one could say that if some of the loss of social structure in this society, and particularly within high impact inner city areas, has removed some of the civilizing evolutionary things that we have built up and that maybe it isn't just a careless use of the word when people call certain areas of certain cities jungles, that we may have gone back to what might be more natural, without all of the social controls that we have imposed upon ourselves as a civilization over thousands of years in our evolution. (237-38)

Again, this scientific racism is nothing new. All that has changed are the dimensions of the racism underpinning projects such as Operation: Fruehmenschen and the federal Violence Initiative. In the past, the Fruehmenschen was the black American. Today, the Fruehmenschen is anybody who does not occupy the same layer of socioeconomic stratum as the elite.

In *Evolution and Ethics*, Darwinian Arthur Keith candidly stated: "The German Fuhrer as I have consistently maintained, is an evolutionist; he has consciously sought to make the practice of Germany conform to the theory of evolution" (*Evolution and Ethics*, 230). As was previously established, the evolutionary Weltanschauung infers a hierarchical societal framework. Such was the case with the scientific dictatorship of Nazi Germany. In *Hitler Speaks*, the Fuehrer claimed that mankind was evolving into two distinct forms: "I might call the two varieties the god-man and the mass animal... Man is becoming God—that is the simple fact. Man is God in the making" (qutd. in Keith, *Casebook on Alternative Three*, 151). In hopes of eugenically regulating the population of the "mass animals," the national security apparatus of Nazi Germany turned in on its own people.

Likewise, the ruling class of today, which wield substantial control over the military establishment, are employing state machinations in a war against those they consider "mass animals." An article in *Parameters Magazine*, the official publication of the Army War College, most painfully illustrates this reality. The article is entitled "The New Warrior Class" and is authored by Ralph Peters, a particularly smug Army Major with a penchant for unabashedly elitist rhetoric.

Peters begins the tract with the following remarks:

> The soldiers of the United States Army are brilliantly prepared to defeat other soldiers. Unfortunately, the enemies we are likely to face through the rest of this decade and beyond will not be "soldiers," with the disciplined modernity that term conveys in Euro-America, but "warriors"—erratic primitives of shifting allegiance, habituated to violence, with no stake in civil order. Unlike soldiers, warriors do not play by our rules, do not respect treaties, and do not obey orders they do not like. Warriors have always been around, but with the rise of professional soldieries their importance was eclipsed. Now, thanks to a unique confluence of breaking empire, overcultivated Western consciences, and a worldwide cultural crisis, the warrior is back, as brutal as ever and distinctly better-armed. (no pagination)

Who are the "erratic primitives" that constitute the "new warrior class?" Peters states: "Most warriors emerge from four social pools which exist in some form in all significant cultures" (no pagination). He proceeds to enumerate the four social pools and their respective warrior offspring:

> First-pool warriors come, as they always have, from the underclass (although their leaders often have fallen from the upper registers of society). The archetype of the new warrior

class is a male who has no stake in peace, a loser with little education, no legal earning power, no abiding attractiveness to women, and no future. With gun in hand and the spittle of nationalist ideology dripping from his mouth, today's warrior murders those who once slighted him, seizes the women who avoided him, and plunders that which he would never otherwise have possessed. (no pagination)

In other words, the "first-pool" of "erratic primitives" is composed of unattractive and patriotic males who suffer the misfortune of occupying a lower layer of socioeconomic stratum. Bear in mind, Peters is serious. Inherent in such a contention is credence to the Darwinian concept of sexual selection. Like male birds that must flaunt their plumage in order to sexually attract potential mates, men must now meet a demanding aesthetic criteria or be deemed unfit to breed. Men who take issue with such a shallow criteria are summarily deemed a "threat" to be expunged through force. Also inherent in this contention is credence to the Malthusian economics of Herbert Spencer. Lower income means a lower form of life and, thus, relegates one to the category of "worthless eater." Finally, Peters' disdain for the "spittle of nationalist ideology" echoes the globalist sentiments of the power elite.

Peters identifies the "second pool warriors" as:

society's preparatory structures such as schools, formal worship systems, communities, and families are disrupted, young males who might otherwise have led productive lives are drawn into the warrior milieu. These form a second pool. For these boys and young men, deprived of education and orientation, the company of warriors provides a powerful behavioral framework. (no pagination)

These younger "anthropomorphic apes" are potential recruits for the "warriors." They, too, must be expunged. Reiterating his globalist contentions, Peters proceeds to identify patriots as the next class of "warrior":

The third pool of warriordom consists of the patriots. These may be men who fight out of strong belief, either in ethnic, religious, or national superiority or endangerment, or those

who have suffered a personal loss in the course of a conflict
that motivates them to take up arms. (no pagination)

This particular variety of "anthropomorphic ape" would probably
oppose the amalgamation of its respective nation-state into a global
government. Therefore, it must be eradicated as well. Finally, Peters
reveals the fourth "pool" of "mass animals":

> Dispossessed, cashiered, or otherwise failed military men
> form the fourth and most dangerous pool of warriors. Officers,
> NCOs, or just charismatic privates who could not function
> in a traditional military environment, these men bring other
> warriors the rudiments of the military art—just enough to
> inspire faith and encourage folly in many cases, although the
> fittest of these men become the warrior chieftains or warlords
> with whom we must finally cope. (no pagination)

These soldiers of the "obsolete military paradigm" have no place
in the new society. The duty of the new soldier no longer involves the
protection of nation, family, or the traditional way of life. Now, the new
soldier's duty is to impose the will of the elite upon the "mass animals."
Gestapo officer Werner Best most succinctly voiced this mandate: "As
long as the police carries out the will of the leadership, it is acting legally"
(Shirer 271). This is the philosophy that is now being promulgated within
America's own armed services.

Although Darwin was clearly a bigot, evolutionists like the deceased
Stephen Jay Gould have commonly argued that evolutionary theory can
be divorced from its racist origins. To be sure, not every evolutionist
is a bigot and many neo-Darwinian theoreticians rightly decry racism.
Yet, can an evolutionary Weltanschauung be separated from the racist
thinking that created it?

In terms of racial specificity, modern evolutionary theory does not
single out any one given race as "inferior." However, it does lower all
men to the level of beasts and this is where a new, more virulent form
of racism takes shape. If all humans are animals, then one must logically
conclude that the concept of "human rights" is obsolete. In the animal
kingdom, one must kill or be killed. Theistic evolutionist William
Windwood Reade, who was a chief inspiration of Cecil Rhodes, flatly
stated, "The law of Nature is the law of death." No one is to be granted
immunity. As animals, humans would be equally susceptible to being
indiscriminately murdered.

The result of such thinking has been the emergence of a culture preoccupied with death. Of all the particular varieties of death exalted by this emergent culture, self-immolation has become the most popular. This morbid preoccupation is exemplified by a hypothetical scenario presented by neo-Darwinian Richard Dawkins:

> Even if we don't eat chimpanzees (and they are eaten in Africa, as bushmeat) we do treat them in otherwise inhuman ways. We incarcerate them for life without trial (in zoos). If they become surplus to requirements, or grow old and miserable, we call the vet to put them down. *I am not objecting to these practices, simply calling attention to the double standard. Much as I'd like the vet to put me down when I'm past it, he'd be tried for murder because I'm human.* (No pagination; emphasis added)

Therefore, Dawkins does not find the practice of euthanasia, consensual or coerced, morally objectionable. In fact, he believes that it should be extended to the whole human race. He even volunteers himself to be "put down," an eventuality he welcomes once he is "past it." At least Dawkins is remaining amorally consistent. However, his rather candid remarks reveal a twisted form of death worship. No doubt, this is a consequence of his adherence to the Gnostic myth of Darwinism. Gnosticism also entertained suicide. The Albigenses are one case in point. This Gnostic sect mandated suicide, favoring starvation as the chief means of dispatching one's self ("Albigenses").

If this is what it means to be human, then one must conclude that humanity is a condition to be reviled and condemned. Herein is the ultimate form of racism. Even the most basic examination of humanity reveals a universal anthropic template according to which people are designed, irrespective of skin hue and other superficial physical variations. Thus, when speaking of the many varieties of mankind, the term "race" becomes utterly meaningless. Following this argument to its logical end, one must conclude that a truly consistent racist hates the whole human race. To put it in Kantian terms, misanthropy is the maxim of racism universalized. Darwinism, which reduces the whole human race to beasts, qualifies as misanthropy. It universalizes racism, thus condemning the whole human race as "inferior." It is, in essence, the doctrine of anti-humanity.

Examining ideologies that have resulted in genocide, Steven Pinker observed that "...Marxism, had no use for race, didn't believe in genes and denied that human nature was a meaningful concept" (no pagination).

Of course Marxism had no need of race. With its emphasis upon the subordination of the individual to the collective and ecumenical wealth confiscation, Marxism elevated the State at the expense of humanity. In this sense, Marxism was, like Darwinism, a doctrine of anti-humanity. Premised as they were upon Darwinism, the communist scientific dictatorships throughout history followed evolutionary racism to its logical conclusion. Communism was an equal opportunity murderer, conducting campaigns of genocide against all people. This was made possible by Darwinism's gradual deconstruction of humanity. During the 60s, a manual enumerating communist brainwashing techniques surfaced. Entitled *Brainwashing: A Synthesis of the Russian Textbook on Psychopolitics*, the manual synopsized the rationale for humanity's perpetual subjugation by the State:

> Basically, Man is an animal. He is an animal which has been given a civilized veneer. Man is a collective animal, grouped together for his own protection before the threat of the environment. Those who so group him must then have in their possession specialized techniques to direct the vagaries and energies of the animal Man toward greater efficiency in the accomplishment of the goals of the State. (No pagination)

Ever-present is the theme of man as an animal. A similar theme pervades the technocratic social sciences, which reduce people to mere cells in a nebulous social organism. Remaining consistent with the technocratic elements of Marxism, the manual emphasizes "animal Man's" role in achieving "greater efficiency in the accomplishment of the goals of the State." As usual, it is efficiency that governs a scientific dictatorship, not objective moral law.

Astute readers will also recognize the text's use of the capital letter M in "Man," connoting his potential to be apotheosized through evolution. However, the connotation invoked by the capital M does not raise man above the level of the animals. It merely signifies his potential to do so by realizing the Nietzschean mandate to overcome humanity itself. After all, if humans are nothing more than animals, then one can only hope to transcend such a condition by becoming something other than human. Man must become a god. This theme underpins Luciferianism, the ruling class religion.

Of course, such a belief system is no less hierarchical. In a world governed by Singer's contention that the differences between man and "nonhuman animals" are simply "differences of degree," it stands to reason that some men more closely approximate Man than others. More importantly, most men will never become gods. According to the power elite's evolutionary Weltanschauung, humanity is a liability.

Throughout the years, this evolutionary Weltanschauung has been disseminated on the popular level in numerous forms. Gnosticism promises Man's escape of his wretched human condition through *gnosis*, which facilitates the "transfiguration of man into divine." The occult institution of Freemasonry practices degrees of initiation, mimicking the "differences of degree" inherent to the purported evolutionary process. Transhumanism and other futurist sects portend a "post-human" era, which will witness the transformation of man into machine. In all cases, Man ceases to be human.

Evolutionary theory, which scientifically dignified several forms of racism, was the beginning of a slippery slope. With the occult doctrine of "becoming" epistemologically exalted in academia and institutionalized through the development of certain societal machinations, it was only a matter of time before racism was universalized. The Africans, Jews, and other historically oppressed peoples were the first victims of an opening salvo in an anti-human crusade. Now, humanity has begun to deconstruct itself.

Philosopher and evolutionist Daniel Dennett characterizes Darwinism as "a universal solvent, capable of cutting right to the heart of everything in sight" (521). Indeed, that "universal solvent" has been employed by the power elite in their ongoing *Weltanschauungkrieg* against humanity. The acid of Darwinian thought has worn away the image of God within man, reducing him to a mere animal that can either be further devalued or, through genetic manipulation and eugenical regimentation, deified. This Weltanschauung is a paradoxical amalgam of both racial self-loathing and anthropocentric hubris. Its progeny is the scientific dictatorship, which has actuated itself in numerous forms over the years. Now, as the "universal solvent" wears away patriotism and nation-states, a global scientific dictatorship is appearing on the horizon.

With the opening of the London Zoo's "human" exhibit, the masses have been further psychologically conditioned to accept the view of man as an animal. As this view becomes more entrenched in the public mind, so does the feasibility of a stratified society governed by a technocratic elite. The cage in which humans were so casually displayed at the London Zoo semiotically gesticulates towards a larger enclosure. That enclosure is global in scope, encompassing the entire world. The "mass animals," "erratic primitives," "warriors," and "anthropomorphic apes" that it was built to hold are all those outside the insular clique of the power elite.

There have been various appellations applied to the common man by the power elite. While Darwin invoked the label of "anthropomorphic ape," the actual derogation he had in mind was even less attractive. It is an ugly word that has poured forth from the lips of racists for years. Although it was once bestowed exclusively upon people of African ancestry, the title has been expanding its borders to include larger portions of the human race. Researcher and PBS journalist Tony Brown comments on the growing elasticity of this obscene racial derogation:

> The new world in which the only color of freedom is green demands a new "nigger." New conditions dictate that the new class of niggers cannot be race based. You are now a nigger when you don't know that you are being robbed of your money and your freedom. Niggers get no respect, die in wars so other people can profit (the Vietnam War produced an $80 billion profit for the companies that sold products to the military), and their human rights confiscated on a daily basis and their property taken from them by the statists every April 15. (156)

In the scientific dictatorship, those with the misfortune of being human are now deemed "niggers." The "masters" occupy corporate boardrooms and regal offices. As their supremacy doctrine metastasizes and subsumes our thoughts, their yoke weighs heavier upon our shoulders. According to their elitist Weltanschauung, all those below them are anthropomorphic apes waiting to be domesticated.

Katrina and the Social Darwinian Politics of Disaster

In the article "Katrina Meets Social Darwinism and its Son, Eugenics," Anisa Abd el Fattah makes some interesting observations concerning the allocation of disaster relief shortly after hurricane Katrina. She writes:

> Through Eugenics we identify who shall live, who shall have chances, and advantages, and who shall be denied, and whose deaths we shall welcome, and advance through humane and civil means if possible, and through other means if not. In the past, Eugenics depended heavily upon race, or rather skin

color and physical attributes as a determinant in such matters. Today, due to DNA studies we now use intellectual aptitude, class, and health status to distinguish between the weak, and the strong, realizing that race is not so easily determined after centuries of human migration, and mixture. Public policies regarding disaster relief, the allocation of natural and other resources, and also social service policies are increasingly predicated upon such distinctions, especially since the world has acquiesced to the United Nation's sustainable development theory and its emphasis on population control. (No pagination)

Evidently, there was a Darwinian rationale underpinning the government's disaster response. Katrina served a critical function for the emergent scientific dictatorship in America. According to Anisa Abd el Fattah, disasters like Katrina are now viewed as "a means by which to accelerate the evolution of the human species" (no pagination). This view is clearly demonstrated by the criminal negligence exhibited by certain officials and organizations shortly after Katrina hit New Orleans.

The negligence begins with, according to former Clinton advisor Sydney Blumenthal, President George W. Bush. Blumenthal wrote:

In 2001, FEMA warned that a hurricane striking New Orleans was one of the three most likely disasters in the U.S. But the Bush administration cut New Orleans flood control funding by 44 percent to pay for the Iraq war. (No pagination)

However, President Bush was not the only one blamed for the Katrina debacle. Former director of the Federal Emergency Management Agency, Michael Brown, was also tagged for his inaction. The Associated Press reported the following:

The top U.S. disaster official waited hours after Hurricane Katrina struck the Gulf Coast before he proposed to his boss sending at least 1,000 Homeland Security workers into the region to support rescuers, internal documents show.
Part of the mission, according to the documents obtained by The Associated Press, was to ``convey a positive image" about the government's response for victims.
Acknowledging that such a move would take two days, Michael

Brown, director of the Federal Emergency Management Agency, sought the approval from Homeland Security Secretary Michael Chertoff roughly five hours after Katrina made landfall on Aug. 29.

Before then, FEMA had positioned smaller rescue and communications teams across the Gulf Coast. But officials acknowledged the first department-wide appeal for help came only as the storm raged.

Brown's memo to Chertoff described Katrina as "this near catastrophic event" but otherwise lacked any urgent language. The memo politely ended, ``Thank you for your consideration in helping us to meet our responsibilities." (No pagination)

An August 29, 2005 FEMA press release also revealed that Brown had urged first responders not to respond to hurricane impact areas unless they were first called in by state and local authorities. The press release states:

Michael D. Brown, Under Secretary of Homeland Security for Emergency Preparedness and Response and head of the Federal Emergency Management Agency (FEMA), today urged all fire and emergency services departments not to respond to counties and states affected by Hurricane Katrina without being requested and lawfully dispatched by state and local authorities under mutual aid agreements and the Emergency Management Assistance Compact. (No pagination)

Expecting the state and local authorities to first request assistance before responding defies all logic. Given the fact that a major hurricane had just taken place, state and local authorities would already be anticipating a federal response. FEMA's primary function is supposed to be dealing with emergencies, and yet the agency was sitting on its hands while a major emergency was unfolding before the nation's very eyes on every major news network. This is tantamount to a fire department expecting a call from a homeowner whose house is on fire before deciding to deploy.

Another shocking detail of the Katrina disaster is the fact that Homeland Security refused to let the Red Cross deliver food. Ann Rodgers reported:

As the National Guard delivered food to the New Orleans convention center yesterday, American Red Cross officials said that federal emergency management authorities would not allow them to do the same.
Other relief agencies say the area is so damaged and dangerous that they doubted they could conduct mass feeding there now.
"The Homeland Security Department has requested and continues to request that the American Red Cross not come back into New Orleans," said Renita Hosler, spokeswoman for the Red Cross.
"Right now access is controlled by the National Guard and local authorities. We have been at the table every single day [asking for access]. We cannot get into New Orleans against their orders."
Calls to the Department of Homeland Security and its subagency, the Federal Emergency Management Agency, were not returned yesterday. (No pagination)

Authorities also did nothing to protect the inhabitants of hurricane impact areas from crime and lawlessness. Jamie Doward reports:

Those trapped inside the two main shelters, the Superdome and the Convention Centre, paint a picture of a city that was subsumed beneath waves of violence, rape and death and accuse the police and National Guard of standing by, ignoring their pleas for help. (No pagination)

One New Orleans citizen, Correll Williams, also described authorities ignoring the need to take action:

"We had to wrap dead people in white sheets and throw them outside while the police stood by and did nothing," said Correll Williams, a 19-year-old meat cutter from the Crowder Road district in the east of the city, who waded two miles through waist-high water to make it to the Convention Centre after hearing on the radio it was being turned into a refuge.
"The police were in boats watching us. They were just laughing at us. Five of them to a boat, not trying to help nobody. Helicopters were riding by just looking at us. They weren't helping. We were pulling people on bits of wood, and

the National Guard would come driving by in their empty military trucks." (No pagination)

Probably the most disturbing aspect described by Williams and another inhabitant, Arineatta Walker, was authorities' plan to intentionally flood parts of the city:

> Williams only left his apartment after the authorities took the decision to flood his district in an apparent attempt to sluice out some of the water that had submerged a neighbouring district. Like hundreds of others he had heard the news of the decision to flood his district on the radio. The authorities had given people in the district until 5pm on Tuesday to get out-- after that they would open the floodgates.
> "We thought we could live without electricity for a few weeks because we had food. But then they told us they were opening the floodgates,' said Arineatta Walker, who fled the area with her daughter and two grandchildren.
> "So about two o'clock we went on to the streets and we asked the army, 'Where can we go?'. And they said, 'Just take off because there's no one going to come back for you.' They kicked my family out of there. If I knew how to hotwire a car I would have," Walker said. (No pagination)

Many in the mainstream media have interpreted these revelations the same way: gross incompetence on the part of the government. Apparently, Uncle Sam cannot get a thing right these days. What the media has completely missed (or ignored) is how certain factions within government could use the Katrina catastrophe to introduce social changes previously unthinkable. There is a discomforting possibility that Americans must consider in light of the fact that there is no one else looking out for their best interest. It is the possibility that warnings were ignored and assistance was intentionally delayed to create a pretext for unprecedented government growth. One supporter of this contention is Paul Craig Roberts, the former Assistant Secretary of the Treasury. On the 5th September 2005 Alex Jones show, Roberts:

> agreed that FEMA has deliberately withheld aid, and cut emergency communication lines, and automatically made the crisis look worse in order to empower the image of a

police state emerging to 'save the day'" (Watson and Jones, no pagination).

Steve Watson and Alex Jones also report:

Roberts further commented "There is no excuse for this, we have never had in our history the federal government take a week to respond to a disaster. . .this is the first time ever that the help was not mobilized in advance. The proper procedure is that everything is mobilized and ready to go." (No pagination)

While Roberts, Watson, and Jones are correct in characterizing the Katrina debacle as a pretext for police state measures, there is another motive that can be discerned from the evidence. New Orleans and other hurricane impact areas have had predominantly black and non-white populations. Is it too farfetched to say that the Katrina disaster relief was predicated upon eugenical regimentation and social Darwinism? Given the fact that sociopolitical Darwinians have the reins of government firmly in their hands, the contention is quite sound.

The Technocratic Agenda

As was previously established, Freemason Aldous Huxley coined the term "scientific dictatorship" and presented an allegorized version of the concept in his famous *roman 'a clef* entitled *Brave New World*. Freemason H.G. Wells, Huxley's mentor, also presented a fictionalized scientific dictatorship engineered through the ideological forces of the technocratic movement. The term "Technocracy" can be used interchangeably with the term "scientific dictatorship." "Technocracy" is an interesting designation for a world government managed by functional elites and scientists. It is derived from the Greek word *techne*, which means "craft." Given Wells' membership in the Craft of Freemasonry, the synchronicity becomes apparent.

Moreover, the term *craft* is associated with witchcraft or *wicca*. From the term *wicca*, one derives the word *wicker* (Hoffman 63). Examining this word a little closer, Michael Hoffman explains: "The word wicker has many denotations and connotations, one of which is 'to bend,' as in the 'bending' of reality'" (63). This is especially interesting when considering the words of Mark Pesce, co-inventor of Virtual Reality Modeling Language. Pesce writes: "The enduring archetype of techne within the pre-Modern era is magic, of an environment that conforms entirely to the will of being." (No pagination)

Techne is also from whence the word "technology" is derived. The significance of this fact becomes evident when Pesce opines:

> Each endpoint of techne has an expression in the modern world as a myth of fundamental direction--the mastery of matter, and the collection of spirit. The myth of matter comes to its end as the absolute expression of will as artifact; in a word, nanotechnology. (No pagination)

Herein is the technocratic agenda: the reconfiguration of reality through the sorcery of technology, specifically the emergent field of nanotechnology. This is the ultimate end of the Masonic doctrine of "becoming," which was disseminated on the popular level as Darwinism.

The Epistemological Pretext for Reality Reconfiguration

As was already established, most of contemporary science is predicated upon empiricism. This is the epistemological stance that all knowledge is derived exclusively through the senses. Lyndon LaRouche explains the inherent flaws of empiricism:

> By the nature of our processes of sense-perception, our direct perception of the world "outside our skins" (so to speak) does not show us that world "outside our skins," but, rather, the impact of that unperceived real world upon the biology of our mental-sensory processes. In other words, the shadows on the wall of Plato's Cave. ("The Pagan Worship of Isaac Newton," no pagination)

Thus, the world becomes little more than an ever-shifting pliancy of impressions. All that a percipient surveys is an amorphous amalgam of "shadows." It comes as little surprise that an exclusively empirical approach relegates causality to the realm of metaphysical fantasy. The obviation of causality holds enormous ramifications for science. In the absence of causality, the results of even the most rigorous experimentation criteria require some investment of faith. No matter how probabilistic, empirical observation can never yield any degree of certainty. Again, the elite merely exchanged one form of mysticism for another. It comes as little surprise that, within certain occult circles, contemporary science is considered sorcery disseminated on the popular level. For instance, Satanic high priest Anton LeVey regarded science and technology as "sanctioned, but ineffectual 'occultism'" (Raschke, *Painted Black* 214).

In fact, science has become a new form of sorcery for the manipulation of matter. According to the epistemology of empiricism, reality is little more than a quagmire of impressions. It is analogous to a holograph, the fabric of which is pliable enough to be manipulated. Thus, reality becomes the ever-shifting canvas upon which scientists paint whatever they wish. The scientist's role in this reconfiguration of reality was delineated in an esoteric tract entitled *The Way of Light*. Authored by Comenius in 1668, the manifesto was dedicated to the British Royal Society. Researcher Michael Hoffman elaborates:

> In it, Comenius addressed the first formal scientists as "illuminati" and outlined their scientific purpose, " . . .*which is to secure...the empire of the human mind over matter.*" (23; emphasis added)

Years later, Bertrand Russell would recapitulate the "illuminati's" (i.e., scientists') role in the establishment of "the empire of the human mind over matter." Redefining science as an instrument of radical empiricism, Russell wrote:

> The way in which science arrives at its beliefs is quite different from that of medieval theology. Experience has shown that it is dangerous to start from general principles and proceed deductively, both because the principles may be untrue and because the reasoning based upon them may be fallacious. Science starts, not from large assumptions, but from particular facts discovered by observation or experiment. From a number of such facts a general rule is arrived at, of which, if it is true, the facts in question are instances. . . .Science thus encourages abandonment of the search for absolute truth, which belongs to any theory that can be successfully employed in inventions or in predicting the future. "Technical" truth is a matter of degree: a theory from which more successful inventions and predictions spring is truer than one which gives rise to fewer. *"Knowledge" ceases to be a mental mirror of the universe, and becomes merely a practical tool in the manipulation of matter*. (*Religion and Society*, 13–15; emphasis added)

In other words, science or "knowledge" becomes the instrument by which the "illuminati" re-sculpts reality. It also becomes an epistemological weapon against the minds of men, wielded by the proverbial Descartean "evil demon." Thus, the technocratic elite become gods, creating their own paradise and keeping the rest of the human herd blinded. This was the central precept of Weishaupt's Illuminati and the conceit of the Technocracy today: God was not in the beginning, but evolved from Man in the end. According to this conceit, Man could recreate Eden without the God. It comes as little surprise that sci-fi predictive programmer and British intelligence asset Arthur C. Clarke commented, "Any sufficiently advanced technology is indistinguishable from magic."

The Global Holodeck

The technocratic agenda of reconfiguring the "holograph of reality" is most clearly delineated by William Sims Bainbridge, sociologist and member of the National Science Foundation. Citing sci-fi predictive programmer Gene Roddenberry, Bainbridge writes:

> An interesting feature of the popular Star Trek universe is that mass-media popular culture is absent from its fictional future world. Several characters play musical instruments and the preferred styles of music are classical, whether European or belonging to some other high culture. Perhaps precisely because the characters are living very future-oriented lives, they turn to historical sources like Mozart for their aesthetic recreation. Presumably, the copyrights have all expired. Instead of passively watching television programmes and movies, they programme their own "holodeck" virtual reality dramas in which they play active roles, often with historical settings. Government is certainly not in the science fiction business, but government-encouraged research is currently developing the technology to realise the Star Trek prophecies. ("Memorials," no pagination)

Evidently, government-sponsored research programs are already dedicated to the Technocracy's vision of re-sculpting reality. Bainbridge is certainly no stranger to this vision, as is evidenced by his association with Scientology. In *Religion and the Social Order*, Bainbridge presented a mandate for scientists to become "religious engineers" in the development of a new world religion ("New Religions, Science, and Secularization"). This new world religion, which Bainbridge calls a "Church of God Galactic," would find its origins with science fiction literature (Bainbridge, "Religions for a Galactic Civilization," no pagination). In the formulation of his "Church," Bainbridge used the scientistic cult of Scientology as a working model:

> Today there exists one highly effective religion actually derived from science fiction, one which fits all the known sociological requirements for a successful Church of God Galactic. I refer, of course, to Scientology. ("Religions for a Galactic Civilization," no pagination)

In the Scientologist bible, L. Ron Hubbard's *Dianetics*, one finds a reiteration of Comenius' mission statement delivered to the "illuminati" (i.e., scientists). Hubbard states:

> Man has something more: some people call it imagination, some call it this or call it that; but whatever it is called, it adds up to the interesting fact that man is not content merely to "face reality" as most other life forms are. Man makes reality face him. Propaganda about "the necessity of facing reality," like propaganda to the effect that a man could be driven mad by a "childhood delusion" (whatever that is), does not face the reality that where the beaver down his ages of evolution built mud dams and keeps on building mud dams, man graduates in a half century from a stone and wood dam to make a mill wheel pond to structures like Grand Coulee Dam, and changes the whole and entire aspect of a respectable portion of nature's real estate from a desert to productive soil, from a flow of water to lightening bolts. (Hubbard 308)

It is very interesting that Hubbard would cite the Grand Coulee Dam as an instant when man "made reality face him." Again, Comenius' mandate for mankind to establish "the empire of the human mind over matter" emerges. Discussing the "Saturnian-masonic" era of erecting megalithic structures, Hoffman observes:

> Actually, with some crucial exceptions, the rise of the megaliths marked the rise of the Hermetic Academy into its dominant

physical phase. The theory is that the megaliths "pin down" natural forces, helping to subdue nature's most savage furies. We marvel today at the Hoover Dam but that symbol laden construction is but a crude parody of the technology of the megaliths which helped to "dam" the wildest forces of nature. (21)

Evidently, men like Hubbard and Bainbridge see something entirely different when they view structures like the Grand Coulee Dam and the Hoover Dam. They are viewing the "endpoint of techne--the mastery of matter." It is the Technocracy's project in consciously shaping the terrain of the global holodeck.

The Promethean Crusade

As has been previously established, Darwin's theory of evolution was cribbed liberally from Freemasonry's occult doctrine of "becoming." According to this doctrine, humanity was gradually evolving towards apotheosis. The architecture of Masonry's evolutionary mythology is a counterpart to the Biblical account of humanity's expulsion from Eden. However, there are some major modifications. In *The Meaning of Masonry*, W.L. Wilmshurst alleges that:

> In all Scriptures and cosmologies the tradition is universal of a "Golden Age," an age of comparative innocence, wisdom and spirituality, in which *racial unity* and individual happiness and enlightenment prevailed; in which there was that open vision for want of which a people perisheth, but in virtue of which men were once in conscious conversation with the unseen world and were shepherded, taught and guided by the "gods" or discarnate superintendents of the infant race, who imparted to them the sure and indefeasible principles upon which their spiritual welfare and *evolution* depended. (173; emphasis added, ed. note: In its Masonic context, "racial unity" means one race, not concord in race relations)

However, Wilmshurst contends that a peregrination of human consciousness away from the "racial mind" caused humanity to fall from its former glory:

> The tradition is also universal of the *collective soul* of the human race having sustained a "fall," a moral declension from its true

path of life and *evolution*, which has severed it almost entirely from its creative source, and which, as the ages advanced, has involved its sinking more and more deeply into physical conditions, its *splitting up from unity employing a single language into a diversity of conflicting races of different speeches and degrees of moral advancement*, accompanied by a progressive densification of the material body and a corresponding darkening of the mind and atrophy of the spiritual consciousness. (173; emphasis added)

Recall Pesce's statement that *techne* was expressed in the modern world as the "collection of spirit." This is precisely the objective of Masonry. . . the "collection of spirit" through the facilitation of evolution! Wilmshurst proceeds to reveal the chief means by which this will be achieved:

And it required something further. It required the application of an orderly and scientific method to effect the restoration of each fallen soul-fragment and bring it back to its primitive pure and perfect condition. I emphasize that the method was necessarily to be not a haphazard, but a scientific one. (174)

A little later, Wilmshurst recapitulates this theme:

Unable to effect its [Man's] own recovery it required skilled *scientific* assistance from other sources to bring about its restoration. Whence could come that skill and *scientific* knowledge if not from the Divine and now invisible world, from those "gods" and angelic guardians of the erring race of whom all ancient traditions and sacred writings tell? Would not that regenerative method be properly described if it were called, as in Masonry it is called, a *"heavenly science,"* and welcomed in the words that Masons in fact use, "Hail, Royal Art!" (175; emphasis added)

Can there be any wonder why Freemasons Aldous Huxley and H.G. Wells were proponents of a "scientific dictatorship?" It is an intrinsic feature of their Masonic heritage. This heritage led them to bestow absolute epistemological primacy upon Science, spelled with a capital "S" to denote its divine role in man's purported ascent towards apotheosis and the reconstitution of the Masonic "collective soul." This is scientism. In a speech before the Royal Institute of International Affairs in 1936, H.G. Wells succinctly expressed the core precept of scientism:

> "At first the realization of the ineffectiveness of our best thought and knowledge struck only a few people, like Mr. Maynard Keynes, for example...*It is science and not men of science that we want to enlighten and animate our politics and rule the world.*"(Qutd. in Keith, *Mind Control, World Control*, 306–307; emphasis added)

In addition to espousing scientism, Wells' speech also mentioned a concept bearing an ominous resemblance to the "collective soul" concept of his Freemasonic heritage:

> I want to suggest that something, a new social organization, a new institution—which for a time I shall call World Encyclopaedia...This World Encyclopaedia would be the mental background of every intelligent man in the world...Such an Encyclopaedia would play the role of an undogmatic Bible to World culture. It would do just what our scattered and disoriented intellectual organizations of today fall short of doing. It would hold the world together mentally...It would compel men to come to terms with one another...It is a super university I am thinking of, a *World Brain*; no less...Ultimately, if our dream is realized, it must exert a very great influence upon everyone who controls administration, makes wars, directs mass behavior, feeds, moves, starves and kills populations...You see how such an Encyclopaedia organization could spread like a nervous network, a system of mental control about the globe, knitting all the intellectual workers of the world through a common interest and cooperating unity and a growing sense of their own dignity, informing without pressure or propaganda, directing without tyranny." (Qutd. in Keith, *Mind Control, World Control*, 306-307; emphasis added)

It is very possible that Wells' "World Encyclopaedia" was derivative of the Masonic "collective soul." Wells also dubbed this cognitive singularity "The Mind of the Race." W. Warren Wagar elaborates on Wells' "racial mind" doctrine:

It was at once the capstone and the mortar of his [Wells'] faith: a belief in the emergence in human evolution of a collective racial being with the collective racial mind, which gathered the results of the individual mental effort into a single fund of racial wisdom and grew gradually toward organic consciousness of itself. Individuals could escape the frustration inherent in the fact of their individuality and mortality only by consecrating their lives to the service of the Mind of the Race. (100-101)

Wells believed that the final coalescence of human consciousness into a "racial mind" would result in the emergence not of a mere man, but of perfected Man with a capitalized M (Wagar 104). The M is capitalized to denote the purported divinity that is dormant within humanity. Wells' Weltanschauung remained consistent with the Masonic themes of a "collective soul" and man's evolutionary ascent towards deification. In *H.G. Wells and the World State*, author Warren Wagar elaborates:

But the transcendent reality Wells actually professed to see emerging here and now was the collective being of humanity, rather than any "God." At the level of the individual the species Homo sapiens might be nothing more than a swarm of unique individuals descended in an unbroken sequence from remote protozoan ancestors; yet Homo sapiens was more than a name. At this moment in cosmic time it also denoted a class of similar if not identical individuals, evolving in ceaseless interaction with one another, and through the unique gift of speech able to pool their experiences and so give birth to a higher order of being entirely: a racial memory, *a collective mind*, the emergent intelligence of an emergent racial being. (104; emphasis added)

According to Wells' Weltanschauung, the ecumenical singularity into which humanity was being compressed by evolution would relegate the individual to obsolescence:

As Wells grew older, he tended to look at life more and more from the synthetic level of racial being and less and less from the analytical level of the individual. At the end of his spiritual pilgrimage he virtually accepted the realist argument that the whole is real and the individual an illusion. (104)

In *The Undying Fire*, Wells distilled this monistic view of humanity in an allegorical form. W. Warren Wagar provides a synopsis of Wells' allegory:

In the symbolic prologue to *The Undying Fire*, he [Wells] even likened the opposition of essence and existence to the interplay of good and evil. God was here represented as the inscrutable creator, who created things perfect and exact, only to allow the intrusion of a marginal inexactness in things through the intervention of Satan. God corrected the marginal uniqueness by creation at a higher level, and Satan upset the equilibrium all over again. Satan's intervention permitted evolution, but the ultimate purpose of God was by implication a perfect and finished and evolved absolute unity. (104)

According to Wells, this "absolute unity" would represent the culmination of the evolutionary process: "The dialectic of good and evil was the method of evolution, from absolute and perfect nonbeing to absolute and perfect all-being" (105).

This monistic Weltanschauung is similar to the elite's view of humanity. Researcher Dee Zahner refers to this paradigm as the "Old World View" and characterizes it as:

. . .a fatalistic view that the individual is helpless to determine his destiny, that he is controlled by forces outside himself and can do nothing to improve his lot in life. Therefore, he needed a king or leader to guide and control him. In this stagnant world, with no hope of progress, men must be herded into a collective mass, a bee hive, and controlled. (59-60)

This vision of human civilization as a "bee hive" is also illustrated by the iconography and language pervading the elite's semiotic lexicon. This semiotic lexicon is based on ancient occult principles and doctrines, which were maintained and perpetuated by numerous mystical secret societies throughout the ages. Deceased researcher Jim Keith makes the following observation concerning elite semiology:

. . .I now see that any number of "insect" metaphors reside in the lore of governments and secret societies, with an emphasis on bees, according to Ordo Templi Orientis head Kenneth

Grant a symbolic representation of a group mind proceeding from the "Queen" goddess Isis, beloved of the Freemasons and many another a mystical sect. Recalling the name of Illuminist Adam Weishaupt's secret society, the *Beenan Orden* (Order of the Bees), recalling the beehive emblem of the Freemasons and the Masonic offshoot Mormon Church's hive symbolism, I have to think that this must be a clue to the philosophy of these mystic Machiavellians. (*Casebook on Alternative Three*, 157)

Eloquently synopsizing the philosophy towards which these various "insect" metaphors semiotically gesticulate, Keith states:

The insect metaphor resident in mystical literature is reminiscent of the basic elitist, anti-human theme of aristocratic "bluebloods" sustaining the hive through reproductive processes and blood lineage. (*Casebook on Alternative Three*, 158)

The numerous collectivist crusades throughout history (i.e., communism, fascism, or any other permutation of Marxism) could be considered crude attempts to tangibly enact just such a vision. The virulent anti-individualism and uniformity of thought intrinsic to historical cases of socialist totalitarian regimes certainly reinforces this contention. According to Wells' monistic Weltanschauung, the species of *Homo sapiens* was analogous to Prometheus (Wagar 76). The mythical character of Prometheus was also central to the Utopian vision of early socialist revolutionaries. James A. Billington explains:

A recurrent mythic theme for revolutionaries-- early romantics, the young Marx, the Russians of Lenin's time-- was Prometheus, who stole fire from the gods for the use of mankind. The Promethean faith of revolutionaries resembled in many respects the general belief that *science would lead men out of darkness into light*. (6; emphasis added)

One immediately discerns the overt scientism of this Promethean faith, which guided the early sociopolitical Utopians in their crusade to establish absolute unity. Given the Masonic origins of these revolutionaries, there is a definite possibility that the absolute unity they desired represented a reconstituted "collective soul." The final goal may not have been a world government, although that was certainly one of the crusade's objectives. The final goal may have been a world mind, the unification of human consciousness into a single entity. Science fiction authors seem to portend a similar outcome in human evolution: Wells' "World Encyclopaedia," Clarke's "Overmind," Lucas' animistic "Force," etc.

Many contemporary movements have also semiotically communicated the theme of a "collective soul." For instance, the technocratic movement of the 1930s sported a geometric representation of the Pythagorean Monad. The word "monad" finds its etymological origins with the Greek word for "one" and "single" ("Monad," no pagination). Not surprisingly, the symbol is prominent in the philosophy of monism, which contends that "the metaphysical and theological view that all is of one essence" ("Monad," no pagination).

Another case in point is Singularitarianism, a futurist religion closely aligned with Transhumanism. According to the techno-Utopian outlook of this scientistic Weltanschauung, man stands just before a defining moment in human history. This pivotal juncture, which has been edified by rapid scientific and technological advancement, is dubbed the Singularity. Scientist and inventor Ray Kurzweil is one of the chief proponents of this religion. He describes the Singularitarian period as follows:

> . . .the Singularity is a future period during which the pace of technological change will be so fast and far-reaching that human existence on this planet will be irreversibly altered. We will combine our brain power—the knowledge, skills, and personality quirks that make us human—with our computer power in order to think, reason, communicate, and create in ways we can scarcely even contemplate today. (39)

This is a rather ambiguous description. However, as Kurzweil continues his explication of the Singularity, a familiar codex becomes discernible: "I teach you the overman. Man is something to be overcome." Kurzweil states:

> This merger of man and machine, coupled with the sudden explosion in machine intelligence and rapid innovation in gene research and nanotechnology, will result in a world where there is no distinction between the biological and the mechanical, or between physical and virtual reality. These technological

revolutions will allow us to transcend our frail bodies with all their limitations. Illness, as we know it, will be eradicated. Through the use of nanotechnology, we will be able to manufacture almost any physical product upon demand, world hunger and poverty will be solved, and pollution will vanish. Human evolution will undergo a quantum leap in evolution. We will be able to live as long as we choose. The coming into being of such a world is, in essence, the Singularity. (39-40)

The "Singularity" is an interesting appellation to assign such a time. Convergent technology programs promise to obliterate "the distinction between individuals and the entirety of humanity" (Cochrane, no pagination). The word "singularity" is defined as the condition of being one. Again, one discerns the occult themes of the Masonic "collective soul," Wells' "Mind of the Race," Durkheim's "collective consciousness," Pierre Teilhard de Chardin's "Noosphere," Jung's "collective unconscious," and other variants of monism. In turn, all of these themes echo a single political doctrine: collectivism.

The scientific dictatorship is not merely a dictatorship of the world, but of the mind as well. The Promethean crusade, whether conducted by the Marxist Utopians of the past or the technocratic neoconservatives of the present, is devoted to the erection of such a dictatorship. Crude though they may be, the communist and fascist campaigns of genocide and eugenical regimentation could be considered variants of a historical Promethean crusade. Their ultimate objective was the eradication of individual thought and the unification of human consciousness.

Pax Narcotica

There have been many subtle and scientifically refined methods employed in the ongoing campaigns to amalgamate individual consciousness into an ecumenical "racial mind." The 60s counterculture could be considered one example. This becomes especially apparent with the counterculture's vigorous circulation of hallucinogenic drugs, particularly LSD. Through the promulgation of such narcotics, human consciousness could be significantly altered and individual wills could be made more tractable. Once rendered more pliable, individual minds could be more comfortably amalgamated into a psychocognitive singularity.

Ostensibly, the 60s counterculture appeared to be a grass roots mobilization against the monopolistic capitalists of the Establishment. However, many of the counterculture's own radicals have suggested quite the reverse. In the radical treatise *Do It!*, revolutionary leader of the Yippies Jerry Rubin writes:

> The hip capitalists have some allies within the revolutionary community: longhairs who work as intermediaries between the kids on the street and the millionaire businessmen.

In his *The Strawberry Statement: Notes of A College Revolutionary*, former revolutionary Kunen gives us the following account of the 1968 S.D.S. (Students for a Democratic Society) national convention:

> Also at the convention, men from Business International Roundtables-the meetings sponsored by the Business International for their client groups and heads of government-tried to buy up a few radicals. These men are the world's leading industrialists and they convene to decide how our lives are going to go. These are the boys who wrote the Alliance for Progress. They're the left wing of the ruling class.
> They agreed with us on black control and student control. . .
> They want McCarthy in. They see fascism as the threat, see it coming from Wallace. The only way McCarthy could win is if the crazies and young radicals act up and make Gene more reasonable. They offered to finance our demonstrations in Chicago.
> We were also offered Esso (Rockefeller) money. They want us to make a lot of radical commotion so they can look more in the center as they move to the left. (116)

Another individual to discover this connection between the elite and the revolutionary community was undercover police intelligence operative David Gumaer. Gumaer took part in SDS demonstrations. Gumaer states that he:

> wondered where the money was coming from for all this activity, and soon discovered it came through radicals via the United Nations, from the Rockefeller Foundation, the Ford Foundation, United Auto Workers, as well as cigar boxes of American money from the Cuban embassy. (Epperson 403)

The evidence indicated that the ruling class financed violence on the part of the counterculture. In 1970, Ohio legislators were startled by a briefing, which included an Illinois commission report that addressed SDS uprisings on Ohio campuses. The report revealed: ". . .that $192,000 in Federal money and $85,000 in Carnegie Foundation funds were paid to [the] Students for a Democratic Society...during the fall of 1969" (Epperson 403). Before the House and Senate Security Committees, former Communist Party member and FBI informant James Kirk made the following statement:

> They (60s radicals) have no idea they are playing into the hands of the Establishment they claim to hate. The radicals think they are fighting the forces of the super-rich, like Rockefeller and Ford, and don't realize that it is precisely such forces which are behind their own revolution, financing it, and using it for their own purposes. (Griffin 107-08)

Simultaneously, the counterculture was systematically infused with mind altering narcotics like LSD. The dissemination of drugs served an alchemical purpose integral to evolution. Before it assumed the "scientific" guise of Darwinism, evolution was an occult doctrine of the Babylonian Mystery religions. As the Mysteries were diffused throughout the East, so was the concept of evolution. Reincarnation, which was the spiritual correlative of evolution, accompanied this doctrine. One of the major Eastern religions resulting from this diffusion was Hinduism. Researchers Paul deParrie and Mary Pride explain:

> Ancient Babylonian and Hindu beliefs included the doctrine of evolution. The goddess Kali was designated, among other things, the goddess of "becoming" or evolution. Reincarnation, the spiritual form of evolution, was part of both of these religions. (27)

In addition to evolution and reincarnation, the East also embraced two other practices: meditation and drug use. Researchers Patricia and Weldon Witters explain the augmentative role of both drugs and meditation in human evolution:

> Experimental psychiatrists, neurophysiologists, psychologists, and physicians are investigating the mind. Some of the most intriguing work is being done on the state of the mind during meditation. Countries like India have long histories linked to people who were able to achieve certain goals through

meditation. The word *yoga* is derived from the Sanskrit word for *union*, or yoking, meaning the process of discipline by which a person attains union with the Absolute. In a sense, it refers to the use of the mind to control itself and the body. Various systems of mind control have been used for thousands of years to find peace and contentment within. . . These effects occur without drugs, but drugs can speed up the process tremendously, and often unpredictably.

The category of people who take drugs as part of their search for the meaning of life eventually look for other methods of maintaining the valuable parts of the drug experience. Such people learn to value the meditation "high" and abandon drugs. They describe their drug experiences as having given them a taste of their potential, as something they grew out of now that they are established in the real thing... (382-87)

Allen Hollub, a protégé of infamous occultist Aleister Crowley, reiterated this augmentative function. Commenting on the magical tract *Book of the Forgotten Ones*, Hollub mandated that "the Mage must invoke his most primal self by the sacramental use of the proper drugs (blood, raw meat, cocaine, etc., and sex)" (Raschke, *Painted Black* 160). In *Diary of a Drug Fiend*, Crowley himself provided a fictionalized account of his own experimentation with narcotics in an effort to augment the evolutionary process:

We obtained the ineffable assurances of the existence of a spiritual energy that worked its wondrous will in ways too strange for the heart of man to understand until the time should be right... we had attained a higher state of *evolution*. (368; emphasis added)

It is interesting that drug use, meditation, ritual magic, and evolutionary thought all intersect within an occultist like Aleister Crowley. This strange conjunction comes into clearer focus, however, when one linguistically dismantles the word "pharmacy." It is derivative of the Greek word *pharmakeia*, which means "sorcery" (Daniel 47). It is most appropriate that the "sorcery" of drug use would be so closely associated with evolution. Cribbed from occult doctrines, the theory of evolution seems to invariably reunite itself with its correlating dogmas of meditation and drug experimentation.

The Nazis were also examples of this occult nexus. During his examination of Heinrich Himmler's Wewelsburg castle in Westphalia, satanic high priest and military intelligence officer Michael Aquino learned that the stronghold was used for Nazi "black magic" rituals (Raschke, *Painted Black* 245). Aquino revealed that the castle's chambers constituted "nothing less than an SS laboratory for experiments in *'conscious evolution'* (*Painted Black*, 245; emphasis added). Evidently, while most Darwinians eschew spiritualism, the adept occultist recognizes the mystical roots of evolutionary thought.

Aleister Crowley's pharmaceutical experimentation in "conscious evolution" was eventually imparted to his protégé, Aldous Huxley. Huxley was probably introduced to Crowley under the guidance of H.G. Wells (*Dope, Inc.* 538-39). It is possible that either Aldous or his brother, Julian, was a member of Crowley's Golden Dawn cult (Daniel 147-48). Huxley subscribed to Crowley's belief that drugs promised instantaneous mystical enlightenment, as is evidenced by his famous tract *The Doors of Perception*. In this treatise for drug use, Huxley provides the following description of the mescaline experience:

[W]hat happens to the majority of the few who have taken mescalin under supervision can be summarized as follows.
1. The ability to remember and to "think straight" is little if at all reduced. (Listening to the recordings of my conversation under the influence of the drug, I cannot discover that I was then any stupider than I am at ordinary times.)
2. Visual impressions are greatly intensified and the eye recovers some of the perceptual innocence of childhood, when the sensum was not immediately and automatically subordinated to the concept. Interest in space is diminished and interest in time falls almost to zero.
3. Though the intellect remains unimpaired and though perception is enormously improved, the will suffers a profound change for the worse. The mescalin taker sees no reason for doing anything in particular and finds most of the causes for which, at ordinary times, he was prepared to act and suffer, profoundly uninteresting. He can't be bothered with them, for the good reason that he has better things to think about.
4. These better things may be experienced (as I experienced them) "out there," or "in here," or in both worlds, the inner and the outer, simultaneously or successively. That they are better seems to be self-evident to all mescalin takers who

come to the drug with a sound liver and an untroubled mind.
(25-27)

Huxley's *The Doors of Perception* became one of the central treatises
on drug experimentation for the 60s counterculture. Another major
manifesto of the 60s radicals was *The Aquarian Conspiracy* by Marilyn
Ferguson. Making its public appearance in the spring of 1980, Ferguson's
book delineated the conspiratorial machinations underpinning the
counterculture movement. Revealing the evolutionary background of
this project, Ferguson states:

> The Aquarian Conspiracy is indeed loose, segmented,
> *evolutionary*, redundant. Its center is everywhere. Although
> many social movements and mutual-help groups are
> represented in its alliances, its life does not hinge on any of
> them (217; emphasis added).

Reiterating the theme of "conscious evolution," Ferguson adds:

> Millennia ago humankind discovered that the brain can
> be teased into profound shifts of awareness. The mind can
> learn to view itself and its own realities in ways that seldom
> occur spontaneously. These systems, tools for serious inner
> exploration, made possible the *conscious evolution of consciousness*.
> The growing worldwide recognition of this capacity and how
> it can be accomplished is the major technological achievement
> of our time. (71; emphasis added)

Seeking to "tease" the brains of counterculture radicals into
"profound shifts of awareness," Huxley initiated a project in the
mass narcotization of America. In October 1960, Huxley encouraged
Timothy Leary to "become a cheerleader for evolution" by flooding
Western democratic states with "brain-drugs, mass-produced in
the laboratories" (44). The experiment in "conscious evolution" was
expanded beyond the finite scope of individual tests and inundated
American society.

Huxley was not alone in this project of mass narcotization. The
Central Intelligence Agency was also involved in this experiment in
"conscious evolution." Marilyn Ferguson elaborates:

> Ironically, the introduction of major psychedelics, like LSD, in

the 1960s was largely attributable to the Central Intelligence Agency's investigation into the substances for possible military use. Experiments on more than eighty college campuses, under various CIA codenames, unintentionally popularized LSD. Thousands of graduate students served as guinea pigs. Soon they were synthesizing their own "acid". By 1973, according to the National Commission on Drug and Marijuana Abuse, nearly 5 percent of all American adults had tried LSD or a similar major psychedelic at least once. (126)

Although Ferguson characterizes the CIA's project of drug popularization as "unintentional," an occult undercurrent permeates the Agency's experiments in "conscious evolution." This undercurrent suggests a darker agenda. According to the authors of *Dope, Inc.*, the OSS, which was the forerunner of the CIA, was merely a subsidiary of British intelligence (540). When the Office of Strategic Services was being organized, William Stephenson, Britain's Special Operations Executive representative in the United States, was brought in for "technical assistance" (418). Stephenson's involvement would lead to the creation of "a British SOE fifth column embedded deeply into the American official intelligence community" (454).

British intelligence, in turn, seems to be little more than a subsidiary of Freemasonry. It is quite possible that occult involvement in British Intelligence goes back to its very beginning. The connection can be found with Sir Francis Walsingham, and advisor to Queen Elizabeth and the individual credited with founding British Secret Service (Howard 52). According to researcher Michael Howard:

It was rumoured that, like Dee (John Dee, the confidant to Elizabeth I), Walsingham was a student of occultism and that he used the underground organization of witch covens in Tudor England to gather material for his intelligence service. (53)

Walsingham would also work very closely on intelligence operations with Elizabeth's confidant, John Dee (53). Dee is alleged to be a Grand Master of the Rosicrucians, the occult forerunner to Freemasonry (51). This occult involvement would continue to the present day through Freemasonry. One individual who noticed the Freemasonic influence over British intelligence was Peter Wright, former Assistant Director of MI5. In his autobiography entitled *Spy Catcher*, Wright records an incident involving Personnel Director John Marriott that reveals a Freemasonic connection:

> After lunch I made my way back along the fifth floor for the routine interview with the Personnel Director, John Marriott. During the war Marriott had served as Secretary to the Double Cross Committee, the body responsible for MI5's outstanding wartime success—the recruitment of dozens of double agents inside Nazi intelligence. After the war he served with Security Intelligence Middle East (SIME) before returning to Leconfield House. He was a trusted bureaucrat. "Just wanted to have a chat—a few personal details, that sort of things," he said, giving me a distinctive Masonic handshake. I realized then why my father, who was also a Mason, had obliquely raised joining the brotherhood when I first discussed with him working for MI5 full-time. (30)

Evidently, membership in the brotherhood was an important factor in the selection of recruits for British intelligence. If nothing else, Masonic membership provided a definite advantage. At any rate, this strong Masonic influence remained within the CIA through the "British SOE fifth column" embedded deep within it and may have been one of the guiding visions for the Agency's mass narcotization projects.

Having established the presence of Masonic sorcerers within the CIA, one may proceed to examine the form of *pharmakeia* (sorcery) employed. The predominant alchemical potion that the CIA inoculated the counterculture with was LSD. The story of LSD begins with its development in 1943 by Albert Hoffman. Hoffman was a chemist in the employ of Sandoz A.B., a pharmaceutical house located in Switzerland that was owned by oligarch S.G. Warburg (*Dope, Inc.*540). During this period, Allen Dulles was in Berne, Switzerland acting as station chief for the OSS, precursor to the CIA. Dulles would go on to be Director of Central Intelligence (DCI) during the period when CIA was beginning MK-Ultra (*Dope, Inc.*540). While station chief in Berne, one of Dulles' OSS assistant was James Warburg, a member of the same oligarchical family that owned Sandoz A.B (*Dope, Inc.*540). This suggests that the OSS, later to become the CIA, may have played a role in the creation of LSD (*Dope, Inc.*540). Aldous himself would play a role in the Agency's project. During a return trip to America from Britain, Aldous would bring with him Dr. Humphrey Osmond, the Huxley's private physician.

Osmond was almost immediately enlisted by Allen Dulles to participate in MK-Ultra (*Dope, Inc.* 540). The "Opium War" against the United States had begun in earnest.

Re-examining Huxley's observations regarding the effects of mescaline, the hidden agenda of the CIA's mass narcotization project becomes clearer. Huxley states:

> Though the intellect remains unimpaired and though perception is enormously improved, the will suffers a profound change for the worse. The mescalin taker sees no reason for doing anything in particular and finds most of the causes for which, at ordinary times, he was prepared to act and suffer, profoundly uninteresting. He can't be bothered with them, for the good reason that he has better things to think about. (25-27)

Mescaline, LSD, and other psychotropic drugs substantially reduce the human impulse to resist coercion. It makes the user more susceptible to external manipulation and control. Herein was one of the chief objectives of the CIA's mass narcotization project: the creation of a tractable and compliant population. Jim Keith elaborates:

> As repugnant as it may be for a liberal audience to consider, the '60s "counter-culture" of LSD may have constituted an action reminiscent of the goals of the earlier British "vitality sapping" assault on China through opium; it may have also provided an Illuminist-derived injection of mysticism into American culture, a "peace pill." (*Casebook on Alternative Three* 67)

Was the project successful? One need only read the words of counterculture revolutionary James Simon Kunen to answer that question. Recounting a discussion with a woman in a restaurant, Kunen writes:

> We're the bridge generation, I continued. We're the product of all the past and we'll determine the future.
> Depressing or what?
> No, it's exciting. It's a challenge. It's up to us to keep future people human, assuming that's desirable.
> Is it?

I don't know. I mean, in *Brave New World* the people were all always happy. They were dehumanized and low but the fact remains they were happy. It was repugnant to the observer, but they couldn't step outside their system to see. They were just happy. That seems all right. (107)

Through the counterculture left, the oligarchs were able to neutralize grass root attempts to resist oligarchy. The ruling class used the movement to induce a paradigm shift. Confidence in America's Constitutional Republican form of government was considerably eroded. In addition, more power was concentrated into the hands of government, an entity the elitists could control. Both outcomes worked in accordance with the elite's criteria for maintaining and strengthening their power.

It is very appropriate that one CIA agent referred to the Haight Ashbury district, an area where 60s radicals commonly congregated, as a "human guinea pig farm" (Keith, *Mind Control, World Control*, 174). As Ferguson made clear, "many social movements and mutual-help groups" comprised the Aquarian Conspiracy, but the life of the project itself "does not hinge on any of them." With their minds effectively eviscerated by drug abuse, the "human guinea pigs" of the counterculture acquiesced to the emergent scientific dictatorship that was now firmly embedded within America.

Emerging concurrently with this population of mental and emotional invalids was an upsurge of New Age spiritualism. Many of the counterculture radicals adhered to some form of New Age mysticism. This may have been the intended result. Keith explains:

There is a line to be drawn. While mysticism perhaps comprises a vital, higher form of perception, in the matter of the real world that perception needs to be checked with critical analysis. A lack of a practical understanding is one reason that the hippie revolution failed, and this perhaps inherent shortcoming of drugged enlightenment may provide a rationale for the injection of drugs and mystical philosophy into a society. It may, in fact, be a technique for "softening up" populations. Hasn't religion and mysticism always been used in this manner? (*Casebook on Alternative Three*, 67)

The CIA's narcotization project might have been designed to augment the elite's program of religious and mystical manipulation.

The Masonic-British SOE "fifth column" within the Agency certainly reinforces this contention. Whatever the case might be, this new theocratic order and its pharmaceutically pacified automatons were the results of an ideological salvo successfully launched by the elite.

The Media Mind-Meld

Electronic media, such as TV and film, are two other means by which the unification of human consciousness could be attempted. The TV's application as a weapon of psychocognitive warfare is historically documented. In her book *The Perfect Machine: TV and the Nuclear Age*, Joyce Nelson offers the following case study:

In November 1969, a researcher named Herbert Krugman, who later became manager of public-opinion research at General Electric headquarters in Connecticut, decided to try to discover what goes on physiologically in the brain of a person watching TV. He elicited the co-operation of a twenty-two-year-old secretary and taped a single electrode to the back of her head. The wire from this electrode connected to a Grass Model 7 Polygraph, which in turn interfaced with a Honeywell 7600 computer and a CAT 400B computer.

Flicking on the TV, Krugman began monitoring the brain-waves of the subject What he found through repeated trials was that within about thirty seconds, the brain-waves switched from predominantly beta waves, indicating alert and conscious attention, to predominantly alpha waves, indicating an unfocused, receptive lack of attention: the state of aimless fantasy and daydreaming below the threshold of consciousness. When Krugman's subject turned to reading through a magazine, beta waves reappeared, indicating that conscious and alert attentiveness had replaced the daydreaming state.

What surprised Krugman, who had set out to test some McLuhanesque hypotheses about the nature of TV-viewing, was how rapidly the alpha-state emerged. Further research revealed that the brain's left hemisphere, which processes information logically and analytically, tunes out while the person is watching TV. This tuning-out allows the right hemisphere of the brain, which processes information emotionally and noncritically, to function unimpeded. "It appears," wrote Krugman in a report of his findings, "that the

mode of response to television is more or less constant and very different from the response to print. That is, the basic electrical response of the brain is clearly to the medium and not to content difference. . .[Television is] a communication medium that effortlessly transmits huge quantities of information not thought about at the time of exposure."
Soon, dozens of agencies were engaged in their own research into the television-brain phenomenon and its implications. The findings led to a complete overhaul in the theories, techniques, and practices that had structured the advertising industry and, to an extent, the entire television industry. The key phrase in Krugman's findings was that TV transmits "information not thought about at the time of exposure." (69-70)

Because the human brain does not engage active critical analysis during viewing, the TV can redefine the percipient's notions of reality. Through visceral imagery, the media creates a surrogate reality where what is presented on the screen is typically disproportionate with genuine reality. What TV presents as reality for one becomes reality for all. Herein is a somewhat effective means by which the ruling class have already unified mass consciousness. Apart from those who profitably eschew electronic media, a vast majority of the world are subconsciously fettered by a glowing screen in the living room. Nelson continues:

As Herbert Krugman noted in the research that transformed the industry, we do not consciously or rationally attend to the material resonating with our unconscious depths at the time of transmission. Later, however, when we encounter a store display, or a real-life situation like one in an ad, or a name on a ballot that conjures up our television experience of the candidate, a wealth of associations is triggered. Schwartz explains: "The function of a display in the store is to recall the consumer's experience of the product in the commercial. . . You don't ask for a product: The product asks for you! That is, a person's recall of a commercial is evoked by the product itself, visible on a shelf or island display, interacting with the stored data in his brain." Just as in Julian Jaynes's ancient cultures, where the internally heard speech of the gods was prompted by props like the corpse of a chieftain or a statue, so, too, our internalized media echoes are triggered by products, props, or situations in the environment.

As real-life experience is increasingly replaced by the mediated "experience" of television-viewing, it becomes easy for politicians and market-researchers of all sorts to rely on a base of mediated mass experience that can be evoked by appropriate triggers. The TV "world" becomes a self-fulfilling prophecy: *the mass mind takes shape, its participants acting according to media-derived impulses and believing them to be their own personal volition arising out of their own desires and needs. In such a situation, whoever controls the screen controls the future, the past, and the present.* (82; emphasis added)

Those who share in the mass media experience comprise the "mass mind," a somewhat diffuse psychocognitive singularity. This "mass mind" is just pliable enough to be semiotically re-sculpted according to whatever designs the media cartel might have. Thus, those who control the media also control vast quantities of percipients. Little do they know that they are being directly exposed to a subtle form of brainwashing. By presenting a selection of Establishment-sanctioned products and programs, viewers develop illusory notions of liberty and self-determinism. After all, are not the people free if they can choose between Pepsi and Coke? Meanwhile, through the alchemical sorcery of electronic media, individual consciousness is immersed within a "mass mind." TV and electronic media have become instrumental in the fulfillment of the elite's evolutionary script for humanity. It synchronizes with the Masonic vision of a unified consciousness, which is the purported outcome of the evolutionary process.

Convergent Technologies for Convergent Consciousness

William Sims Bainbridge may prove to be instrumental in the demise of the individual. In an article entitled "US report foretells of brave new world," journalist Nathan Cochrane examines *Converging Technologies for Improving Human Performance*, a report edited and contributed to by Bainbridge:

A draft government report says we will alter human *evolution* within 20 years by combining what we know of nanotechnology, biotechnology, IT and cognitive sciences. The 405-page report sponsored by the US National Science Foundation and Commerce Department, Converging Technologies for

Improving Human Performance, calls for a broad-based research program to improve human performance leading to telepathy, machine-to-human communication, amplified personal sensory devices and enhanced intellectual capacity. (Cochrane; emphasis added)

Elaborating on this research program, Cochrane explains how this convergent-technologies plan would be instrumental in the unification of mass consciousness:

People may download their consciousnesses into computers or other bodies even on the other side of the solar system, or participate in a giant "hive mind," a network of intelligences connected through ultra-fast communications networks. "With knowledge no longer encapsulated in individuals, the distinction between individuals and the entirety of humanity would blur," the report says. "Think Vulcan mind-meld. We would perhaps become more of a hive mind—an enormous, single, intelligent entity." (Cochrane, no pagination)

Of course, preparations must be made for the humanity's comfortable acclimation to this new "hive mind." Cochrane writes:

The report says the abilities are within our grasp but will require an intense public-relations effort to "prepare key organisations and societal activities for the changes made possible by converging technologies", and to counter concern over "ethical, legal and moral" issues. Education should be overhauled down to the primary-school level to bridge curriculum gaps between disparate subject areas. (Cochrane, no pagination)

The "endpoint of techne" may be drawing nigh as the Technocracy constructs its global holodeck. The nadir will be the reconstitution of the Masonic collective soul and the engineering of a "hive mind" through the sorcery of nanotechnology.

Digital Transformation

Semiotic intimations of this "hive mind" may have already been presented to the public eye. This becomes evident with one of the strange murals at Denver University. Pictures of this mural are available at the official website of Michael Corbin's outstanding radio program, *A Closer Look*. The wall painting, entitled "Digital Transformation," portrays:

> a goddess with her hands outstretched, palms up. Issuing forth from her palms are the double helix of the human DNA. As the helix moves upward, it dissolves into binary numbers. The numbers appear to be going skyward, as depicted by the back drop of stars and planets. (No pagination)

Encapsulated within this image is the Transhumanist concept of "cybersurvival." According to cybersurvival, man can achieve technological immortality through the marriage of the human mind with machines. Thus, man becomes a god who lives forever in cyberspace. This is accomplished through "mind uploading," which is defined as follows:

> Uploading is the transfer of the brain's mindpattern onto a different substrate (such as an advanced computer) which better facilitates said entity's ends. Uploading is a central concept in our vision of technological ascension. . . (*Mind Uploading: An Introduction*, no pagination)

Within the framework of this neo-Gnostic concept, the corporeal body is a mere prosthetic to be manipulated. Meanwhile, the human spirit, which Transhumanist ideologues reduce to an information pattern, is free to selectively incarnate itself in any number of physical shells. This is merely an updated version of Adam Weishaupt's "inner Areopagites: man made perfect as a god-without-God" (Billington 97). It is also the newest permutation of an older ideational contagion: the Gnostic myth of Darwinism.

Arthur C. Clarke prophesied such a digital transformation in the novel *The City and the Stars* (Bainbridge, "Technological Immortality," no pagination). It is also semiotically communicated through popular sci-fi films like *The Matrix*. Now, as is the case with all sci-fi predictive programming, fiction is struggling to become fact. In the very influential book *The Age of Spiritual Machines*, Ray Kurzweil asserts that this digital transformation could be achieved through magnetic resonance imaging or some technique of reading and replicating the human brain's neural structure within a computer (no pagination). Through the merger of computers and humans, Kurzweil believes that man will "become god-like spirits inhabiting cyberspace as well as the material universe" (no pagination). Kurzweil is hardly some marginalized cult leader. He has been involved in the founding and development of nine businesses, one of which dealing with virtual reality. He has also addressed the Council on Foreign Relations, one of the most prominent globalist organizations. Individuals such as he could be amassing the political and financial capital to tangibly enact the "prophecies" of science fiction.

However, Kurzweil is not alone. As William Sims Bainbridge previously stated: ". . .government-encouraged research is currently developing the technology to realise the Star Trek prophecies" ("Memorials," no pagination). Bainbridge himself is intimately involved in similar research, as is evidenced by his membership in the National Science Foundation. This independent government agency sponsored the report *Converging Technologies for Improving Human Performance*, which advocated the amalgamation of human consciousness into "a network of intelligences connected through ultra-fast communications networks" or a "hive mind" (Cochrane, no pagination). The National Science Foundation boasts an "annual budget of about $5.5 billion, and provides approximately 20 percent of all federally supported basic research conducted by America's colleges and universities" (*Wikipedia: The Free Encyclopedia*, "National Science Foundation," no pagination). Evidently, there are already powerful government and independent entities devoted to realizing the scientistic vision depicted by the "Digital Transformation" mural.

The theme of the Denver University mural is inherently religious. Goddesses are a common motif encountered throughout pagan mythology. The stars and the planets also semiotically gesticulate towards occult astrology. Yet, the DNA double helix adds a scientistic dimension to the mural's religious theme. It depicts the reduction of the human mind (i.e., soul) to a digital code. The ascension of the binary numbers into the heavens represents humanity's apotheosis via some form of mind uploading. Man is digitally unified with "God," who reflects the totality of human consciousness. The "all-seeing eye" of the Masonic Great Seal has become the "all-seeing Consciousness" of the computerized "hive mind." This is the "Digital Transformation," inspired by the power elite's evolutionary religion and semiotically communicated by the Denver University mural.

Evolutionary Pantheism
The ostensibly futuristic notion of man joining a computerized hive consciousness is really derivative of an ancient religion. It is the

latest incarnation of pantheism, albeit of a more scientistic character. Gnosticism and the ancient Mystery religions, which promised the "transfiguration of human into divine," both exhibited pantheistic features. Father Clarence Kelly comments on pantheism:

> Pantheism is a favorite doctrine of collectivists because.
> . .it offers a concept of man which, on religious grounds, subordinates the individual to the collective. (179)

The pantheistic concept of man teaches that humanity constitutes but a fragment of a greater whole. This greater whole, in turn, constitutes "God." However, "God" is not the personal Heavenly Father of the Scriptures. Instead, "God" is an immanent energy that channels itself through all things—-rocks, trees, oceans, clouds, animals, humans, and, ultimately, the universe itself. In rejecting the inherent uniqueness of the Creator, pantheism rejects the inherent uniqueness of humanity. No longer a creature made in the image of his Creator, man cannot lay claim to any God-given inalienable rights. According to pantheism, the immanent energy of "God" must flow unabated. This stipulates the individual's subordination to the collective. Kelly reiterates, stating that pantheism:

> functions as an effective tool in the subversion of God-centered religion by making religion man-centered, and thereby giving religious sanction to the doctrines and programs of political collectivism. (179)

Collectivistic as it is in nature, pantheism has had a historical appeal to oligarchs. It was the most frequently invoked religious doctrine of ancient rulers, especially Nimrod and his successors in Babylon. Of course, such theocratic power structures would eventually vanish with antiquity, giving way to the age of scientific dictatorships. However, not only did pantheism manage to survive this shift, but it may have been integral to it as well. Kelly observes that pantheism:

> . . .can be used as a stage in bringing people from theism to atheistic materialism. In religion, pantheism is most often expressed as Naturalism-—"the doctrine that religious truth is derived from nature, not revelation. . ." (17)

With its depiction of God as an immanent force and its metaphysical emphasis upon the ontological plane of the corporeal universe, pantheism presaged the dialectical materialism of contemporary scientific dictatorships. The religion also synchronized comfortably with evolutionary theory. No doubt, the metaphysical harmony between pantheism and evolutionism is attributable to Darwinism's Gnostic features. In a Darwinian context, the immanent force of "God" becomes the guiding hand of Nature, a self-created golem directing all evolutionary development. Such was the belief of Freemason Erasmus Darwin, Charles' grandfather. Erasmus developed "every important idea that has since appeared in evolutionary theory" (Darlington, "The Origin of Darwinism," 62). Thus, an undercurrent of occult pantheism may have been embedded within Darwinism from the very beginning. Of course, Darwinism would act as the edifying science for all the scientific dictatorships of modernity. Likewise, an evolutionary form of pantheism would become the galvanizing religious doctrine of the contemporary power elite.

With the widespread dissemination of Darwinism, several philosophers, cults, and subcultures adopted the elite's evolutionary pantheism. Of course, the diffusion of an idea leads to almost countless variations. Evolutionary pantheism was certainly no exception. The numerous ideological beneficiaries of evolutionary pantheism engaged in religious engineering, refining the doctrine according to their own theological dispositions. An ideational contagion had been unleashed and new permutations of evolutionary pantheism began to sprout.

One such permutation is Cosmotheism ("Pantheism," *Wikipedia: The Free Encyclopedia*, no pagination). Cosmotheism contends that "God was something created by man, perhaps even an end state of human evolution, through social planning, eugenics and other forms of genetic engineering" (no pagination). This belief in evolutionary pantheism and the emergent deity of man has resurfaced throughout history in various forms. It constituted the core doctrine of Adam Weishaupt's Illuminati. Researcher Dee Zahner explains:

> They [the Illuminati] taught that, rather than God creating the universe, the universe is creating God and that man is himself god and therefore unaccountable to a higher power. This is similar to the New Age doctrine of the 20th century. (30)

Fabian socialist and Freemason H.G. Wells would call the emergent deity of Cosmotheism the "world brain" ("Pantheism," *Wikipedia: The*

Free Encyclopedia no pagination). Evolutionist and Marxist liberation theologian Pierre Teilhard de Chardin promulgated the doctrine of the "Noosphere," which qualified as a form of Cosmotheism (no pagination). Likewise, Gnostic psychologist Carl Jung's "collective unconscious" and sociologist Emile Durkheim's "Collective consciousness" reflect the paradigmatic character of the Cosmotheist deity (no pagination). The "Overmind" in Arthur C. Clarke's *Childhood's End* is a "possible reference to the Cosmotheist Noosphere" (no pagination). With such noteworthy ideologues carrying its banner, the religion of Cosmotheism successfully transported pantheism into modernity.

The religion was also adopted by William Luther Pierce, a founder of the white supremacist organization called the National Alliance (Gardell 135). Mattias Gardell characterizes Pierce's variety of Cosmotheism as "racist pantheism" (135). Frighteningly enough, this new incarnation of pantheism is also the religious inspiration of some sects of Transhumanism. The official website of the Cosmotheist Community Church features several links to transhumanist sites. Given Transhumanism's considerable diffusion throughout academia and other cultural institutions, such connections are disturbing. To be sure, not every transhumanist may subscribe to Cosmotheism. However, that the precepts of Transhumanism are so easily united with the racialist principles of Pierce's Cosmotheism certainly prompts philosophical misgivings, if not outright suspicions.

Like its progenitor, pantheism, Cosmotheism is the "belief in the oneness of all material and spiritual elements" (135). Expanding on this monistic outlook, Gardell writes: "Monism bridges the illusory separation between God and man, creator and creation, mind and matter suggested by the Judea-Christian tradition" (135). A corollary of Cosmotheism's monism is the religion's overt advocacy of collectivism. *The Path*, which is the first of Cosmotheism's so-called "Holy Books," states:

> There is an essential unity, or consciousness that binds all living beings and all of the inorganic cosmos, as one. And what our true identity is this: we are the cosmos, made self-aware and self-conscious by evolution. Our undeniable human purpose, is to know and to complete ourselves as conscious individuals, and also as a self-aware species, and thereby to co-evolve with the cosmos towards total and universal awareness, and towards the ever-higher perfection of consciousness and being. (No pagination)

Examining the evolutionary pantheism of Cosmotheism, Gardell writes:

> The self-created whole is constantly evolving along the path toward perfection, propelled by a pre-biological force that animates all things in the tangible universe. Man is part of nature and subject to nature's laws, principal of which is the law of inequality and survival of the fittest. Evolving through a succession of states, the purpose of man is to ascend into godhood. (135)

Of course, evolutionary theory is premised upon metaphysical naturalism and materialism. Cosmotheism emphasizes evolution as the chief means of attaining "ever-higher perfection of consciousness and being." Thus, metaphysical naturalism and materialism remain firmly embedded within this new scientistic faith. Cosmotheism rejects the "notion of a transcendent divinity of a nature that is fundamentally Other than man" (Gardell 136). With this outright rejection of a "transcendent divinity," Cosmotheism reiterates materialism's claim that matter holds primacy and metaphysical naturalism's claim that all living things created themselves. In *The Path*, adherents are told:

> We give you the Truth, which is this: There is but one Reality, and that Reality is the Whole. It is the Creator, the Self-Created.
> The meaning of the Truth is this: Man, the world, and the Creator are not separate things, but man is a part of the world, which is a part of the Whole, which is the Creator. (No pagination)

Inherent to the Cosmotheistic version of "Truth" is the concept of self-creation, which comprises the metaphysical core of Darwinism. Of course, this metaphysical claim is closely aligned with the Gnostic claim of "self-salvation." The Cosmotheistic "Truth" also reiterates the realist argument, which was thoroughly accepted by H.G. Wells (Wagar 104). This position contends "that the whole is real and the individual an illusion"(104). This argument would reinforce Wells' belief in:

> the emergence in human evolution of a collective racial being

with the collective racial mind, which gathered the results of the individual mental effort into a single fund of racial wisdom and grew gradually toward organic consciousness of itself. (100-01)

Wells assigned this "collective racial mind" several appellations: the "World Encyclopaedia," the "World Brain," and the "Mind of the Race." Given Wells' possible Masonic heritage, this concept may be a derivation of the Freemasonic "Group Soul" or "collective soul." It was Wells' religious conviction that:

[i]ndividuals could escape the frustration inherent in the fact of their individuality and mortality only by consecrating their lives to the service of the Mind of the Race. (100-101)

Cosmotheism presents a virtually identical mandate:

And the meaning of the second way in which man serves the Creator's Purpose is this: The evolution of the Whole toward Self-completion is an evolution in spirit as well as in matter. Self-completion, which is Self-realization, is the attainment of perfect Self-consciousness. The Creator's Urge, which is immanent in the Universe, evolves toward an all-seeing Consciousness. (*The Path*, no pagination)

As an outgrowth of the power elite's own evolutionary pantheism, Cosmotheism reflects the paradigmatic character of the ruling class religion. It reiterates the Promethean aspirations of establishing a collectivist society and eugenically guiding man towards apotheosis.

The Neo-Eugenic State

Like the evolutionary outlook of Hitler, Cosmotheism contends that human evolution is producing two distinct varieties of man. Although Cosmotheism does not employ the same terms that Hitler invoked, its categories are no less elitist than the Nazi categories of the "god-man" and the "mass animal." *The Path* reveals the two races:

Man stands between sub-man and higher man, between immanent consciousness and awakened consciousness, between unawareness of his identity and his mission and a state of Divine Consciousness. Some men will cross the threshold, and some will not. (No pagination)

According to Cosmotheism, "higher man" stands to inherit divinity:

> Those who attain Divine Consciousness will ascend the Path of Life toward their Destiny, which is Godhood; which is to say, the Path of Life leads upward through a never-ending succession of states, the next of which is that of higher man, and the ultimate that of the Self-realized Creator. True reason will illuminate the Path for them and give them foresight; it will be a mighty aid to the Creator's Urge within them. (No pagination)

Meanwhile, the "sub-man" will find himself fettered by evolutionary stultification:

> And those who do not attain Divine Consciousness will continue groping in the darkness, and their feet will be tripped by the snares of false reason, and they will stumble from the Path, and they will fall into the depths. (No pagination)

Just as Hitler foresaw a dialectical struggle between the Aryans and the Jews, Cosmotheism foresees a struggle between "higher man" and "sub-man." Within this Cosmotheistic Armageddon, the "higher man" will defeat "sub-man" and consciously guide humanity's evolution toward immanent consciousness:

> And those who do not attain Divine Consciousness will continue groping in the darkness, and their feet will be tripped by the snares of false reason, and they will stumble from the Path, and they will fall into the depths. (No pagination)

Like the early revolutionary faith, Cosmotheism is preoccupied with the recurrent mythic theme of Prometheus. This becomes evident when one examines its corresponding philosophy, which is dubbed "Prometheism." The adherents of this philosophy seek to create the "First Sovereign Transhuman and Neo-Eugenic Libertarian Religious-State" ("Prometheism," no pagination). Neo-eugenics is defined as:

conscious evolution (these words are interchangeable). Purposefully directed evolution via voluntary positive neo-eugenics (including voluntary selective breeding), cloning, genetic engineering and ultimately any and all transhuman technologies. Neo-Eugenics means harnessing all science, technology and knowledge available now or in the future, guiding it with spirituality, ethical considerations and higher consciousness,ultimatelytowardsachievingtotalandunlimited self transformation. The term Neo-Eugenics embodies the sciences and philosophies involved in Biotechnology, Extropy and Transhumanism all merged in a philosophy of spiritual Conscious Evolution. (No pagination)

Just as Wells believed man to be analogous to Prometheus, this transhuman philosophy characterizes the human race as a "Promethean species" (no pagination). Man, who embodies Prometheus, now guides his own evolution through neo-eugenics:

Our immediate aim is to create a neo-eugenically enhanced race that will eventually become a new, superior species with whatever scientific means are available at the present time. In the short-term, this will be achieved via neo-eugenics, ie. voluntary positive eugenics, human cloning, germ-line engineering, gene therapy and genetic engineering.
In the long-term, when the science becomes available we intend to utilize transhuman technologies: nanotechnology, mind uploading, A/I and other variations of ultra exo-tech.
Our goal is to enable total and unlimited self-transformation, consciousness and expansion across the universe of our species. (No pagination)

Like the Utopian society envisioned by neo-Jacobins, the "First Sovereign Transhuman and Neo-eugenic Libertarian Religious-State" is a plebiscitary democracy:

Eventually, the goal of a neo-eugenically equalized society is to displace representative democracy with direct democracy. Only in this way can the corruption inherent in democracy be eliminated. This direct democracy requires that all members of society be highly intelligent and capable of understanding the issues as well as our elite representatives do now. In a neo-eugenically direct democracy the people and not corrupt politicians make decisions. (No pagination)

Following the Malthusian principles of Darwinism to their logical ends, the Promethean transhumanist subscribes to the eschatological belief in an overpopulation crisis. To combat this crisis, the transhumanist advocates the neo-eugenical regimentation of society:

Potential children are in abundant supply and the world is overpopulated with people without a future. Every child brought into this world should be of the finest intellect possible, and free of genetic diseases or abnormalities. Every generation needs to be an incremental step in the evolution to a new species. The only traits to be altered during the first genesis shall be an increase in overall intelligence, typical intellectual engagement (TIE), and patriotism. Other behavioral traits must only be altered when there is no longer a danger from competitive species and our knowledge of our species has progressed to a state of understanding that makes behavioral traits modification beyond question. Until then, we must retain the full spectrum of human variation for the sake of higher adaptability and survivability. (No pagination)

It should be noted that human variation is tolerated only as a matter of evolutionary expediency. Yet, what will happen once the neo-eugenic state has successfully achieved behavioral trait modification on a societal level? Will human variation remain desirable? Perhaps the question has already been answered by Dr. Richard Lynn, emeritus professor at the University of Ulster. Lynn candidly states:

"What is called for here is not genocide, the killing off of the population of incompetent cultures. But we do need to think realistically in terms of the 'phasing out' of such peoples. . .Evolutionary progress means the extinction of the less competent." (Hayes, no pagination)

Again, such thinking reflects the paradigmatic character of the power elite's religious doctrine. Concepts like Transhumanism, Cosmotheism, Prometheism, and neo-eugenics are all derivative of elite thought. They are theological variations of the Luciferian evangel, promulgated by occult secret societies and practiced by oligarchs for centuries.

Luciferianism: The Religion of Apotheosis

Luciferianism constitutes the nucleus of the ruling class religion. While there are definitely political and economic rationales for elite criminality, Luciferianism can account for the longevity of many of the oligarchs' projects. Many of the longest and most brutal human endeavors have been underpinned by some form of religious zealotry. The Crusades testify to this historical fact. Likewise, the power elite's ongoing campaign to establish a socialist totalitarian global government has Luciferianism to thank for both its longevity and frequently violent character. In the mind of the modern oligarch, Luciferianism provides religious legitimacy for otherwise morally questionable plans.

Luciferianism is the product of religious engineering, a tradition that even precedes Bainbridge's coining of the term itself. It has been the practice of Freemasonry for years. It was also the practice of Masonry's religious and philosophical progenitors, the ancient pagan Mystery cults. The inner doctrines of the Mesopotamian secret societies provided the theological foundations for the Christian and Judaic heresies, Kabbalism and Gnosticism. All modern Luciferian philosophy finds "scientific" legitimacy in the Gnostic myth of Darwinism. As evolutionary thought was popularized, variants of Luciferianism were popularized along with it. A historical corollary of this popularization has been the rise of several cults and mass movements, exemplified by the various mystical sects and gurus of the sixties counterculture. The metastasis of Luciferian thinking continues to this very day.

Luciferianism represents a radical revaluation of humanity's ageless adversary: Satan. It is the ultimate inversion of good and evil. The formula for this inversion is reflected by the narrative paradigm of the Gnostic *Hypostasis* myth. Like the *Hypostasis*, the binary opposition of Luciferian mythology caricatures Jehovah as an oppressive tyrant. He becomes the "archon of arrogance," the embodiment of ignorance and religious superstition. Satan, who retains his heavenly title of Lucifer, is the liberator of humanity.

Like some varieties of Satanism, Luciferianism does not depict the devil as a literal metaphysical entity. Lucifer only symbolizes the cognitive powers of man. He is the embodiment of science and reason. It is the Luciferian's religious conviction that these two facilitative forces will dethrone God and apotheosize man. It comes as little surprise that the radicals of the early revolutionary faith celebrated the arrival of Darwinism. Evolutionary theory was the edifying "science" of Promethean zealotry and the new secular religion of the scientific dictatorship.

Of course, modern scholars and Masonic apologists reject the concept of Lucifer and Satan being one in the same. They contend that this concept was purely the invention of St. Jerome. According to this argument, Jerome developed the name "Lucifer" by combining the Latin words *lucis* (meaning "light") and *ferre* (meaning "bearer"). Within the context of this position, the famous segment from the fourteenth chapter of Isaiah no longer refers to the spiritual entity of the Devil before his fall. Instead, it is nothing more than a reference to the king of Babylon, whose "golden city ceased" (Isaiah 14:4). However, this contention exhibits many significant weaknesses.

There is no consensus over which Babylonian king is being referred to in Isaiah. Some have argued that it is Nebuchadnezzar. However, the Lucifer depicted by Isaiah has been "cut down to the ground." If this is a figurative depiction of a king losing his kingdom, then it could not be Nebuchadnezzar. Nebuchadnezzar did not lose Babylon. It was Belshazzar who lost the "golden city" to the Persians and Medes. Meanwhile, others contend that Isaiah is referring to Tiglath-pileser III. However, Tiglath-pileser was never deposed. He merely died, leaving a power vacuum to be filled by the governor of Babylon, Ululai. Therefore, the only Babylonian king to whom Isaiah could be referring is Belshazzar.

Yet, even if Belshazzar is the object of Isaiah 14:4 and 14:12-15, these verses describe certain qualities that are not characteristic of mere mortals. In reference to the Babylonian king, Isaiah uses the term *Helel Ben Shachar*. Translated from Hebrew, this means "shining one." Unless Belshazzar's skin was naturally luminescent, this term hardly describes a human. It is, however, strangely similar to another personage from the Scriptures.

Hannachash is the Hebrew word used for the serpent that approaches Eve in the third chapter of Genesis (Heiser, no pagination). Yet, the "ha" portion of *hannachash* is a prefixed article meaning "the" in the original Hebrew (no pagination). In actuality, the base word, *nachash*, is an adjective meaning "bright" or "brazen"(no pagination). Michael S. Heiser, Ph.D. explains: "In Hebrew grammar, it is not unusual for an adjective to be 'converted' for use as a noun (the proper word is 'substantivized')" (no pagination).

As a noun, *nachash* can mean "snake/serpent or one who practices of divination" (no pagination). However, as a substantivized adjective, *nachash* is translated "shining one" (no pagination). Heiser concludes that:

> . . .Eve was not talking to a snake. She was speaking to a bright, shining upright being who was serpentine in appearance, and who was trying to bewitch her with lies. (No pagination)

Scripture is quite clear about the identity of this "bright, shining upright being who was serpentine in appearance." Revelation 20:2 reads: "that old serpent, which is the devil, and Satan . . ." In the context of the *nachash* interpretation, the curse bestowed upon the Devil by the Lord becomes more meaningful. Heiser states:

> My view of the curse in simple terms is that, as the *nachash* desired to vaunt himself above all created things on earth (and above the other created elohim, the "stars of God"—cf. Isa. 14:12-15; Job 38:7-10), and above the apex of that creation, humanity, so God turns the tables on him. He will now be placed under humanity's authority, who also governs the animals. He's going to be put on the bottom of the barrel of created things, so to speak. I view this as eschatological—it is not true now, since the kingdom of God—-which will be administered by a HUMAN, God incarnated in Jesus Christ, and by HUMANS—those of us who will, in Paul's words "rule over angels" (I Cor. 6). When the kingdom comes, the *nachash* will be put in his place. (No pagination)

The *nachash* interpretation also makes much more sense than the traditional interpretations. Heiser comments:

> I also think this is a preferable understanding to a literal curse on a   snake primarily because the curse on the woman—-whose seed is at odds with the seed of the *nachash*—-is the messiah (cf. Gal.3). The ultimate outcome of the curse is tied to the messiah's reign—and OUR reign with him as messiah's seed.

My view also makes more sense than the traditional view in that:

1) Not all women fear snakes.

2) There is no indication that snakes had limbs and walked upright and talked prior to the fall. This has been read into the passage "of necessity" by some who are not aware of the many divine council terms and motifs that clearly demonstrate the Eden incident concerns a confrontation against Yahweh and his human imagers from within the divine council.

3) What would the *nachash's* "seed" be if this referred to a literal snake? If this is in fact a shining divine being, then it makes eminent sense that the "counter seed" that would arise to oppose and kill off the seed of the woman (both in terms of godly humans and the ultimate seed, the messiah) would be an actual genealogical line that would be evil in origin—which is exactly what we see in Genesis 6:1-4. The activity of the fallen sons of God is juxtaposed with the thorough corruption of all human lineages (except Noah's), and the slate must be wiped clean. It's all about preserving bloodlines and "hybrid" bloodlines. . .the offspring of the sons of God—-the nephilim and other giant clans—-show up in very interesting places in the OT, and references to them in other texts (like that from Ugarit) show inextricably link them to the netherworld, the Canaanite hell. (No pagination)

The nephilims' origin with the nachash illustrates the inextricable link between the term "shining one" and supernatural entities. The nephilim represented an attempt by the "shining one" to sabotage the Lord's plan for humanity's redemption. Herein is the battle between God and Satan as it is to unfold on earth. Shortly after the *nachash's* deception in Eden and mankind's subsequent fall, the Lord revealed his plan of salvation for humanity: "I will put enmity between you and the woman, and between your seed and her seed; he shall bruise your head, and you shall bruise his heel" (Genesis 3:15). Encapsulated within this prognostication is a vivid portrait of Jesus Christ, who is the divinely implanted seed of the woman, defeating the seed of the *nachash*. This victory was accomplished through the sacrifice of Christ, whose bloodline remained unadulterated by the nephilim.

In fact, this purity of Christ's blood is etymologically communicated through the names of those who would comprise His lineage. In *Alien Encounters*, Chuck Missler and Mark Eastman linguistically dismantle the names of Adam, Seth, Enosh, Kenan, Mahalalel, Jared, Enoch, Methuselah, Lamech, and Noah. Translating the original Hebrew into English, Missler and Eastman reveal an amazing message embedded within this genealogy:

> Now let's put it all together:
> HEBREW   ENGLISH
> Adam   Man
> Seth   Appointed
> Enosh   Mortal
> Kenan   Sorrow;
> Mahalalel The blessed God
> Jared   Shall come down
> Enoch   Teaching
> Methuselah   His death shall bring
> Lamech The Despairing
> Noah   Rest, or comfort.
> That's rather remarkable:
> "Man [is] appointed mortal sorrow; [but] the blessed God shall come down teaching [that] his death shall bring [the] despairing rest."
> This is, of course, a summary of God's plan of redemption for mankind (called the "gospel" in the New Testament) hidden in a genealogy in Genesis. (220)

Yet, the genealogical lineage of evil—-"the nephilim and other giant clans"—-may not have been the only seeds of the *nachash*. Although the giants of antiquity were eventually wiped out, the seed of the *nachash* still manifest itself in far more pervasive form. Researcher John Daniel elaborates:

> Through the "seed" of woman, God would provide a Redeemer. The serpent, representing Satan, would also have a "seed," a counterfeit redeemer. Conflict would break out between the serpent's seed and the woman's seed. (102)

Therefore, the new "seed" of the *nachash* could be a succession of false messiahs. Such a succession would probably culminate with a final counterfeit Christ. Of course, this last anti-Christ would be vanquished with the return of the Lord's true Messiah. Daniel explains:

To understand how this conflict between God and Satan is to be played out in human history, we must consider the key Hebrew words in the statement, "he [Christ] shall bruise your [the serpent's] head, and you shall bruise his heel." The Hebrew primitive root word for *heel* means to "supplant, circumvent, or trip up." It suggests that the Serpent or Satan shall set up a religion which becomes a stumbling block to supplant or circumvent the plane of God for our redemption; the Adversary will attempt as well as to "trip up," or "circumvent" the Redeemer. The Redeemer, on the other hand, would bruise the head of the serpent. The Hebrew word for *head* means "ruler," and the word for bruise means "overwhelm." In other words, Satan is the "head" or "ruler" of this present world, but in the end Jesus Christ, the Redeemer, shall ultimately bruise, or "overwhelm" Satan (Rev. 19:11-20:15). (102)

Since it no longer has a biological lineage through which it can propagate itself, the "seed" of the *nachash* could be manifesting through the false messianic doctrines that are currently challenging Christianity. Given the numerous false messiahs resulting from the religious engineering of modern scientistic cults, this scenario seems highly likely. It is further reinforced by the resurgence of Gnosticism, which has been scientifically edified by its corresponding myth of Darwinism. Gnosticism holds that humanity is fettered by the curse of the archons and must wait for the arrival of the "perfect man," the "Gnostic adept" who will break the spell (Raschke, *The Interruption of Eternity* 27).

Contemporary scholars and Masonic apologists automatically assume that Lucifer was some sort of fiction concocted by St. Jerome because he substituted a Latin word for the Hebrew appellation of "shining one." From this assumption, they conclude that the idea of an intelligent evil, namely Satan, is merely some meme that gradually evolved over time. Such a contention synchronizes comfortably with the moral relativism of the postmodern era. After all, an invariant criterion for determining evil presupposes the existence of a defining principle of evil. This defining principle would have to be embodied by some sentient entity, which tangibly enacts the otherwise abstract precepts of evil. The secular and Masonic arguments reject Satan, who constitutes just such a sentient entity. The rejection of Satan is invariably accompanied by the rejection of evil itself.

However, the etymological continuity of the appellation of "shining one" suggests that St. Jerome had good reason to encapsulate Satan's former glory as the "light bearer" within the Latin name, "Lucifer." 2 Corinthians 11:14 warns that Satan can appear as an "angel of light." Perhaps he can assume this appearance because he never ceased to "shine," even after his fall from heaven. As an angel, albeit a fallen one, Satan would remain the "shining one."

The "shining one" is precisely the same designation assigned to the supposed "king of Babylon" in Isaiah 14:12-15 (no pagination). This etymological synchronicity suggests that the purely historicized interpretation of Isaiah's words concerning the "shining one" is incomplete. It ignores the obvious spiritual characteristics intrinsic to the "shining one" mentioned by Isaiah. Even if Belshazzar is the object of Isaiah 14:4 and 14:12-15, the spiritual implications of these verses are inescapable. Therefore, Isaiah 14:4 and 14:12-15 must have both a historical context and a spiritual context.

Isaiah 14:4 and 14:12-15 present what some Biblical scholars call a "type." That is, an actual event that "typifies" a spiritual reality. This becomes evident when one examines the forth verse:

That thou shalt take up this *proverb* against the king of Babylon, and say, How hath the oppressor ceased! the golden city ceased! (Emphasis added)

The word "proverb" is derived from the Hebrew root *mashal*, which can be translated as "allegory" or "similitude." Allegories are symbolic representations of ideas, principles, and personages of significance. Certainly, Belshazzar voices the same ambitions of the "shining one," making him the ideal symbol of the fallen *nachash*. A "similitude" is an appreciable likeness, which closely resembles a paradigmatic counterpart. Indeed, the paradigmatic character of Isaiah's parable closely resembles the account of the "shining one's" fall from heaven. Thus, Belshazzar typifies Satan. Historically, he is the king of Babylon. Symbolically, he is the "shining one" who fell from glory. Like the oligarchs of the present world system, Belshazzar merely represented a stronger, supra-sensible source of evil. Isaiah's metaphorical message is beautifully reiterated in Ephesians 6:11-12:

For we wrestle not against flesh and blood, but against principalities, against powers, against the rulers of the darkness of this world, against spiritual wickedness in high places.

Behind the flesh and blood of Belshazzar was a "spiritual wickedness." As a vessel for this "spiritual wickedness," the king of Babylon provided the "shining one" with access to the "high places" of the dominant political and social structure. The same holds true for Belshazzar's contemporaries, the architects of the global scientific dictatorship.

Transhumanism: The Cult of Techno-Luciferianism

Transhumanism offers an updated, hi-tech variety of Luciferianism. Following the Biblical revisionist tradition of the Gnostic *Hypostasis* myth, Transhumanists invert the roles of God and Satan. In an essay entitled "In Praise of the Devil," Transhumanist ideologue Max More depicts Lucifer as a heroic rebel against a tyrannical God:

> The Devil—-Lucifer--is a force for good (where I define 'good' simply as that which I value, not wanting to imply any universal validity or necessity to the orientation). 'Lucifer' means 'light-bringer' and this should begin to clue us in to his symbolic importance. The story is that God threw Lucifer out of Heaven because Lucifer had started to question God and was spreading dissension among the angels. We must remember that this story is told from the point of view of the Godists (if I may coin a term) and not from that of the Luciferians (I will use this term to distinguish us from the official Satanists with whom I have fundamental differences). The truth may just as easily be that Lucifer resigned from heaven. (No pagination)

According to More, Lucifer probably exiled himself out of moral outrage towards the oppressive Jehovah:

> God, being the well-documented sadist that he is, no doubt wanted to keep Lucifer around so that he could punish him and try to get him back under his (God's) power. Probably what really happened was that Lucifer came to hate God's kingdom, his sadism, his demand for slavish conformity and obedience, his psychotic rage at any display of independent thinking and behavior. Lucifer realized that he could never fully think for himself and could certainly not act on his independent thinking so long as he was under God's control. Therefore he left Heaven, that terrible spiritual-State ruled by the cosmic sadist Jehovah, and was accompanied by some

of the angels who had had enough courage to question God's authority and his value-perspective. (No pagination)

More proceeds to reiterate 33rd Degree Mason Albert Pike's depiction of Lucifer:

Lucifer is the embodiment of reason, of intelligence, of critical thought. He stands against the dogma of God and all other dogmas. He stands for the exploration of new ideas and new perspectives in the pursuit of truth. (No pagination)

Lucifer is even considered a patron saint by some Transhumanists ("Transtopian Symbolism," no pagination). Transhumanism retains the paradigmatic character of Luciferianism, albeit in a futurist context. Worse still, Transhumanism is hardly some marginalized cult. Richard Hayes, executive director of the Center for Genetics and Society, elaborates:

Last June at Yale University, the World Transhumanist Association held its first national conference. The Transhumanists have chapters in more than 20 countries and advocate the breeding of "genetically enriched" forms of "post-human" beings. Other advocates of the new techno-eugenics, such as Princeton University professor Lee Silver, predict that by the end of this century, "All aspects of the economy, the media, the entertainment industry, and the knowledge industry [will be] controlled by members of the GenRich class...Naturals [will] work as low-paid service providers or as laborers. . ." (No pagination)

Max More is another case in point. He came to the United States from Oxford University in England("Extropianism," *Wikipedia: The Free Encyclopedia*, no pagination). While in England, he founded Mizar Limited, which was the first European cryonics organization (no pagination). He also co-founded the Extropy Institute, which began organizing the first transhumanist conferences in 1992 (no pagination). With a growing body of academic luminaries and a techno-eugenical vision for the future, Transhumanism is carrying the banner of Luciferianism into the 21st century. Through genetic engineering and biotechnological augmentation of the physical body, Transhumanists are attempting to achieve the very same objective of their patron saint:

I will ascend into heaven, I will exalt my throne above the stars of God: I will sit also upon the mount of the congregation, in the sides of the north: I will ascend above the heights of the clouds; I will be like the most High. (Isaiah 14:13-14)

Return of the Sun God

In *Fires in the Minds of Men*, James H. Billington observes that the scientistic Promethean faith was accompanied by "the more pointed, millennial assumption that, on the new day that was dawning, the sun would never set" (6). Billington states that the tumult of the "French upheaval" birthed a "solar myth of the revolution" (6). This "solar myth" contended that "the sun was rising on a new era in which darkness would vanish forever" (6). This ideational contagion became embedded "at a level of consciousness that simultaneously interpreted something real and produced a new reality" (6). Indeed, the sun motif seems to be part of another recurring metaphorical icon pervading the crusade to create this "new reality," which Billington describes as follows:

> The new reality they sought was radically secular and stridently simple. The ideal was not the balanced complexity of the new American federation, but the occult simplicity of its great seal: an all-seeing eye atop a pyramid over the words *Novus Ordo Seclorum*. In search of primal, natural truths, revolutionaries looked back to pre-Christian antiquity-—adopting pagan names like "Anaxagoras" Chaumette and "Anacharsis" Cloots, idealizing above all the semimythic Pythagoras as the model intellect-turned-revolutionary and the Pythagorean belief in prime numbers, geometric forms, and the higher harmonies of music. (6)

It is very interesting that such a "radically secular" reality would be so preoccupied with the "occult simplicity" and "pagan names" of "pre-Christian antiquity." Again, it becomes evident that secularization is merely a philosophical segue. Once the religious institutions of a culture have been effectively eviscerated by atheism, "cults and occultism explode to fill the spiritual vacuum" (Bainbridge, no pagination, "Religions for a Galactic Civilization"). The recurring sun motif of the ancient Mystery cults and the Promethean faith may provide a fragmentary glimpse of the emergent world religion's paradigmatic character.

Within esoteric circles, the sun is inextricably linked with alchemical magic:

Ashmun is the Egyptian name for Hermopolis, the home of the sacred eight, the site of the world's creation out of chaos, the cosmic egg. Hermopolis, believed to be city "where the sun first rose on earth," was also the center of the god of wisdom and learning, the ibis-headed moon-god Thoth. The Greeks called him Hermes Trismegistrus, the god of time and "conductor of the dead," the inventor of alchemy and magic, and a traditional source for Western ritual magic. (409)

Underpinning the concept of alchemical magic is the theme of man's inherent mutability. This was also the theme of transformism, a belief in the progressive development of humanity:

Transformism was a rather Lamarckian view of the mutability of species that preceded Darwinian evolution in Germany, France, and elsewhere. What connected the two theories was the essential belief that life-forms had changed over time; what separated them was the proposed mechanism, which for Darwin was natural selection and for transformists was a vaguely described will or yearning of the organism for self-improvement. However, transformism was the scientific equivalent of the French Revolution: a dangerous doctrine of the possibility of change in social as well as biological spheres. (Shipman 91-92)

Such beliefs are nothing new. The religious belief in man's potential for "transfiguration into the divine," typified by Gnosticism and the ancient Mysteries, qualified as a form of transformism. This ancient belief system remained preserved as the Masonic doctrine of "becoming," which would provide the foundation for the early evolutionary theories of John Locke and Erasmus Darwin. Of course, these precursory theories would inspire Darwinism, which was vigorously promulgated by the Masonic Royal Society. Commensurate with the popularization of Darwinism were social Darwinism, eugenics, population control, and other forms of self-directed evolution through societal intervention. In a sense, all of these comprised an alchemical agenda to realize the aspiration to achieve apotheosis, which was characteristic of the ancient occult theocracies. As modern societies implemented policies of self-directed evolution, scientific dictatorships began to rise.

The theocratic forerunners to the scientific dictatorship were largely governed by sun god cults (Keith *Saucers of the Illuminati* 78). The mystical significance of the sun was reiterated by the "solar mythology" of the sociopolitical Utopians and secular Gnostics of the late 19th century. However, the early Promethean revolutionaries would jettison a majority of the metaphysical concepts inherent to the sun god myth. Now, the sun was merely a symbol for the "new reality" that they were determined to create. Yet, the theme of man's inherent mutability remained intact. In addition, an even older promise continued to kindle the Promethean fire: ". . .ye shall be as gods."

Freemason Albert Pike states: ". . .Osiris, himself symbolized the Sun. . ." (15). Pike also reveals that Osiris had a rival: "Long known as. . .Adonai [another name for Jehovah, the Lord of the Bible]; . . .the Rival of Bal and Osiris. . ." (697). In fact, Bal and Osiris were one in the same, representing the "invisible God" worshipped "beyond the orb [sun]" (77). This is the reason for Pike's capitalization of the word "Sun." He is not referring to the corporeal "orb" that provides earth with daylight, but an "invisible God" whose identity was known only to a few.

Pike provides a hint regarding the identity of the "invisible God" lurking "beyond the orb." Referring to the Egyptians, one of the many ancient peoples that worshipped the sun god, Pike explains: "The horned serpent was the hieroglyphic for a God" (495). Of course, the Bible also speaks of a serpent that opposed Adonai and promulgated the conceit that he was a god. Deceased researcher William Cooper elaborates:

> The *snake* and the dragon are both symbols of wisdom. Lucifer is the personification of the symbol. It was Lucifer who tempted Eve to entice Adam to eat of the tree of knowledge and thus free man from the bonds of ignorance. (70; emphasis added)

At this juncture, it is important to recall Wilmshurst's contention that the process of evolution is to culminate with the unification of human consciousness with the "Omniscient" (94). In light of this fact, Pike's later statements concerning the sun come into painful focus: ". . .the Blazing Star has been regarded as an emblem of Omniscience, or the All-Seeing Eye, which to the ancients was the Sun" (506).

The symbol of this counterfeit god adorns the halls of Freemasonry.

Pike states: "The Sun. . .his is the All-Seeing Eye in our Lodges" (477). Of course, the All-Seeing Eye was placed on America's dollar during the presidency of Freemason Franklin D. Roosevelt. Interestingly enough, this period would also witness the rise of the technocratic movement and the enshrinement of the New Deal's socialist policies. This period of American history was just the beginning. America's technocratic restructuring continues under the guidance of the sociopolitical Darwinians of today.

Aldous Huxley, who coined the term "scientific dictatorship," retained membership in what author Martin Green dubbed the "Children of the Sun" (3). Green recognized the group as a revival of an older Egyptian model, which also called itself the "Children of the Sun" (437). Green provides a description of this earlier model:

> This culture was diffused by migration, and in the farther-off lands the new king-gods it brought were said to have come "from the sky" because they came from abroad. These kings, and sometimes one section of the ruling class, called themselves Children of the Sun; they claimed the sun as their father, and expected to go up to the sky when they died (it was the normal expectation that one would go underground). . .This class, then, felt themselves to be an elite within their own culture, and felt their culture to be an elite in relation to other cultures. (437)

Of course, this was the creed of John Ruskin and his protégé, Cecil Rhodes. . .the superiority of the British ruling class culture. In fact, the offspring of Round Table members peopled the modern Children of the Sun. According to Green, this cult held "prominence within" and "partial dominance over" the "English culture after 1918" (3). It is possible that the Children of the Sun's belief system partially constituted the doctrinal foundation of the anglophile "scientific dictatorship," *Pax Britannia*.

Initially, Osiris was the locus of the ancient Egyptian cult's worship (438-39). However, through cultural and religious development, the cult's locus of praise relocated itself within the pagan deity of Dionysus (439). Green explains the changes that accompanied this shift in worship:

> Dionysus, when he comes, is dependent on no one and responsible for no one. He is neither son nor father, and the culture built around him--perhaps we can get some idea of this

from recent rock festivals--must be orgiastic and solipsistic, defiant of all responsibility and all relationship. Osiris is always a son, even though he has no father. Most typically, gods like him were born from an egg, or from a lotus blossom, or from a divine cow--in other words, from the divine mother-earth, without any ordinary impregnation. (439)

Osiris was begat by the golem of "mother-earth" or Gaia. Dionysus, however, was sovereign and claimed no progenitor. While the Osiric model was clearly desirable, it was merely a transitional phase for the processing of the masses into the Dionysian theocracy (439). The banner of the Dionysian model:

was taken up by Bachofen's Munich disciples, Alfred Schuler and Ludwig Klages, who for a time advocated it as the moist perfect of all cultural phases. We might look briefly at Schuler's essay "Die Sonnenkinder," the fifth of his seven lectures "Vom Leben der Ewigen Stadt," given in Munich in 1917, in which he recreated, highly imaginatively, the religious culture of ancient Rome. He described Osiris as passive toward Isis, but as giving light to mankind by virtue of containing the two poles of boyhood within himself. He is both Castor and Pollux, and their love for each other, in him, makes him a radiant god. (439-40)

According to Schuler, the rebirth of the cult of Osiris could break the grip of the:

evil Apollonian forces of progress and mechanization, of increase by production and reproduction, of self-justification in one's children rather than in oneself, that he saw ruining his own Germany. (440)

Green reveals the institution that most succinctly embodied Schuler's accursed Apollonian forces:

Christianity was for Schuler the enemy of all life, being spiritual, ascetic, mental--Apollonian. What he and his disciples loved--what, less ideologically, Oscar Wilde in contemporary England loved--was the culture of the mother goddess in her decadence, when the radiant son was triumphant over her. (440)

The British Children of the Sun, of which Aldous Huxley was a member, represented a revival of this belief system. Over the years, this pagan Weltanschauung has resurfaced under numerous appellations and in numerous forms. However, several features have remained consistent. First, it is a thinly veiled form of Luciferianism that is diametrically opposed to Christianity. Second, it depicts man as an inherently mutable organism whose development is facilitated through the alchemy of science. Third, it venerates the sun as a symbol of the alchemical forces that promise humanity's apotheosis. In spite of its modifications through consistent religious engineering, the institution of sun worship has retained these traits and remains a permanent fixture of the numerous conspiracies to erect a scientific dictatorship.

Consider the words of scientism's high priest, Darwinian Carl Sagan:

Our ancestors worshiped the Sun, and they were far from foolish. And yet the Sun is an ordinary, even a mediocre star. If we must worship a power greater than ourselves, does it not make sense to revere the Sun and stars? (243)

Evidently, the modern sociopolitical Darwinians are of the same opinion.

Thanatos Universalized: The Ruling Class Suicide Cult
One of the many dialectics that has divided humanity throughout history is spiritualism against materialism. As is typically the case with dialectics, the competing ideational entities involved in the conflict are not dichotomously related. Instead, they represent variants of metaphysical irrationality. One elevates the soul to the detriment of the physical body. The other elevates the physical body to the detriment of the soul. Invariably, the synthesis of these two results in cynical nihilism and the primacy of some form of authoritarian Gnosticism. The emergent world religion of the scientific dictatorship is being birthed by just such a dialectical climate.

Michel Foucault asserted that "the soul is the prison of the body" (30). This assertion is somewhat true in relation to spiritualism. Spiritualism tends to bestow absolute metaphysical and epistemological primacy upon the soul. Meanwhile, the body is treated like an unfortunate and annoying afterthought. Although the soul is obviously important, spiritualism virtually detaches it from the physical body and depicts its relationship with corporeality as some sort of accident. Within the spiritualist conceptual framework, the body becomes an impediment to humanity's acquisition of knowledge and is regarded with increasing derision. This seemed to be a commonly held view among the early Greek philosophers. Such was the contention of Socrates and his protégé, Plato. In *Phaedo*, Plato writes:

[T]he body provides us with innumerable distractions in the pursuit of our necessary sustenance, and any diseases which attack us hinder our quest for reality. Besides, the body fills us with loves and desires and fears and all sorts of fancies and a great deal of nonsense, with the result that we literally never get an opportunity to think at all about anything. Wars and revolutions and battles are due simply and solely to the body and its desires. All wars are undertaken for the acquisition of wealth, and the reason why we have to acquire wealth is the body, because we are slaves in its service. That is why, on all these accounts, we have so little time for philosophy. Worst of all, if we do obtain any leisure from the body's claims and turn to some line of inquiry, the body intrudes once more into our investigations. Interrupting, disturbing, distracting, and preventing us from getting a glimpse of the truth. We are in fact convinced that if we are ever to have pure knowledge of anything, we must get rid of the body and contemplate things by themselves with the soul itself. (65A-66E)

Invariably, such a dismal Weltanschauung promulgates a morbid preoccupation with death. After all, given the physical body's alleged inconvenience, is not death a welcome liberator? Again, this was a shared contention of both Socrates and Plato. As the discourse in *Phaedo* proceeds, Plato exalts death:

It seems, to judge from the argument, that the wisdom which we desire and upon which we profess to have set our hearts will be attainable only when we are dead, and not in our lifetime. If no pure knowledge is possible in the company of the body, then either it is totally impossible to acquire knowledge, or it is only possible after death, because it is only then that the soul will be separate and independent of the body. It seems that so long as we are alive, we shall continue closest to knowledge if we avoid as much as we can all contact and association with the body, except when they are absolutely necessary, and instead of allowing ourselves to become infected with its nature, purify ourselves from it until God himself gives us deliverance. In this way, by keeping ourselves uncontaminated by the follies of the body, we shall probably reach the company of others like ourselves and gain direct knowledge of all that is pure and uncontaminated—that is, presumably, of truth. For one who is not pure himself to attain to the realm of purity would no doubt be a breach of universal justice. (66E-67B)

Obviously, such a Weltanschauung is hardly life affirming. In such a world, self-immolation would constitute a meritorious duty. Eventually, this way of thinking was married to Christianity, birthing the aberration of Gnosticism. According to Gnosticism, the physical universe is hell. Corporeal existence is a prison that fetters man through the demonic agents of space and time. Remaining consistent with the Greeks' death worship, Gnosticism also entertained suicide. The Albigenses are one case in point. This Gnostic sect mandated suicide, favoring starvation as the chief means of dispatching one's self ("Albigenses").

On the other pole of this dialectic resides materialism, which bestows absolute primacy upon matter. Commensurate with the ascendancy of this metaphysical doctrine has been the dogmas of both Karl Marx and Charles Darwin. Ironically, both of these ideologues were merely reiterating the precepts of Gnosticism within the context of metaphysical naturalism. The *Encyclopedia of Religion* reveals that "both Hegel and his materialist disciple Marx might be considered direct descendants of gnosticism" (576). Darwinism, which acted as the legitimizing science for Marxism, amounts to little more than a Gnostic myth. Dr. Wolfgang Smith elaborates:

As a scientific theory, Darwinism would have been jettisoned long ago. The point, however, is that the doctrine of evolution has swept the world, not on the strength of its scientific merits, but precisely in its capacity as a Gnostic myth. It affirms, in effect, that living beings created themselves, which is in essence a *metaphysical* claim... Thus, in the final analysis, evolutionism is in truth a metaphysical doctrine decked out in scientific garb. In other words, it is a scientistic myth. And the myth is Gnostic, because it implicitly denies the transcendent origin of being; for indeed, only after the living creature has been speculatively reduced to an aggregate of particles does Darwinist transformism become conceivable. Darwinism, therefore, continues the ancient Gnostic practice of depreciating "God, the Father Almighty, Creator of Heaven and earth." It perpetuates, if you will, the venerable Gnostic tradition of "Jehovah bashing." And while this in itself may gladden Gnostic hearts, one should not fail to observe that the doctrine plays a vital role in the economy of Neo-Gnostic thought, for only under the auspices of Darwinist "self-creation" does the Good News of "self-salvation" acquire a semblance of sense. (242-43)

Edified by the Gnostic myth of Darwinism, Marxism was disseminated on the popular level as both communism and fascism. Both of these ideological camps (which are not diametrically opposed, as some political scientists would have one believe) qualified as forms of secular Gnosticism. Of course, both were also scientific dictatorships. Following the Gnostic tradition of the Albigenses, the social contracts underpinning communism and fascism amounted to enormous suicide pacts. Most of the socialist totalitarian regimes of the 20th century self-destructed, either by financial stultification resulting from Marxist planned economies or by violent revolution.

The dialectic of spiritualism against materialism may witness a Hegelian synthesis in the emergent philosophies of Transhumanism, Singularitarianism, and other futurist variants of the ruling class *Weltanschauung*. Transhumanism serves as a prime example. This philosophy not only expresses an inherently Gnostic derision for the physical body, but an overall disdain for humanity as well. Christian philosopher C. Christopher Hook expands on the continuity of Gnostic thinking within Transhumanism:

Transhumanism is in some ways a new incarnation of gnosticism. It sees the body as simply the first prosthesis we all learn to manipulate. As Christians, we have long rejected the gnostic claims that the human body is evil. Embodiment is fundamental to our identity, designed by God, and sanctified by the Incarnation and bodily resurrection of our Lord. Unlike gnostics, transhumanists reject the notion of the soul and substitute for it the idea of an information pattern. (no pagination)

Transhumanism carries on the secular Gnostic tradition, attempting to "immanetize the Eschaton" on this ontological plane through the technological transformation of mankind. The Transhumanist contends that biotechnology, nanotechnology, and neurotechnology will propel humanity into a "posthuman" condition. Once he has arrived at this condition, man will cease to be man. He will become a machine. Like spiritualism, Transhumanism portrays the relationship between the soul (which Transhumanists reduce to a mere information pattern) and the physical body as one enormous cosmic mistake. Transhumanist ideologue Bart Kosko flatly declares: "Biology is not destiny. It was never more than tendency. It was just nature's first quick and dirty way to compute with meat. Chips are destiny" (qutd. in Hook, no pagination). Again, the Gnostic proclivity towards suicide becomes evident. According to Transhumanism, man can only become more once he has killed himself.

This perpetually metastasizing and mutating form of irrationality owes its existence to the dialectic of spiritualism against materialism. It was probably this dialectic that Thomas Aquinas sought to overcome. As a medieval Christian philosopher, Aquinas had, no doubt, been exposed to more than a few Gnostic thinkers. Endowed with a certain degree of philosophical foresight, Aquinas would have recognized the logical ends of Gnostic thought and its ramifications for the future (although it's doubtful that even Aquinas would have prognosticated where it has led today).

Aquinas argued that Plato's substance dualism went too far. Aquinas agreed with Plato's contention that the soul is a *per se* subsistent entity, which is possible given its ability of abstraction apart from the body. However, Aquinas disagreed with Plato's contention that the soul represents an entirely independent species and genus of substance. Such a contention makes the relation of the soul and the body accidental. Thus, death would not qualify as a substantial corruption. This contention is obviously false. The problem stems from a categorical misallocation. Categorically, the soul qualifies as a subsistent form of a substance, not a primary substance. The soul is the energizing mechanism of the corporeal body. It animates the hylomorphic composite of man. This is an appropriate portrait of the soul, especially given its etymological origin. "Soul" is derived from *anima*, which also provides the root for the word "animation."

The body is tailored according to the parameters of empirical utility. It serves a teleological function for the soul. The body to which the soul is united must equilibrate all of its biological qualities. Otherwise, it would be prone to consistent empirical miscalculations. Although empirical miscalculations do occur, they are not quite so frequent because the body maintains a delicate balance among its various biological dispositions. Moreover, this equilibrium allows the body to correct empirical miscalculations. Viewing the physical body from this vantage point, one automatically dismisses Gnosticism's derisive notion of all things corporeal. It demolishes the portrayal of the body as an impediment to the soul and reverences it as wonderfully designed machine. This is a position reiterated by King David in the Scriptures: "I am fearfully and wonderfully made" (Ps. 139:14). Christian philosopher Ravi Zacharias recapitulates:

> Jesus made it clear that the body is not just informationally different from other quantities; it is purposefully different. That is why the resurrection is a physical one at its core. The body matters in the eternal sense, not just the temporal. (71)

In emphasizing the physical body's purposeful difference, Zacharias reinforces the teleological function that it serves for the soul. In *Summa Theologiae*, Aquinas illustrates the centrality of the physical body to humanity's acquisition of knowledge. Through the body, humanity obtains sensory data concerning individual things. Thus, the body supplies the passive component of knowledge. This is the fundamental stage of knowledge acquisition. It is the bedrock upon which the active component of knowledge, which is supplied by the mind, rests. Aquinas elaborates:

> Our intellect cannot know the singular in material things directly and primarily. The reason for this is that the principle of singularity in material things is individual matter; whereas our intellect understands by abstracting the intelligible species from such matter. Now what is abstracted from individual matter is universal. Hence our intellect knows directly only universals. But indirectly, however, and as it were by a kind of reflexion, it can know the singular, because. . .even after abstracting the intelligible species, the intellect, in order to understand actually, needs to turn to the phantasms in which it understands the species... Therefore it understands the universal directly through the intelligible species, and indirectly the singular represented by the phantasm. And thus

it forms the proposition, "Socrates is a man." (Pt. I, Qu. 86, Art. I)

Man's innate ability of abstraction is clearly demonstrated by the observations of the Apostle Paul:

For the invisible things of him from the creation of the world are clearly seen, being understood by the things that are made, even his eternal power and Godhead; so that they are without excuse. (Romans 1:20)

Through abstraction, Paul recognizes the intrinsic finality of creation. The axiomatic nature of such finality leaves man with no excuse for doubt and unbelief. Knowledge of this finality is only possible through the possession of the physical human body, which was divinely designed by Almighty God Himself. Thus, the body serves a definite teleological purpose.

However, the Christian role for the body is not merely a sensory instrument for the soul. In fact, according to the Christian Weltanschauung, the physical body is integral to man's communion with God. Zacharias eloquently expands on this theme: ". . .in His [Christ's] incarnation He exalts the body, first by being conceived in the womb of a virgin, then by taking on human form and giving it the glorious expression of God in the flesh" (72).

Thus, the physical body is not simply analogous to a temple. *It is the temple itself!* Zacharias explains:

The Christian does not go to the temple to worship. The Christian takes the temple with him or her. Jesus lifts us beyond the building and pays the human body the highest compliment by making it His dwelling place, the place where He meets with us. (73)

The apostle Paul argued this position several times in his first letter to the Corinthians. In 1 Corinthians 6:15, Paul writes: "Know ye not that your bodies are the members of Christ?" Later, in 1 Corinthians 6:19, Paul recapitulates this message: "What? know ye not that your body is the temple of the Holy Ghost which is in you, which ye have of God, and ye are not your own?"

Evidently, the physical body is extremely important to the Christian. It plays an integral role in providing humanity with knowledge of God. Once acquainted with this truth, man can begin to commune with the Creator. In communion, the mysteries of the Lord's supra-rational mind can be gradually explicated through revelation. There is no *gnosis* or transformation of humanity's sensate being. There is merely the acknowledgement of what has been evident in creation from the beginning. This was the position that Aquinas philosophically dignified in *Summa Theologiae*. His mission was two-fold: reconcile the body with the soul and re-establish the temple of humanity.

However, Christianity was corrupted by nominalism during the Lutheran Reformation. Developed by English Franciscan William of Ockham, nominalism: "rejected Aquinas' high assessment of human powers and his confident belief in the ordered and knowable structure of the natural world" (Chambers, Hanawalt, et al. 424). In rejecting the knowable structure of creation, nominalism also rejected a knowable God (424). Of course, an unknowable God may as well be an absentee landlord. Thus, nominalism was a precursor to deism, which was, in turn, a precursor to atheism. Secularization had begun.

Martin Luther was the unconscious agent of secularization. Understandably, Luther sought to reform the Church. Under Catholicism, the truth had become the province of priests and other self-proclaimed "mediators of God." However, Luther made the mistake of adopting nominalism as one of the chief philosophical foundations for his doctrines. In *The Western Experience*, the authors write:

[S]ome of Luther's positions had roots in nominalism, the most influential philosophical and theological movement of the fourteenth and fifteenth centuries, which had flourished at his old monastery. (450)

By the time Luther's ideas were codified in the Augsburg Confession, nominalism was already beginning to co-opt Christianity. Nominalism's rejection of a knowable God harmonized with the superstitious notions of the time. Misunderstanding the troubles that beset them, many peasants made the anthropic attribution of the Black Death to God's will. Following this baseless assumption to its logical conclusion, many surmised that God was neither merciful nor knowable. Such inferences clearly overlooked the sacrifice of Jesus Christ, which represented the ultimate act of grace on God's part. Nevertheless, the superstitious populace were beginning to accept the new portrayal of God as an indifferent deistic spirit. Nominalism merely edified such beliefs. Invariably, nominalism would seduce those who would eventually convert to Protestantism.

Christians should have had more than a few philosophical misgivings with nominalism, especially in light of its commonalities with humanism:

> Although nominalists and humanists were frequently at odds, they did share a dissatisfaction with aspects of the medieval intellectual tradition, especially the speculative abstractions of medieval thought; and both advocated approaches to reality that concentrated on the concrete and the present and demanded a strict awareness of method. (424)

Suddenly, Christianity was infused with materialism and radical empiricism. The occult character of both of these philosophical positions has already been established. Radical empiricism rejects causality, thereby abolishing any epistemological certainty and reducing reality to a holograph that can be potentially manipulated through the "sorcery" of science. Materialism emphasizes the primacy of matter, inferring that the physical universe is a veritable golem that created itself. Despite their clearly anti-theistic nature, these ideas began to insinuate themselves within Christianity.

With abstraction thoroughly rejected, man was ontologically isolated from his Creator. Knowledge was purely the province of the senses and the physical universe constituted the totality of reality itself. Increasingly, theologians invoked naturalistic interpretations of the Scriptures, thereby negating the miraculous and supernatural nature of God. The spiritual elements that remained embedded in Christianity assumed more of a Gnostic character, depicting the physical body as an impediment to man's knowledge of God and venerating death as a welcome release from a corporeal prison. Gradually, a Hegelian synthesis between spiritualism and materialism was occurring. The result was a paganized Christianity, which hardly promised the abundant life offered by its Savior.

Luther's unwitting role in the popularization of such thinking suggests an occult manipulation. There is already a body of evidence supporting the contention that occult elements had penetrated Christendom and were working towards its demise. Malachi Martin states:

> As we know, some of the chief architects of the Reformation-

—Martin Luther, Philip Melanchthon, Johannes Reuchlin, Jan Amos Komensky—-belonged to occult societies. (521)

Author William Bramley presents evidence that backs Martin's contention:

Luther's seal consisted of his initials on either side of two Brotherhood symbols: the rose and the cross. The rose and cross are the chief symbols of the Rosicrucian Order. The word "Rosicrucian" itself comes from the Latin words "rose"("rose") and "cruces" ("cross"). (205)

Luther's involvement in the Rosicrucian Order made him an ideal instrument of secret societies. Michael Howard reveals explains the motive for this manipulation:

The Order had good political reasons for initially supporting the Protestant cause. On the surface, as heirs to the pre-Christian Ancient Wisdom, the secret societies would have gained little from religious reform. However, by supporting the Protestant dissidents they helped to weaken the political power of the Roman Catholic Church, the traditional enemy of the Cathars, the Templars and the Freemasons. (54)

However, occultism was not the only belief system benefiting from the Reformation. Elitism and oligarchy would also receive a boost from Luther's activities. It should be recalled that many of the secret societies supporting Luther acted as elite conduits. While Luther was already ideologically aligned with the elites in many ways, he officially became their property in 1521. In this year, the papacy's secular representative, Emperor Charles V, summoned Luther to a Diet at the city known as Worms (Chambers, Hanawalt, et al. 449). Luther was to defend himself against a papal decree that excommunicated him from the Church (449).

At the Diet, Luther refused to recant any of his beliefs (450). This led to the Emperor issuing an imperial edict for the monk's arrest (450). However, Luther was rescued by the Elector Frederick III of Saxony (450). Frederick staged a kidnapping of the monk and hid him away in Wartburg Castle (450). The regional warlord of Saxony had much to gain by protecting Luther. Frederick represented a group of German princes that opposed the influence of the Church and its secular representative, the Emperor (450). These elites would use Luther's teachings to justify breaking with the ecclesiastical authorities and establishing their own secular systems. When the peasants wished to use the Luther's teachings as justification for challenging unjust nobles, Luther wrote a vicious pamphlet entitled *Against the Rapacious and Murdering Peasants* (454). In the end, the Reformation reformed nothing at all. It caused a division in Christendom and led Europe down the path of secularization. Howard states:

Indirectly the Reformation gave the impetus for the Scientific Revolution of the seventeenth century, which centred on Newton, and led to the founding of the Royal Society after the English Civil War. (148)

The "Scientific Revolution" facilitated by the Reformation led to the popularization of Baconian concepts, which were radically scientistic and occult in character. Commensurate with this paradigm shift was the rise of the elite's first secular epistemological cartel and the acculturation of the masses to technocratic ideas.

Out of this era would emerge many of the theoreticians responsible for the development and augmentation of technocratic concepts. Comte was one of the most significant. His "science" of sociology provided a theoretical apparatus for the maintenance of bio-power and his philosophy of Positivism provided the edifying theology for a new theocracy. Not surprisingly, Comte's scientistic Weltanschauung was accompanied by the same morbid veneration for death that was infecting Christianity. Positivism would act as an ideational segue for movements devoted to the "right to die." Dowbiggin states:

Comte's positivism served as a kind of bridge between the naturalism of the late nineteenth century and the liberalism of the Progressive era, a discourse that played a large role in the origins of the euthanasia movement. (11)

In essence, Positivism was basically the scientistic incarnation of Calvinism, a movement birthed by the Reformation. This theological aberration was a direct result of Protestantism's nominalist theology, which rejected man's ability of abstraction and the intelligibility of God.

Calvinism promoted a doctrine of predestination, which presented the following contentions:

- That all humans are, inherently wicked and offend God;
- That there is an elect that God chose to be saved regardless of their actions and how deserving;
- That Jesus died just for those special elect, not for everyone;
- That once God has chosen an elect they are saved by irresistible grace no matter what;
- That these elect or Saints cannot fall from grace once saved. (Millegan 405)

According to Calvinism, there is only abundant life for some. Jesus Christ did not "set the captives free." He merely affirmed the elitist pedigree of a few. In Haeckelian terminology, supernatural selection is "aristocratic in the strictest sense of the word." The vast majority of humanity can only expect death, both physical and spiritual, irrespective of the individual's capacity for accepting for Christ as savior. Apparently, the Gnostic preoccupation with death was becoming increasingly evident within the early Protestant movement.

This hideous bowdlerization of Christianity synchronized comfortably the Positivist notion of a cosmos whose destiny was completely predetermined by impersonal forces. Positivism was further edified by Darwin's evolutionary determinism (Desmond and Moore 261). In this sense, Calvinism acted as a precursor to Positivism and Darwinian concepts of biological predestination. In fact, Desmond and Moore characterize Comte's scientistic Weltanschauung as a sort of naturalistic Calvinism (261). The secularizing forces embedded within the Reformation were paving the way for a technocratic society.

Concurrently, a culture of death was beginning to emerge. Intimations of Thanatos pervade the fabric of almost every movement, organization, and synarchical machination associated with the scientific dictatorship. Examples of this preoccupation with death abound. Adam Weishaupt's infamous Illuminati, which contributed to erection of an abortive scientific dictatorship in revolutionary France, exemplifies the principle of Thanatos exalted. The becomes evident when one examines the initiatory rites of this subversive organization:

The candidate to the Illuminati initiation was conducted

through a long, dark tunnel into a great vestibule adorned with black drapings and genuine corpses wrapped in shrouds. In the center of the hall the initiate would behold an altar composed of human skeletons. Two men dressed as ghosts would then appear and would tie a pink ribbon, which had been dipped in blood and bore the image of Our Lady of Loretto, around the forehead. A crucifix would be a funeral pyre. Crosses of blood would be painted on his body. (*Painted Black* 143)

Corpses, human skeletons, ghosts, blood, and a funeral pyre are all symbols of death. The dark theme communicated by these semiotic items was actuated with chilling results during the last bloody days of the French Revolution. The Illuminati was a suicide pact and its self-immolating propensities were channeled through Jacobinism. It was not long before the same guillotine that beheaded so many innocent souls fell upon the necks of the conspirators themselves. Yet, the meme of death was not eradicated. Not long after, the Bavarian Illuminati was discovered by the local authorities and driven from their headquarters. As they spread throughout Europe, so did the pandemic of Thanatos.

Illuminism eventually begat the Promethean faith, which would birth the scientific dictatorships of communism and fascism. These governmental aberrations would all be built atop the bedrock of the Gnostic myth called Darwinism. Massive democides would follow, claiming more lives than all of the world's wars combined. In some instances, the regime would expire along with the very population that it decimated. Perhaps the "fire in the minds of men" was actually a funeral pyre fueled by the irrationality of sociopolitical Utopianism and the religious fanaticism of secular Gnosticism. The flames of that pyre continued to burn further into the 20th century.

The 60s counterculture, which was partially a project of the intelligence community, sported the Teutonic rune of death. It was during this turbulent era that the euthanasia movement began to gain momentum (Dowbiggin 120). Shortly after this period of American history, Roe v. Wade made abortion legal. Planned Parenthood, which was closely aligned with the race scientists of the vanquished Nazi scientific dictatorship, could now continue the eugenics agenda with the sanction of American jurisprudence. As a result, millions of children have been murdered before they could even emerge from the womb. Meanwhile, America has continued her steady descent towards demographic implosion.

After September 11th, 2001, Thanatos began to spread abroad. The

Western incarnation of a global scientific dictatorship, *Pax Americana*, initiated several military expeditions in hopes of restructuring the world according to the technocratic designs of neoconservatives. Thousands have died in Afghanistan and Iraq. A large number of those casualties have been civilians. Interestingly enough, this bloody campaign has taken place under the presidency of George W. Bush. Of course, Bush is a member of Skull and Bones, a secret society headquartered at Yale University. Yale was established and administrated by Calvinist clerics (Millegan 417). This prompts a few disturbing questions. Could Bush's secret Order and Calvinism be bound by the same ideational thread of death? Worse still, could that same self-immolating mode of thought be guiding America's foreign and domestic policies?

Of course, the Order's title and iconography both echo the theme of death. If death is a governing theme of Skull and Bones, then its members tangibly enact it with frightening cognizance. Kris Millegan provides some horrifying examples:

As Governor of Texas, George W. Bush "hung" all prisoners sentenced to death that he had an opportunity to kill, *except one*, Henry Lee Lucas—-a confessed serial killer for a shadowy cult. Bonesman and US political power player Henry L. Stimson takes credit for talking President Truman into dropping *the bombs* on Hiroshima and Nagasaki. These are just two of the Order's death-dealing activities. (418)

In light of these examples, Millegan postulates:

The primary families of the Order are all inter-related and related to European royalty. In a private letter from one Bonesman to another, a member does treat another of its members as royalty. When one takes into consideration that George W. Bush has more royal relations than any other US President, along with royalty's heavy involvement in Masonry and other western ritual magical circles, one wonders about the mindset created from the influences of the Hyper-Calvinist beliefs of Hell, predestination, and infallible salvation mixed with potent duality of Western Ritual Magic tradition. Is our republic being undermined by a fervent multi-generational death-magic cult that is trying to bring about an apocalyptic New World Order through synarchical means—-with

themselves playing the leading roles—-no matter the cost to
the rest of us? Do *they* know? (419)

One must wonder if those involved in the conspiracy to establish
world oligarchy are even aware of their own self-immolating proclivities.
The sociopolitical Darwinians of globalism have endeavored to create a
world premised upon a single codex: "Survival of the fittest." According
to the elitist criterion for superiority, a vast majority of humanity
qualifies as "unfit." Malthusian theoreticians and population control
advocates further this agenda of mass extermination. It is interesting to
note that Masonry, which played a significant role in the popularization
of Darwinism, features ceremonies involving coffins and other items
semiotically related to death. Indeed, the principle of Thanatos exalted
seems to pervade the very ideational fabric of the conspiracy.

Meanwhile, it would appear as though the ideational contagion
is metastasizing at an alarming rate. Nominalist Christians, who
constitute a sizable portion of the evangelical establishment, continue
to support the murderous policies of neoconservativism. Liberals
remain steadfast in their promotion of birth control and abortion.
Futurist movements like Transhumanism and Singularitarianism eagerly
await man's extinction in some "post-human" era. Science fiction
entertainment, which typically presents humanity as an increasingly
impersonal organism, provides the inspiration for numerous scientistic
cults. Amidst this cacophony, man's universal position of *imago viva Dei*
is being either forgotten or forthrightly rejected.

When the Lord gave the law to Moses, He presented it with a
choice:

I call heaven and earth to record this day against you, *that*
I have set before you life and death, blessing and cursing:
therefore choose life, that both thou and thy seed may live. . .
(Deuteronomy 30:19)

Israel had a choice between life and death. The Lord urged the
people to choose the former. The exact same choice is extended to
the nations of the world today and, once more, the Lord is hoping that
people will choose life. It seems like a simple choice. God has made it
abundantly clear where life can be found:
<begin excerpting?
The thief cometh not, but for to steal, and to kill, and to
destroy: I am come that they might have life, and that they

might have *it* more abundantly. I am the good shepherd: the good shepherd giveth his life for the sheep. (John 10:10-11) Jesus saith unto him, I am the way, the truth, and the life: no man cometh unto the Father, but by me. (John 14:16) For the law of the Spirit of life in Christ Jesus hath made me free from the law of sin and death. (Romans 8:2) And this is the record, that God hath given to us eternal life, and this life is in his Son. He that hath the Son hath life; and he that hath not the Son of God hath not life. (1 John 5: 11-12)

Life is to be found in God, physically expressed through Jesus Christ. Yet, the conspiracy has rejected this option in the vain hope that its members will become gods. They are convinced that their own cognitive powers will tangibly realize this conceit. In Isaiah 1:18, God makes an invitation: "Come now, and let us reason together, saith the LORD." Meanwhile, the power elite apotheosizes reason and embraces God's adversary as a symbol of their alleged cognitive superiority. Yet, in so doing, they have willingly embraced utter darkness: "Professing themselves to be wise, they became fools" (Romans 1:22). Worse still, they have embraced death.

Yes, the scientific dictatorship is dying. Since the invasion of Iraq, the lack of elite consensus has become far worse. On May 15, 2003, a Bilderberg meeting was convened in Versailles, France (Tucker, no pagination). Bilderberg meetings have been held annually since the first meeting at the Bilderberg Hotel in Holland in 1954. The purpose of these meetings is to create harmony between the various factions of the elite. At this meeting, Bush was urged, "to share the spoils of war on Iraq" (no pagination). This pressure was brought to bear on Bush to silence concerns expressed by European elites. James Tucker reports those concerns:

> The Europeans are cynical about the United States urging the United Nations to approve the "coalition of the willing" controlling Iraqi oil for the "benefit of the Iraqis" and using the revenues to rebuild what was destroyed. Effectively, this gives control of Iraq to the United States and Britain, with a tip of the hat to Poland and Spain.
> Several Europeans suggested the "coalition" would generate huge profits by rebuilding Iraq with its oil money and asked: what European companies would get fat contracts. (No pagination)

The European elites have a good reason to desire involvement in the reconstruction of Iraq. Like the American and British elites, the European elites are trying to alleviate the effects of a global financial collapse. A report from the influential and respected French Institute of International Relations has predicted financial doomsday. James Tucker shares the details of this report:

> By 2050, said the report, *World Trade in the 21st Century,* Europe's share of the world economy will be only 12 percent, compared with 20 percent today.
> "The enlargement of the European Union won't suffice to guarantee parity with the United States," the report said. "The EU will weigh less heavily on the process of globalization and a slow but inexorable movement onto 'history's exit ramp' is foreseeable." (No pagination)

Substantial action is required to prevent the EU's movement into history's dustbin. Major contracts in Iraq would certainly help Europe maintain its share of the world economy. However, the Bush Administration has decided to ignore pressure from Bilderberg and to move forward with plans that exclude many European countries, most of which were opposed to the conflict. The *China Daily* was one of many media sources that reported on this move:

> Citing national security reasons, U.S. Deputy Defense Secretary Paul Wolfowitz has ruled that prime contracts to rebuild Iraq will exclude firms from nations such as France and Germany that opposed the U.S. war.
> In a policy document released on Tuesday, Wolfowitz said he was limiting competition for 26 reconstruction contracts worth up to $18.6 billion that will be advertised in coming days.
> "It is necessary for the protection of the essential security interests of the United States to limit competition for the prime contracts of these procurements to companies from the United States, Iraq, coalition partners and force contributing nations," Wolfowitz said in a notice published on the web site www.rebuilding-iraq.net.
> The move is likely to anger France and Germany and other traditional allies in NATO and the U.N. Security Council

who are being blocked out of prime contracts after their opposition to the war. (No pagination)

The move was not only likely to anger France, Germany, and other nations. It made discontent among the various elite factions a certainty. Bluebloods in China, Russia, Germany, France, and other nations left out of the loop are all lining up against the current Administration. Unless the United States government begins to appease these factions, plutocratic warfare seems to be on the horizon. The *China Daily* pointed out the opinion of one expert:

> Procurement specialist Prof. Steven Schooner from George Washington University said it was "disingenuous" to use national security as an excuse and predicted an angry reaction from those nations excluded.
> "This kind of decision just begs for retaliation and a tit-for-tat response from countries (such as Germany, France and Russia)," said Schooner. (No pagination)

Internal conflict among the bluebloods presents a unique problem for the would-be gods of the global scientific dictatorship. There has never been perfect harmony within the ranks of the elites. However, there has always been enough cohesion to keep the rabble in line. After all, the one thing that all the ruling class parties agree upon is that the commoners must be treated like children and kept in the playpen. As the kings of the "new world order" become preoccupied with shaking their fists at one another, the global financial system is failing and the natives are growing restless. Plagued by enemies from without and within, the elite are facing the ultimate juggling contest. It is doubtful that the "experts" of the technocratic paradigm can provide the solutions necessary for victory.

In *Morals and Dogma*, 33rd degree Mason Albert Pike made an interesting observation concerning the patron deity of the secret societies involved in the scientific dictatorship's ascendancy: "...the Sun God...created nothing" (254). Of course this false god has created nothing. . Creation is a thoroughly foreign concept for him. He only knows how "to steal, and to kill, and to destroy" (John 10:10). He is Thanatos universalized, the exalted principle of death worship. Although his adherents do not regard him as a literal metaphysical entity, the reality of his existence is asserting itself through the scientific dictatorship's own self-immolating proclivities. His name is Satan and his time is short.

Yes, the scientific dictatorship is dying. As quickly as it forms, it dissolves. It is, in short, a suicide pact. In opposing the Lord, the oligarchs oppose life itself. Appropriately enough, they shall expire with last enemy of mankind: "The last enemy *that* shall be destroyed *is* death" (1 Corinthians 15:26). To avoid the oligarchs' fate, the rest of humanity must begin now building their arks, starting with the source of life, Jesus Christ, as their foundation.

Works Cited

Abd el Fattah, Anisa. "Katrina Meets Social Darwinism and its son, Eugenics." *Media Monitors*. 04 September 2005 <http://usa. mediamonitors.net/content/view/full/19005

"Abiogenic petroleum origin." *Wikipedia: The Free Encyclopedia* 3 April 2006 <http://en.wikipedia.org/wiki/Abiogenic_petroleum_ origin>

Akin, William E. *Technocracy and the American Dream: The Technocrat Movement, 1900-1941*. Berkeley and Los Angeles: California UP, 1977.

"Albigenses." *World Book Encyclopedia*. 1980 ed.

Alexander, David. *Star Trek Creator*. New York: Dutton Signet, 1994.

Allen, Gary. *None Dare Call It Conspiracy*. Rossmoor, Calif.: Concord Press, 1971.

Allen-Mills, Tony. "Let Bin Laden stay free, says CIA man." *The Sunday Times*. 9 January 2005. <http://www.timesonline.co.uk/ article/0,,2089-1431539,00.html>

Anderson, Kevin. "Revelations and Gaps on Nixon Tapes." *BBC Online*. 1 March 2002 <http://news.bbc.co.uk/2/hi/americas/1848157. stm>

Angebert, Jean-Michel. *The Occult and the Third Reich*. New York: McGraw-Hill Book Company, 1971.

Angus, S. *The Mystery-Religions: A Study in the Religious Background of Early Christianity*. New York: Dover Publications, 1975.

"Anthropocentricism." *Wikipedia: The Free Encyclopedia*. 6 April 2006 <http://en.wikipedia.org/wiki/Anthropocentric>

Aquinas, Thomas. *Summa Theologiae*. Pt. I, Qu. 86, Art. I, in *Basic Writings of Saint Thomas Aquinas*. Ed. Anton C. Pegis (New York: Random House, 1945), I.

Bainbridge, William Sims. "Religions for a Galactic Civilization." Excerpted from *Science Fiction and Space Futures*, edited by Eugene M. Emme. San Diego: American Astronautical Society, pages 187-201, 1982. <http://mysite.verizon.net/ william.bainbridge/dl/relgal.htm>

---. "Social Construction from Within: Satan's Process." Excerpted

from *The Satanism Scare*, edited by James T. Richardson, Joel
Best, and David G. Bromley, New York: Aldine de Gruyter,
pages 297-310, 1991. <http://mysite.verizon.net/william.
bainbridge/dl/satansp.htm>

---. "New Religions, Science, and Secularization." Excerpted from
Religion and the Social Order, 1993, Volume 3A, pages 277-
292, 1993. <http://mysite.verizon.net/william.bainbridge/dl/
newrel.htm>

---. "Memorials." Excerpted from *Social Sciences for a Digital World.*
Edited by Marc Renaud. Paris: Organisation for Economic
Co-Operation and Development, 2000. <http://mysite.
verizon.net/william.bainbridge/dl/newtech.htm>

---. "Technological Immortality." *William Sims Bainbridge: A Personal
Convergenist Website.* 12 November 2005 <http://mysite.
verizon.net/wsbainbridge/dl/techimm.htm>

Baker, Jeffrey. *Cheque Mate: The Game of Princes* Springdale, PA: Whitaker
House, 1995.

Baigent, Michael, Richard Leigh, and Henry Lincoln. *Holy Blood, Holy
Grail.* New York: Delacorte, 1982.

Bainerman, Joel. *The Crimes of a President.* New York: S.P.I. Books,
1992.

Bamford, James. "The Agency That Could be Big Brother." *New
York Times* 25 December 2005 <http://www.nytimes.
com/2005/12/25/weekinreview/25bamford.html?oref=login
&pagewanted=print>

Bannister, Robert. *Sociology and Scientism.* London: North Carolina UP,
1987.

Barzun, Jacques. *Darwin, Marx, and Wagner.* Boston: Little, Brown and
Company, 1941.

Billington, James H. *Fire in the Minds of Men: Origins of the Revolutionary
Faith.* New York: Basic, 1980.

Blumenthal, Sydney. "Former Clinton Advisor: 'No One Can Say
they Didn't See it Coming'." *Spiegel Online* 31 August 2005
<http://service.spiegel.de/cache/international/0,1518,372455
,00.html>

Bobbit, Philip. "Get Ready for the Next Long War." *Time* 1 September
2002   <http://www.time.com/time/covers/1101020909/
abobbit.html>

Bomford, Andrew. "Echelon spy network revealed." *BBC.* 3 November
1999 <http://news.bbc.co.uk/2/hi/503224.stm>

Boot, Max. "Think Again: Neocons" *Foreign Policy* January/February

2004.   <http://www.cfr.org/pub7592/max_boot/think_ again_neocons.php>

Boutros-Ghali, Boutros. "New world order marginalizes UN." *New Zealand Herald Online* 15 April 2003 <http://www.nzherald. co.nz/storyprint.cfm?storyID=3400839>

Brainwashing: A Synthesis of the Russian Textbook on Psychopolitics <http:// www.4acloserlook.com/Brainwashing.pdf>

Bramley, William. *The Gods of Eden.* 1989. New York: Avon Books, 1990.

Brzezinski, Zbigniew. *Between Two Ages: America's Role in the Technetronic Era.* New York: Penguin Books, 1976.

---. *The Grand Chessboard: American Primacy and Geostrategic Objectives.* New York: Basic Books, 1997.

Brown, Tony. *Empower the People.* New York: William Morrow & Company, 1998.

"Brown Berets." *Wikipedia: The Free Encyclopedia.* 19 October 2005. <http://en.wikipedia.org/wiki/Brown_Berets>

Buck, J.D. *Mystic Masonry or the Symbols of Freemasonry and the Greater Mysteries of Antiquity.* Whitefish, MT: Kessinger Publishing, 1990.

Burns, James MacGregor. *John Kennedy: A Political Profile.* New York: Harcourt, Brace, and World, 1961.

Burns, Robert. "Russia gave Saddam intelligence during invasion, Pentagon says." *Seattle Times* 24 March 2006   <http://seattletimes.nwsource.com/html/ nationworld/2002887057_webrussiaintel24.html>

Burrow, John W. *Evolution and Society: A Study in Victorian Social Theory.* 1966. London: Cambridge UP, 1968.

Burstein, Daniel and Arne J. De Keijzer. *Big Dragon.* New York: Simon and Schuster, 1998.

Carlson, Ron, Ed Decker. *Fast Facts on False Teachings.* Eugene, Oregon: Harvest House Publishers, 1994.

Carr, E.H. *Studies in Revolution.* New York: Grossett and Dunlap, 1964.

Carr, Joseph. *The Twisted Cross.* Louisiana: Huntington House Publishers, 1985.

Carr, William Guy. *Pawns in the Game.* Palmdale, CA: Omni/Christian Book Club, 1958.

Carrico, David L. *The Occult Meaning of the Great Seal of the United States.* 1965. Evansville, Ind.: Followers of Jesus Christ Ministries, 1995.

Chambers, Claire. *The SIECUS Circle: A Humanist Revolution*. Appleton, Wisconsin: Western Islands, 1977.

Chambers, Mortimer and Barbara Hanawalt et al. *The Western Experience*. 1974. New York: McGraw-Hill, 2003.

Chase, Allan. *The Legacy of Malthus: The Social Costs of the New Scientific Racism*. New York: Alfred A. Knopf, 1977.

Chaitkin, Anton. *Treason in America*, New York: New Benjamin Franklin House, 1985.

"China, Russia start joint military exercises." *China Daily* 18 August 2005 <http://www.chinadaily.com.cn/english/doc/2005-08/18/content_470125.htm >

"China, Russia Issue Beijing Declaration." *The People's Daily* 18 July 2000 <http://fpeng.peopledaily.com.cn/200007/18/print20000718_45780.html>

Clarke, Arthur C. *Childhood's End*. New York: Ballantine Books, 1953.

Cohn, Werner. "Partners in Hate." 1985, 1995. *The Official Werner Cohn Website*. 6 October 2005. U of British Columbia. <http://www.wernercohn.com/Chomsky.html#anchor22784>

Coleman, Dr. John. *The Conspirator's Hierarchy: The Story of the Committee of 300*. Bozeman, MT.: American West Publishers, 1992.

Coomaraswamy, Rama. "The Fundamental Nature of the Conflict Between Modern and Traditional Man--Often Called the Conflict Between Science and Faith." 2001. *Coomaraswamy Catholic Writings*. 26 August 2005. <http://www.coomaraswamy-catholic-writings.com/Conflict%20Science%20and%20Religion.htm>

Cooper, William. *Behold a Pale Horse*. Sedona, AZ: Light Technology Publishing, 1991.

Cosby, Rita. "'MS-13' is one of nation's most dangerous gangs." *MSNBC* 13 February 2006 <http://msnbc.msn.com/id/11240718/>

"Cosmotheism." *Wikipedia: The Free Encyclopedia* 5 January 2006 <http://en.wikipedia.org/wiki/Cosmotheism>

Courtney, Phoebe and Kent Courtney. *America's Unelected Rulers: The Council on Foreign Relations*. New Orleans: Conservative Society of America, 1962.

Crabtree, James Dowell. *Progressivism, the New Deal, and the Technocratic Movement of the 1930s*. Dayton, Ohio: Wright State University, 1995.

Cuddy, Dennis L. *The Globalists: The Power Elite Exposed*. Oklahoma: Hearthstone Publishing, 2001.

Daniel, John. *Scarlet and the Beast.* Vol. 2 Tyler, Tex.: JKI Publishing, 1995.

Darlington, C.D. "The Origin of Darwinism." *Scientific America* 200 (1959): 60-66.

Darrah, Delmar Duane. *History and Evolution of Freemasonry.* Chicago: Charles T. Powers, 1979.

Darwin, Charles. *The Descent of Man.* 1874. *Evolutionary Classics.* U of Bergen Department of Zoology. 16 April 1998. <http://www.zoo.uib.no/classics/descent.html>

Dawkins, Richard. "The Word Made Flesh." *World of Dawkins* 28 November 2001 <http://www.simonyi.ox.ac.uk/dawkins/WorldOfDawkins-archive/Dawkins/Work/Articles/2001-12-27word_made_flesh.shtml>

DeCamp, John W. *The Franklin Cover-Up: Child Abuse, Satanism, and Murder in Nebraska.* Nebraska: AWT Inc., 1996.

Dennett, Daniel. *Darwin's Dangerous Idea: Evolution and the Meanings of Life.* New York: Simon and Schuster, 1995.

Desmond, Adrian and James Moore. *Darwin: The Life of a Tormented Evolutionist.* 1947. New York, NY: Warner Books, 1991.

deParrie, Paul and Mary Pride. *Unholy Sacrifices of the New Age.* Westchester, Illinois: Crossway Books, 1988.

Douglas Jr., Joseph D. *Red Cocaine: The Drugging of America.* Georgia: Clarion House, 1990.

Dowbiggin, Ian. *A Merciful End: The Euthanasia Movement in Modern America.* New York: Oxford UP, 2003.

Doward, Jamie. "They're not giving us what we need to survive." *The Observer* 4 September 2005 <http://observer.guardian.co.uk/international/story/0,6903,1562415,00.html>

Downs, Anthony. *Opening Up the Suburbs: An Urban Strategy for America.* New Haven: Yale UP, 1973.

Dreyfus, Hubert L. and Paul Rabinow. *Michel Foucault: Beyond Structuralism and Hermeneutics.* Chicago: Chicago UP, 1983.

Dubos, Rene J. *Louis Pasteur: Freelance of Science.* New York: Charles Scribner's Sons, 1976.

Editors of Executive Intelligence Review. *EIR Special Report, Global 2000: Blueprint for Genocide.* Washington D.C.: Executive Intelligence Review, 1982.

---. *Dope Inc.* Washington, D.C.: Executive Intelligence Review, 1992.

"Eisenhower's Farewell Address to the Nation, January 16, 1961." <http://mcadams.posc.mu.edu/ike.htm>

Eliade, Mircea (ed.). *The Encyclopedia of Religion*. England: Macmillan, 1987.

Elsner, Henry. *The Technocrats: Prophets of Automation*. New York: Syracuse UP, 1967.

Epperson, Ralph. *The Unseen Hand*. Tucson, AZ: Publius Press, 1985.

---. *The New World Order*. 1990. Tucson, AZ: Publius Press, 1995.

"Extropianism." *Wikipedia: The Free Encyclopedia* 10 January 2006 <http://en.wikipedia.org/wiki/Extropianism>

"FEMA Chief Sent Help Only When Storm Ended." *Associated Press* 7 September 2005 <http://cnn.netscape.cnn.com/news/story.jsp?idq=/ff/story/0001%252F20050907%252F0218843615.htm&photoid=20050904LADMI20&ewp=ewp_news_0905fema>

"First Responders Urged Not To Respond To Hurricane Impact Areas Unless Dispatched By State, Local Authorities." *FEMA* 29 August 2005 <http://www.fema.gov/news/newsrelease.fema?id=18470>

Fischer, Frank. *Technocracy and the Politics of Expertise*. Newbury Park, California: Sage Publications, 1990.

Fiske, John. *Television Culture*. London: Routledge, 1987.

Flynn, John T. *While You Slept*. Boston: Western Islanders, 1965.

Foucault, Michel. *Discipline and Punish: The Birth of the Prison*. Trans. Alan Sheridan. New York: Pantheon, 1977.

Francis, David R. "Now, dangers of a population implosion." *Christian Science Monitor* 7 October 2004. <http://www.csmonitor.com/2004/1007/p16s02-cogn.html>

Freeman, Christopher. "Malthus with a Computer," *Models of Doom: A Critique of the Limits to Growth*. New York: Universe Books, 1975.

Gaines, Elliot. "The Semiotics of Media Images From Independence Day and September 11th 2001." *The American Journal of Semiotics* 17 (2001): 117-131.

Galton, Francis. *Inquiries Into Human Faculty*. London: Macmillan, 1883.

Gardell, Mattias. *Gods of the Blood: The Pagan Revival and White Separatism*. London: Duke UP, 2003.

---. *Memories of My Life*. London: Methuen & Co., 1908.

Golitsyn, Anatoliy. *New Lies For Old*. New York: Dodd, Mead, and Company, 1984.

Good, Timothy. *Above Top Secret: The Worldwide UFO Cover-Up*. New York: William Morrow, 1988.

Goodrick-Clarke, Nicolas. *The Occult Roots of Nazism: The Ariosophists of Austria and Germany, 1890-1935.* England: Aquarian Press, 1985.

Green, Martin. *Children of the Sun: A Narrative of "Decadence" in England After 1918.* New York: Basic Books, 1976.

Griffin, Des. *Fourth Reich of the Rich.* 1976. Clackamas, OR: Emissary Publications 1995.

Grigg, William Norman. "Revolution in America." *The New American* Vol. 12, No. 4 (1996): 4-10.

---. "Battle Lines in the Drug War." *The New American* Vol. 13, No. 22 (1997): <http://www.thenewamerican.com/tna/1997/vo13no22/vo13no22_battle.htm>

Guenon, Rene. *The Reign of Quantity and the Signs of the Times.* Trans. Lord Northbourne. Baltimore, Maryland: Penguin Books Inc, 1953.

Hall, Manly Palmer. *The Lost Keys of Freemasonry.* 1976. Las Angeles, CA: The Philosophical Research Society, 1996.

Harris, Paul. "Vigilantes gather for Arizona round-up of illegal migrants." *The Observer.* 27 March 2005. <http://observer.guardian.co.uk/international/story/0,6903,1446342,00.html?gusrc=rss>

Hayakawa, Norio. "Why I am a Neo-Conservative." *Civilian Intelligence Network.* 21 March 2005. <http://www.hometown.aol.com/noriomusic/page1.html>

Hayes, Richard. "Selective Science." *TomPaine.commonsense* 12 February 2004. <http://www.tompaine.com/feature2.cfm/ID/9937/view/print>

Heiser, Michael S. Ph.D. *The Nachash and His Seed.* <http://www.thedivinecouncil.com/nachashnotes.pdf>

Henry, Patrick. "Berezovsky Says Putin Knew About FSB Role." *Moscow Times.* 6 March 2002 <http://www.infowars.com/saved%20pages/Hegelian/Berezovsky_Moscow_times.htm>

Hersh, Seymour. "We've Been Taken Over by a Cult." *Democracy NOW!* 26 January 2005. <http://www.democracynow.org/article.pl?sid=05/01/26/1450204>

Hickman, R. *Biocreation.* Worthington, Ohio: Science Press, 1983.

Hitler, Adolf. *Mein Kampf.* 1925 <http://www.mondopolitico.com/library/meinkampf/introduction.htm>

Hoar, William P. *Architects of Conspiracy.* Appleton, WI: Western Islands, 1984.

Hoffman, David. *The Oklahoma City Bombing and the Politics of Terror.*

1998. *The Constitution Society*. <http://www.constitution.org/ocbpt/ocbpt.htm>

Hoffman, Michael. *Secret Societies and Psychological Warfare*. Coeur d'Alene, Idaho: Independent History & Research, 2001.

Hook, C. Christopher. "The Techno Sapiens Are Coming." *Christianity Today* 19 December 2003. <http://www.christianitytoday.com/ct/2004/001/1.36.html>

Hooper, John. "Neo-Nazi leader 'was MI6 agent.'" *The Guardian*. 13 August 2002. <http://www.guardian.co.uk/international/story/0,,773568,00.html#article_continue>

Hooykaas, Reijer. *Religion and the Rise of Modern Science*. London: Chatto and Windus, 1972.

Howard, Michael. *The Occult Conspiracy*. Rochester, Vermont: Destiny Books, 1989.

Howe, Linda Moulton. *An Alien Harvest*. 1989. Huntingdon Valley, PA: Linda Moulton Howe Productions, 1995.

de Hoyos, Linda. "The Enlightenment's Crusade Against Reason." *The New Federalist* 8 Feb. 1993. <http://members.tripod.com/%7eamerican_almanac/dehoyos.htm>

Hubbard, L. Ron. *Dianetics*. Las Angeles, CA: Bridge Publications Inc., 1986.

Hudson, Audrey. "A Supersnoop's Dream." *The Washington Times* 15 November 2002. <http://www.washingtontimes.com/national/20021115-70231.htm>

Hunter, George William. *A Civic Biology Presented in Problems*. New York: American Book Company, 1914.

Huxley, Aldous. *Ends and Means*. New York: Harper and Brothers, 1937.

---. *Brave New World Revisited*. New York: Bantam Books, 1958.

---. *The Doors of Perception*. New York: Harper & Row, 1970.

Huxley, Julian. *UNESCO: Its Purpose and Its Philosophy*. Washington D.C.: Public Affairs Press, 1947.

Huxley, T.H. "Emancipation-Black and White." *Lectures and Lay Sermons*, 1871, Rhys E., ed. London: Everyman's Library, J.M. Dent & Co, 1926, 115.

"Internal Texaco Memo Showing Effort to Reduce Refining Capacity" *The Foundation for Taxpayer and Consumer Rights* <http://www.consumerwatchdog.org/energy/fs/5104.pdf>

"Interview of Zbigniew Brzezinski." Online posting. 15-21 January 1998. Konformist Yahoo Discussion Group. July 2002 <http://groups.yahoo.com/group/konformist/message/2429>

Isindag, Selami. *Masonluktan Esinlenmeler.* Istanbul, Turkey: The Turkish Grand Lodge of Free and Accepted Masons, 1977.

---. *Evrim Yolu.* Istanbul, Turkey: The Turkish Grand Lodge of Free and Accepted Masons, 1979.

Jasper, William F. *Global Tyranny...Step by Step: The United Nations and the Emerging New World Order.* Appleton, Wisconsin: Western Islands Publishers, 1992.

---. "Silk Hats and Brown Berets." *The New American* Vol. 12, No. 4 (1996): 33-36.

---. "Beijing Bailout." The New American Vol. 15, No. 4 (1999): 9-13.

Jeffery, Simon. "Powell: A Dove Among Hawks." *The Guardian.* 15 November 2004. <http://www.guardian.co.uk/usa/story/0,,1351934,00.html>

"John Tanton's Network." *SPLC Intelligence Report.* Summer 2002. <http://www.splcenter.org/intel/intelreport/article.jsp?sid=72>

Jones, E. Michael. *Libido Dominandi: Sexual Liberation and Political Control.* South Bend, Indiana: St. Augustine's Press, 2000.

Jones, John Paul. "What Evil Is and Why It Matters." *Paranoia Magazine* Issue 33 (2003): 62-64.

Jung, C. G. *Psychology and Alchemy.* London: Routledge & Kegan Paul, 1957.

Kadmon, Adam. "Mind Uploading: An Introduction." *Transtopia: Transhumanism Evolved.* 2003 <http://www.transtopia.org/uploading.html>

Kaiser, Robert G. and Ira Chinoy. "Scaife: Funding Father of the Right." *Washington Post.* 2 May 1999. <http://www.washingtonpost.com/wpsrv/politics/special/clinton/stories/scaifemain050299.htm>

Kasun, Jacqueline. *The War Against Population.* San Francisco, California: Ignatius Press, 1988.

Keith, Arthur. *Evolution and Ethics.* New York: G.P. Putnam's Sons, 1947.

Keith, Jim. *Casebook on Alternative Three.* Lilburn, GA: Illuminet Press, 1994.

---. *Mind Control, World Control.* Kempton, Illinois: Adventures Unlimited Press, 1997.

---. *Saucers of the Illuminati.* Lilburn, GA: Illuminet Press, 1999.

Kelly, Rev. Clarence. *Conspiracy Against God and Man.* Appleton, WI: Western Islands, 1974.

Kendall, Bridget. "Who is Putin?" *BBC News* 9 February 2001. <http://

news.bbc.co.uk/2/hi/programmes/correspondent/1156020.
stm>

Keynes, John. *Essays in Biography*. Toronto, Canada: Macmillan, 1933.

King, Alexander and Bertrand Schneider. *The First Global Revolution: A Report by the Council of the Club of Rome*. New York: Pantheon Books, 1991.

Kleiner, Art. *The Age of Heretics: Heroes, Outlaws and the Forerunners of Corporate Change*. New York: Currency Doubleday, 1996.

Kristol, Irving. *Neoconservativism: The Autobiography of an Idea*. New York: The Free Press, 1995.

---. "The Neocon Persuasion." The Weekly Standard. 25 August 2003. <http://weeklystandard.com/Content/Public/Articles/000/000/003/000tzmlw.asp?pg=2>

Kunen, James. *The Strawberry Statement*. New York: Random House, 1968.

Kurzweil, Ray. "Reinventing Humanity: The Future of Machine-Human Intelligence." *The Futurist* March-April 2006: 39-46.

LaRouche, Lyndon. "The Pagan Worship of Isaac Newton." *Executive Intelligence Review*. 21 Nov. 2003 <http://www.larouchepub.com/lar/2003/3045pagan_isaac.html>

Lee, Robert W. *The United Nations Conspiracy*. Appleton, Wisconsin: Western Islands, 1981.

Leigh, Wendy. *Arnold: The Unauthorized Biography*. Congdon and Weed, 1990.

Leopold, Jason. "The NSA Spy Engine: Echelon." *Truthout*. 9 January 2006   <http://www.truthout.org/docs_2006/010806A.shtml>

Letter to President Clinton on Iraq. 26 January 1998. <http://www.newamericancentury.org/iraqclintonletter.htm>

Levenda, Peter. *Unholy Alliance: A History of Nazi Involvement with the Occult*. New York: Avon Books, 1995.

Lewin, Leonard, ed., *The Report from Iron Mountain on the Possibility and Desirability of Peace*, New York: Dell Publishing, 1967.

Lewis, C.S. *Christian Reflections*. Grand Rapids, MI: Eerdmans, 1967.

Lind, Michael. "A Tragedy of Errors." *The Nation* 5 February 2004 <http://www.thenation.com/doc/20040223/lind/2>

Liversidge, Anthony. "Interview with R.D. Laing." *OMNI Magazine* (1988): 60-61

Loeb, Harold. *Life in a Technocracy*. New York: Viking Press, 1933.

Loftus, John and Mark Aarons. *The Secret War Against the Jews*. New York: St. Martin's Press, 1994.

Lovelock, James, *Ages of Gaia* NY: Norton Co. 1988.
"M. King Hubbert." *Wikipedia: The Free Encyclopedia* 3 April 2006 <http://en.wikipedia.org/wiki/M._King_Hubbert>
MacArthur, Douglas. *Reminiscences*. New York: McGraw-Hill, 1964.
Mackay, Neil. "Revealed: US plan to 'own' space." *Sunday Herald* 22 June 2003 <http://www.sundayherald.com/34768>
Mackenzie, Kenneth. *The Royal Masonic Cyclopaedia*. Wellingborough, England: Aquarian Press, 1987.
Makimaa, Julie. "Wolf in 'Humanitarian' Clothing." *New American* 3 July 2000: 35-37.
"Malthusian catastrophe." *Wikipedia: The Free Encyclopedia*. 17 Oct. 2005. <http://en.wikipedia.org/wiki/Malthusian_catastrophe#Neo-Malthusian_theory>
Marks, John. *The Search for the Manchurian Candidate: The CIA and Mind Control*. New York: Time Books, 1979.
Marrs, Jim. *Rule by Secrecy*. New York: HarperCollins, 2000.
Marrs, Texe. *Dark Majesty*. Austin, TX: Living Truth Publishers, 1992.
—. *Circle of Intrigue*. Austin, TX: Living Truth Publishers, 1995.
Martin, Malachi. *The Keys of this Blood*. New York: Simon and Schuster, 1991.
"Mary Parker Lewis." *Wikipedia: The Free Encyclopedia*. 1 December 2004. <http://en.wikipedia.org/wiki/Mary_Parker_Lewis>
"Max More." *Wikipedia: The Free Encyclopedia* 28 December 2005 <http://en.wikipedia.org/wiki/Max_More>
McAlvany, Donald S. *Toward a New World Order*. Phoenix, AZ: Western Pacific Publishing Co., 1992.
McGrath, Roger D. "Remaking America." *The New American Magazine* 13 June 2005: 12-18.
McRae, Donald G. "Darwinism and the Social Sciences." *A Century of Darwin*, 1958, Barnett S.A., ed., London: Mercury Books, 1962, 304.
Meadows, Donnela H. and Dennis L. *The Limits to Growth: A report for the Club of Rome's Project on the Predicament of Mankind*. New York: Universe Books Publishers, 1972.
Missler, Chuck and Mark Eastman. *Alien Encounters*. Coeur d'Alene, Idaho: Koinonia House, 1997.
"Monad." *Wikipedia: The Free Encyclopedia* 28 March 2006 <http://en.wikipedia.org/wiki/Monad>
"Monad (Technocracy)." *Wikipedia: The Free Encyclopedia* 22 March 2006 <http://en.wikipedia.org/wiki/Monad_%28Technocracy%29>

Monteith, Stanley. "The Peak Oil Myth: Part 2." *News With Views*. 10 December 2005. <http://www.newswithviews.com/Monteith/stanley1.htm>

Morahan, Lawrence. "China's weapons testing validates findings of Cox Report." *Capitol Hill Blue* 11 August 1999. <http://www.capitolhillblue.com/Aug1999/081199/coxreport081199.htm>

Moran, Michael. "Bin Laden comes home to roost." *MSNBC*. 24 August 1998. <http//www.msnbc.com/news/190144.asp?cp1=1>

More, Max. "In Praise of the Devil." *Lucifer.com* 1999 <http://www.lucifer.com/lucifer.html>

Morgan, Oliver and Faisal Islam. "Saudi dove in the oil slick." *London Observer*. 14 January 2001. <http://observer.guardian.co.uk/business/story/0,6903,421888,00.html>

"Mysterious monolith marks 2001." *BBC News*. 3 January 2001. <http://news.bbc.co.uk/1/hi/world/americas/1098419.stm>

O Meara, Kelly Patricia. "Money and Madness," *Insight on the News*. 3 June 2002. <http://www.insightmag.com/news/2002/06/24/National/Cover.Storymoney.And.Madness-254286.shtml>

Millegan, Kris. "God and Man and Magic at Yale." *Fleshing Out Skull and Bones*, Kris Millegan, ed., Walterville, OR: Trine Day, 2003, 403-19.

Mills, C. Wright. *The Power Elite*. London: Oxford UP, 1956.

Mumford, Lewis. *The Transformation of Man*. New York: Harper and Brothers, 1956.

"National Science Foundation." *Wikipedia: The Free Encyclopedia*. 12 December 2005 <http://en.wikipedia.org/wiki/National_Science_Foundation>

National Security Study Memorandum 200. April 1974. <http://www.population-security.org/28-APP2.html>

Nelson, Joyce. *The Perfect Machine: TV in the Nuclear Age*. Toronto: Canada, Between the Lines, 1987.

Newman, J.R. *What is Science?* New York: Simon and Schuster, 1955.

Palast, Greg. *The Best Democracy Money Can Buy*. London: Pluto Press, 2002.

"Pantheism." *Wikipedia: The Free Encyclopedia* 6 February 2006 <http://en.wikipedia.org/wiki/Cosmotheism_%28classical%29>

Pesce, Mark. "Ontos and Techne." *Computer-Medicated Magazine*, April 1997 <http://www.december.com/cmc/mag/1997/apr/pesce.html>

Peters, Ralph. "The New Warrior Class." *Parameters* Summer 1994.

<http://carlisle-www.army.mil/usawc/Parameters/1994/
peters.htm>

Phillips, James. "Renovation of the International Economic Order:
Trilateralism, the IMF, and Jamaica." *Trilateralism: The
Trilateral Commission and Elite Planning for World Management*,
Holly Sklar, ed., Boston: South End Press, 1980, 468-91.

Pike, Albert. *Morals and Dogma.* 1871. Richmond, Virginia: L.H. Jenkins,
Inc., 1942.

Pittenger, Mark. *American Socialists and Evolutionary Thought, 1870-1920.*
Madison: Wisconsin UP, 1993.

"Planned neo-Nazi march sparks violence." *CNN.* 15 October 2005.
<http://www.cnn.com/2005/US/10/15/nazi.march/>

Plato. *Phaedo.* Trans: Hugh Tredennick, in *Plato: The Collect Dialogues.*
eds. Edith Hamilton and Huntington Cairns (New York:
Pantheon Books, 1961).

Porter, Henry A. *Roosevelt and Technocracy.* Los Angeles: Wetzel
Publishing Company, Inc., 1932.

Porteus, Liza. "Status of Mental Health Rises in Senate," *Fox
News* 14 November 2002. <http://www.foxnews.com/
story/0,2933,70278,00.html>

Pouzzner, Daniel. *The Architecture of Modern Political Power: The New
Feudalism.* 17 May 2002 <http://www.mega.nu/ampp/>

"President Bush Sworn-in to Second Term." *The White House Official
Website* 20 January 2005 <http://www.whitehouse.gov/
inaugural/>

"Prometheism." *The Official Prometheism Website.* 13 March 2003 <http://
www.prometheism.net/principles.htm>

Prouty, L. Fletcher. *JFK: The CIA, Vietnam and the Plot to Assassinate John
F. Kennedy.* New York: Birch Lane P, 1992.

Quigley, Carroll. *Tragedy and Hope: A History of the World in our Time.* New
York: Macmillan, 1966.

Rand, Ayn. *Capitalism: The Unknown Idea.* New York: Signet Books,
1967.

Raschke, Carl A. *The Interruption of Eternity: Modern Gnosticism and the
Origins of the New Religious Consciousness.* Chicago: Nelson-
Hall, 1980.

---. *Painted Black.* New York: Harper Collins Publishers, 1990.

Rashid, Ahmed. "The Taliban: Exporting Extremism." *Foreign
Affairs Online.* November/December 1999 <http://www.
foreignaffairs.org/19991101faessay1017/ahmed-rashid/the-
taliban-exporting-extremism.html>

Ravenscroft, Trevor. *The Spear of Destiny*. Maine: Samuel Weiser, Inc, 1973.

Rebuilding America's Defenses: Strategy, Forces, and Resources for a New Century. September 2000 <http://www.newamericancentury. org/RebuildingAmericasDefenses.pdf>

Reed, Douglas. *The Controversy of Zion*, South Africa: Dolphin Press, 1978.

Robbins, Alexandra. "White House Bonesman leads nation into the dark," *USA Today Online* 25 September 2002 <www. USATODAY.com>

Robison, John J. *Born in Blood: The Lost Secrets of Freemasonry*. New York: M. Evans, 1989.

Robertson, Pat. *The New World Order*. Dallas, Texas: Word Publishing, 1991.

Rodgers, Ann. "Homeland Security won't let Red Cross deliver food." *Pittsburgh Post-Gazette* 03 September 2005 <http://www. post-gazette.com/pg/05246/565143.stm>

Roosevelt, Franklin D. *Looking Forward*. New York: The John Day Company, 1933.

"Royal Society." *World Book Encyclopedia*. 1980 ed.

Rose, Seraphim. *Orthodoxy and the Religion of the Future*. 1975. Platina, CA: Saint Herman of Alaska Brotherhood, 1996.

Rotberg, Robert. *The Founder: Cecil Rhodes and the Pursuit of Power*. New York: Oxford UP, 1988.

Rubin, Jerry. *Do It!* New York: Ballantine Books, 1970.

Rubin, Michael. "You Must be Likud!" *National Review Online*, 19 May 2004. <http://www.nationalreview.com/rubin/ rubin200405190844.asp>

Rudin, Ernst. "Eugenic Sterilization: An Urgent Need." *Birth Control Review*, Volume XVII, Number 4, April 1933: 102-04.

Rummel, R.J. *Freedom, Democide, War*. 13 March 2000. U of Hawaii. 19 September 2003. <http://www.hawaii.edu/powerkills/>.

Russell, Bertrand. *Religion and Society*. London: Oxford UP, 1947.

—. *The Impact of Science on Society*. New York: Simon & Schuster, 1953.

—. *The Autobiography of Bertrand Russell*. Boston: Little, Brown and Co., 1967.

Ryn, Claes G. *America the Virtuous: The Crisis of Democracy and the Quest for Empire*. News Brunswick, New Jersey: Transaction Publishers, 2003.

Sagan, Carl. *Cosmos*. New York: Random House, 1980.

Sailer, Steve. "Q&A: Steven Pinker of 'Blank State.'" *United Press*

International 30 October 2002 <http://pinker.wjh.harvard.
edu/books/tbs/media_articles/2002_10_30_upi.html>

Sanger, Margaret. *The Pivot of Civilization.* New York: Brentano's Press,
1922.

---. "Plan for Peace." *Birth Control Review*, Volume XVI, Number 4,
April 1932: 107-08.

Saussure, Ferdinand de. *Course in General Linguistics.* 1916. Trans. Wade
Baskin. London: Fontana/Collins, 1974.

---. *Course in General Linguistics.* 1916. Trans. Roy Harris. London:
Duckworth, 1983.

Segal, Howard P. *Technological Utopianism in American Cultural.* Chicago:
Chicago UP, 1985.

Skolnick, Sherman. "The Red Chinese Secret Police In The United
States, Part Two." Skolnick's Report <http://www.
skolnicksreport.com/chinesesp2.html>

Schorr, Daniel. "Poindexter Redux." *The Christian Science Monitor* 30
November 2002. <http://csmonitor.com/2002/1129/p11s01-
coop.html>

Scott, Walter. *The Life of Napoleon Bonaparte,* Vol. 2, Edinburgh:
Ballantyne, 1827.

"Seattle's mystery monolith disappears." *BBC News.* 4 January 2001.
<http://news.bbc.co.uk/1/hi/world/americas/1100524.stm>

Sharlet, Jeffrey. "Jesus Plus Nothing." *Harper's* March 2003: 53-64.

Shaw, Jim and Tom C. McKenney. *The Deadly Deception.* Lafayette:
Huntington House, 1988.

Sheen, Fulton J. *Life of Christ.* New York: McGraw-Hill,1958.

Shermer, Michael. "The Shamans of Scientism." *Scientific
America.* 13 May 2002. <http://www.sciam.com/
article.cfm?articleID=000AA74F-FF5F-1CDB-
B4A8809EC588EEDF>

Shipman, Pat. *The Evolution of Racism.* New York: Simon and Schuster,
1994.

Shirer, William L. *The Rise and Fall of the Third Reich: A History of Nazi
Germany.* New York: Simon and Schuster, 1960.

Shoup, Lawrence H. and William Minter. "Shaping a New World Order:
The Council on Foreign Relations' Blueprint for World
Hegemony." *Trilateralism: The Trilateral Commission and Elite
Planning for World Management,* Holly Sklar, ed., Boston:
South End Press, 1980, 135-56.

"Silent Weapons for Quiet Wars." *Secret and Suppressed.* Ed. Jim Keith.
Los Angeles: Feral House, 1993. 201-14.

Simon, Julian. *The Ultimate Resource*. Princeton, NJ: Princeton UP, 1981.

Simpson, Christopher. *Science of Coercion*. New York: Oxford UP, 1994.

Singer, Peter. *Rethinking Life and Death*. New York: St. Martin's Griffin, 1994.

Skinner, B.F. *Beyond Freedom and Dignity*. New York: Bantam Books, 1972.

---. *Walden Two*. 1948. New York: Macmillan, 1976.

Sklar, Holly. "Trilateralism: Managing Dependence and Democracy." *Trilateralism: The Trilateral Commission and Elite Planning for World Management*, Holly Sklar, ed., Boston: South End Press, 1980, 1-57.

Sloan, Sam. "Pamela Harriman, the Woman who made Bill Clinton President." *Homepage by Sam Sloan* <http://www.samsloan.com/pamela-h.htm>

Smith, C. William. "God's Plan in America," *New Age Magazine*, September 1950.

"Smith Richardson Foundation." *Wikipedia: The Free Encyclopedia*. 10 September 2005. <http://en.wikipedia.org/wiki/Smith_Richardson_Foundation>

Smith, Wolfgang. *Teilhardism and the New Religion: A Thorough Analysis of the Teachings of Pierre Teilhard de Chardin*. Illinois: TAN Books, 1988.

Social Contract Press. Ed. Robert Kyser. 2000-2005. <http://www.thesocialcontract.com/>

Spannaus, Edward. "Richard Mellon Scaife: Who Is He Really?" *The Executive Intelligence Review*. 21 March 1997. <http://members.tripod.com/-american_almanac/scaife.htm>

Spenser, Robert Keith. *The Cult of the All-Seeing Eye*. N.C.: Monte Cristo Press, 1964.

Stanton, William. "Oil and People." *ASPO Newsletter 55*. July 2005. http://www.peakoil.ie/newsletters/588

Stephens, Joe and David B. Ottaway. "From the U.S.A., the ABCs of jihad." *MSNBC*. 2002 <http://stacks.msnbc.com/news/728439.asp>

Stich, Rodney. *Defrauding America*. Alamo, California: Diablo Western Press, 1994.

"Strategic Energy Policy Challenges for the 21st Century: Report of an Independent Task Force Sponsored by the James A. Baker III Institute for Public Policy of Rice University and the Council on Foreign Relations." 2001 April <www.rice.

edu/projects/baker/Pubs/workingpapers/cfrbipp_energy/
energytf.htm>

Strong, Josiah. *Our Country: Its Possible Future and Its Present Crisis*. New York: The American Home Missionary Society, 1885.

Stross, Charles. "The Panopticon Singularity." *Whole Earth Review*. 2002 <http://www.antipope.org/charlie/rant/panopticon-essay.html>

Sutcliffe, Richard J. *The Fourth Civilization: Technology Society and Ethics*. 2003 <http://www.arjay.bc.ca/EthTech/Text/contents.html>

Sutton, Antony. *The Secret Cult of the Order*. Cranbrook, Australia: Veritas, 1983.

---. *America's Secret Establishment*. Billings, Mont.: Liberty House Press, 1986.

Suvin, Darko and Robert M. Philmus. *H.G. Wells and Modern Fiction*. New Jersey: Associated UP, 1977.

Tarpley, Webster Griffin and Anton Chaitkin. *George Bush: The Unauthorized Biography*. Washington, D.C.: Executive Intelligence Review, 1992.

Tarpley, Webster. "How the Venetian System Was Transplanted into England." *The American Almanac* 3 June 1996. <www.tarpley.net/venesys.htm>

Taylor, Ian T. *In the Minds of Men: Darwin and the New World Order*. Toronto: TFE Publishing, 1999.

"Technocratic movement." *Wikipedia: The Free Encyclopedia* 2 April 2006 <http://en.wikipedia.org/wiki/Technocratic_movement>

Tennenbaum, Jonathan. "Towards a New Science of Life." *Executive Intelligence Review* 7 September 2001. <http://www.larouchepub.com/eirtoc/confpres/2001/aug18-19_oberwesel/neo-darwinism/2834life_science.html>

Thio, Alex. *Sociology: A Brief Introduction*. 2000. New York: Pearson Education, 2005.

"The Murals at Denver University." *A Closer Look*. 2004 <http://www.4acloserlook.com/du.htm>

"The National Security Strategy of the United States." <www.whitehouse.gov/nsc/print/nssall.html>

The Path. The Cosmotheist Community Church Website. 2003 <http://www.cosmotheism.net/thepath.shtml>

"The Puppeteer." *SPLC Intelligence Report*. Summer 2002. <http://www.splcenter.org/intel/intelreport/article.jsp?pid=180>

"Thomas Gold." *Wikipedia: The Free Encyclopedia* <http://en.wikipedia. org/wiki/Abiogenic_petroleum_origin>

Timperlake, Edward and William Triplett. *Year of the Rat.* Massachusetts: Regnery Publishing, 1998.

---. *Red Dragon Rising.* Massachusetts: Regnery Publishing, 1998.

"Transtopian Symbolism." *Transtopia: Transhumanism Evolved* 2003-2005 <http://www.transtopia.org/symbolism.html>

Tucker, James. "Bilderberg Puts Heat on 'Loose Cannon' Bush over Mideast Policy." *American Free Press* 12 April 2004 <http:// www.americanfreepress.net/05_24_03/Bilderberg_Puts_ Heat/bilderberg_puts_heat.html>

"U.S.-Concocted 'Cox Report' a Farce to Instigate Anti-China Feelings, Undermine Sino-U.S. Relations: Zhao Qizheng." *People's Daily* 6 January 1999. <http://english1.peopledaily.com.cn/ english/199906/01/enc_990601001032_TopNews.html>

"US shuts out France, Germany for Iraq work." *China Daily* 12 December 2003   <http://www.chinadaily.com.cn/en/doc/2003-12/10/ content_288927.htm>

Utley, Freda. *The China Story.* Chicago: Henry Regnery, 1951.

Vernier, J.P. "Evolution as a Literary Theme in H.G. Wells's Science Fiction." *H.G. Wells and Modern Fiction.* Ed. Darko Suvin and Robert M. Philmus. New Jersey: Associated UP, 1977.

Vinograd, Cassandra. "Humans Are Ones on Display at London Zoo." *Associated Press* 26 August 2005 <http://news.yahoo.com/s/ ap/20050826/ap_on_fe_st/britain_human_zoo_3>

"Voz de Aztlán." *Wikipedia: The Free Encyclopedia.* 19 September 2005. <http://en.wikipedia.org/wiki/Nation_of_Aztl%C3%A1n>

Wagar, W. Warren. *H.G. Wells and the World State.* New Haven, CT.: Yale UP, 1961.

Watson, Steve and Alex Jones. "Former Assistant Secretary of the Treasury on New Orleans: 'Americans are Being Brainwashed.'" *Prison Planet* 6 September 2005 <http:// prisonplanet.com/Pages/Sept05/060905brainwashed.htm>

Webb, James. *The Occult Establishment.* Open Court, 1976.

Webster, Nesta H. *Secret Societies and Subversive Movements.* London: Britons Publishing Society, 1924.

Weikart, Richard. *From Darwin to Hitler: Evolutionary Ethics, Eugenics, and Racism in Germany.* New York: Palgrave Macmillan, 2004.

Wells, Herbert George. *The Outline of History: Being a Plain History of Life and Mankind.* London: Cassell and Company Ltd., 1925.

—. *The Shape of Things to Come*. 1933. *Electronic Text Collection*. Ed. Steve Thomas. U of Adelaide Library. 29 Oct. 2003 <http://etext. library.adelaide.edu.au/w/wells/hg/w45th/index.html>.

—. *Experiments in Autobiography*. New York: Macmillan Co., 1934.

—. *The Open Conspiracy: H.G. Wells on World Revolution*. 1928. Westport, Connecticut: Praeger, 2002.

Westerman, Toby. "The Chinese-Russian Alliance—Birth of a Superstate." *International News Analysis—Today* 11 June 2003 <http://www.inatoday.com/alliance.htm>

Weston, Warren. *Father of Lies*. London: n.p., 1930.

Whalen, William J. *Christianity and American Freemasonry*. Huntington, Indiana: Our Sunday Visitor, 1987.

White, Carol. *The New Dark Ages Conspiracy: Britain's Plot to Destroy Civilization*. New York: New Benjamin Franklin House, 1980.

Whitehall, Geoffrey. "The Problem of the 'World and Beyond': Encountering 'the Other' in Science Fiction." *To Seek Out New Worlds: Science Fiction and World Politics*, Jutta Weldes, ed. NY: Palgrave, 2003, 169-193.

Wilder-Smith, B. *The Day Nazi Germany Died*. San Diego, CA: Master Books, 1982.

Willoughby, Charles and John Chamberlain. *MacArthur: 1941-1951*. New York: McGraw-Hill, 1954.

Wilmshurst, W.L. *The Meaning of Masonry*. New York: Gramercy, 1980.

Witters, Patricia Jones-Witters and Weldon Witters. *Drugs & Society: A Biological Perspective*. Boston: Jones and Bartlett Publishers, 1986.

Wright, Peter. *Spy Catcher: The Candid Autobiography of a Senior Intelligence Officer*. New York: Viking Penguin, 1987.

Zacharias, Ravi. *Jesus Among Other Gods*. Nashville, Tennessee: Word Publishing, 2000.

Zahner, Dee. *The Secret Side of History: Mystery Babylon and the New World Order*. Hesperia, California: LTAA Communications Publishers, 1994.

About the Authors

Paul D. Collins has studied suppressed history and the shadowy undercurrents of world political dynamics for roughly eleven years. In 1999, he earned his Associate of Arts and Science degree. In 2006, he completed his bachelors degree with a major in liberal studies and a minor political science. Paul has authored another book entitled *The Hidden Face of Terrorism: The Dark Side of Social Engineering, From Antiquity to September 11*. Published in November 2002, the book is available online from www.1stbooks.com, barnesandnoble.com, and also amazon.com. It can be purchased as an e-book (ISBN 1-4033-6798-1) or in paperback format (ISBN 1-4033-6799-X).

Phillip D. Collins acted as the editor for *The Hidden Face of Terrorism*. He has also written articles for *Paranoia Magazine*, *MKzine*, *NewsWithViews*, *B.I.P.E.D.: The Official Website of Darwinian Dissent*, the *ACL Report*, *Namaste Magazine*, and *Conspiracy Archive*. In 1999, he earned an Associate degree of Arts and Science. In 2006, he earned a bachelors degree with a major in communication studies and a minor in philosophy. During the course of his seven-year college career, Phillip has studied philosophy, religion, and classic literature.

Read a comprehensive collection of Collins essays at:
http://www.conspiracyarchive.com/Commentary/Collins.php

About the Cover Designer

Terry Melanson is the owner, Developer and Webmaster of the Illuminati Conspiracy Archive. With an eclectic and insatiable reading routine since early childhood, he has set out to archive and compile a unique repository of hidden history and the manipulations of the elite. A Musician and a Web Professional, Terry Melanson lives and works in New Brunswick, Canada where he designs, develops and implements web solutions for a variety of clients worldwide. Visit him on the World Wide Web at: www.conspiracyarchive.com

4896397

Made in the USA
Lexington, KY
12 March 2010